Y0-DJO-575

ESSENTIAL MATHEMATICS

FOR COLLEGE STUDENTS

FRANCIS J. MUELLER

ESSENTIAL MATHEMATICS

FOR COLLEGE STUDENTS

third edition

PRENTICE-HALL, INC. Englewood Cliffs, New Jersey

Library of Congress Cataloging in Publication Data

Mueller, Francis J
Essential mathematics for college students.

Includes index.
1. Mathematics—1961- I. Title
QA39.2.M79 510 75-22216
ISBN 0-13-286518-1

18 17 16 15

Printed in the United States of America

PRENTICE-HALL INTERNATIONAL, INC., London
PRENTICE-HALL OF AUSTRALIA, PTY. LTD., Sydney
PRENTICE-HALL OF CANADA, LTD., Toronto
PRENTICE-HALL OF INDIA PRIVATE LTD., New Delhi
PRENTICE-HALL OF JAPAN, INC., Tokyo
PRENTICE-HALL OF SOUTHEAST ASIA (PTE.) LTD., Singapore

CONTENTS

PART I ESSENTIALS OF ARITHMETIC

PART II ESSENTIALS OF ALGEBRA

PREFACE

The genius of American education lies in its breadth. Nowhere is this more evident than at the post-secondary level, where an ever-increasing variety of courses is matched by an even wider span of student talent and preparation. Maintaining a range of appropriate educational materials that is correspondingly broad and diverse is an endless challenge.

Unfortunately there are no simple common denominators. To be successful amid this diversity, a textbook must limit its focus. The author must decide which students he hopes to serve, study carefully their particular needs, and develop his material accordingly.

In this instance two types of students have been kept in mind: those who wish to acquire or refresh a basic literacy in mathematics, and those who are preparing for certain technical curricula where functional competence with the basic tools of mathematics is expected. No doubt both types of student will often be found in the same person.

The content of the book is organized in such a way as to make it useful for individual self-instruction or for use as a class text. In either case, no prior mathematical background beyond the barest rudiments of arithmetic is necessary—and opportunity to review even that is found in the first three chapters.

The important computational phases of the subject, usually the first victims of disuse, are presented clearly as functional step-by-step procedures. In a sense they are analogous to computer programs, and in fact may be readily expressed in flow-chart form. Each procedure is followed immediately by a variety of relevant, fully explained examples or solved problems. Those who are well acquainted with students operating at this level of mathematical sophistication know the effectiveness of such an instructional pattern.

At the end of each chapter, on perforated pages, is a set of exercises keyed to parts of the chapter. It is expected that most students will attempt to work all the exercises. Only in this way is adequate coverage fully assured. In addition, extensive supplementary exercises are offered at the end of each chapter, including review exercises to help the student maintain understandings and skills.

Answers to all the keyed exercises and half of the supplementary and review exercises appear in the back of the book. Thus a student may check his work immediately. Research has noted that learning and understanding is reinforced when verification is immediate.

Periodic achievement tests are interspersed throughout the book, and a final comprehensive 100 problem test is included. These may be used for review and as a check on retention. The periodic achievement tests may also be used in advance, to determine whether or not one needs to study certain chapters, or which parts need study and which may be bypassed.

For ease of handling, it is suggested that each exercise page be detached from the book before beginning work. The punched holes in the margin make it easy to retain the torn-out pages in an ordinary loose-leaf binder. The exercise pages may be completed as single assignments, or partially as the topics are discussed in the text. The student or the instructor can best determine the optimum use of these pages.

As with any book, the choice of topics—what to include, what to leave out—and how deeply to go rests inescapably with the author. In this edition new material has been added on determinants, relations, and functions. Those choices and lesser modifications were heavily influenced by feedback from the very successful previous edition, and by comments and suggestions made by competent and perceptive instructors in the field who reviewed the current edition when in preparation. My gratitude for their valued assistance is considerable and heartfelt, though responsibility for what finally appears, and how it is presented, is entirely mine.

San Diego, California FRANCIS J. MUELLER

PART I

ESSENTIALS OF ARITHMETIC

1

WHOLE NUMBERS

1. HINDU-ARABIC SYSTEM OF NUMBER NOTATION

Number exists only in our minds. It is an abstraction, a property, something like color, or honesty, or beauty, only much more precise. To express and record his number ideas, man has invented *numerals*, which are arbitrary symbols to represent those ideas. We use the symbol 5, for example, to express the number of fingers on one hand; but an ancient Roman would have used V, and a modern Japanese uses 五. The number, or idea, is the same in each case, but the numerals, or symbols, differ.

The most widespread system of numerals today is the one we use in this country, the Hindu-Arabic. In a way it is a code for naming numbers. It consists of ten basic symbols, or *digits* (0, 1, 2, 3, 4, 5, 6, 7, 8, 9) and a pattern for combining them to represent many other numbers. In this system, the digit in the extreme right position of the numeral represents the number of *ones*, or single units. A digit next to it on the left, when there is one, represents tens of ones, or simply *tens*. If there is a digit to the left of the tens digit, it represents tens of tens, or *hundreds*; the next digit to the left represents tens of hundreds, or *thousands*; the next, *ten-thousands*; and so on. In these numerals the location of each digit has a *place value* ten times that of the digit to its right Thus:

4376 = 4 *thousands* + 3 *hundreds* + 7 *tens* + 6 *ones*
(read four thousand, three hundred seventy-six)

36,824 = 3 *ten-thousands* + 6 *thousands* + 8 *hundreds* + 2 *tens* + 4 *ones*
(read thirty-six thousand, eight hundred twenty-four)

1,642,307 = 1 *million* + 6 *hundred-thousands* + 4 *ten-thousands* + 2 *thousands* + 3 *hundreds* + no *tens* + 7 *ones*
(read one million, six hundred forty-two thousand, three hundred seven)

These three numerals are also shown in the following table. Note that in reading and writing Hindu-Arabic numerals we usually break them into *numeration periods* of three digits each, and set them off by commas in the written numeral, e.g., 1,642,307.

←	Billions	Hundred-millions	Ten-millions	Millions	Hundred-thousands	Ten-thousands	Thousands	Hundreds	Tens	Ones
							4	3	7	6
						3	6	8	2	4
				1	6	4	2	3	0	7
billions		*millions*			*thousands*			*ones*		

NUMERATION PERIODS

2. ADDITION

Addition is one of the four basic operations by which we combine numbers to produce other numbers. We usually compute the result of these operations by processing numerals in a certain way. Most computational procedures are piecemeal techniques—the numeral for the result is put together digit by digit. For example, when adding whole numbers (i.e., 0, 1, 2, 3, 4, . . .) we arrange the numerals for the numbers to be added (called *addends*) in such a way that the ones digits fall in the same column, the tens digits in the next column to the left, the hundreds digits in the next column, and so on. The result (*sum*) is then computed column by column.

EXAMPLE

Add: 456 + 3274 + 67 + 820

		Thousands Hundreds Tens Ones					
addend	⟶	456	=		400 +	50 +	6
addend	⟶	3274	=	3000 +	200 +	70 +	4
addend	⟶	67	=			60 +	7
addend	⟶	820	=		800 +	20 +	0
sum	⟶	4617	=	3000 +	1400 +	200 +	17

3. CHECKS FOR ADDITION

(a) Reverse addend check. The most frequent type of error in computing sums of whole numbers is caused by a mental lapse in forming one of the combinations, such as saying 6 + 8 = 13. In this check, reversing the order of the addends in each column produces a new sequence of combinations, thereby lessening the chance of repeating the same error.

EXAMPLE

The computation		The check	
add downward ↓	642	add upward ↑	642
	337		337
	984		984
	1963		1963

(b) Withheld addend check. Another check, in which one of the addends is withheld temporarily, is outlined in Program 1.1.

(1.1) *To check addition by the withheld addend check:*
Step 1. *Omit one addend (usually the first or last) and find the sum of the remaining addends.*
Step 2. *Add the omitted addend to the sum of Step 1.*
Step 3. *Compare to see that the result of Step 2 is the same as the original sum.*

EXAMPLE

The computation	The check	
463		
872	Step 1. Withhold 463 temporarily and add the remaining addends:	872
946		946
327		327
		2145
2608 ←	Step 2. Add in the withheld addend:	463
	Step 3. Compare with the original sum →	2608

(c) Accountant's check. The check given in Program 1.2 represents yet another way in which to organize the addition process.

(1.2) *To check addition by the accountant's check:*
Step 1. *Find the sum for each column separately without carrying to the next column (columns may be taken from right to left, or left to right).*
Step 2. *Find the sum of the staggered sub-totals of Step 1.*
Step 3. *Compare to see that the result of Step 2 is the same as the original sum.*

EXAMPLE

The computation	The checks (Right to left)	The checks (Left to right)
636	636	636
542	542	542
783	783	783
227	227	227
136	136	136
2324	24	21
	20	20
	21	24
	2324	2324

(d) Excess of nines check. In the following division-by-nine examples, note that the remainders could have been found *by adding the digits of the dividend* (i.e., the number divided).

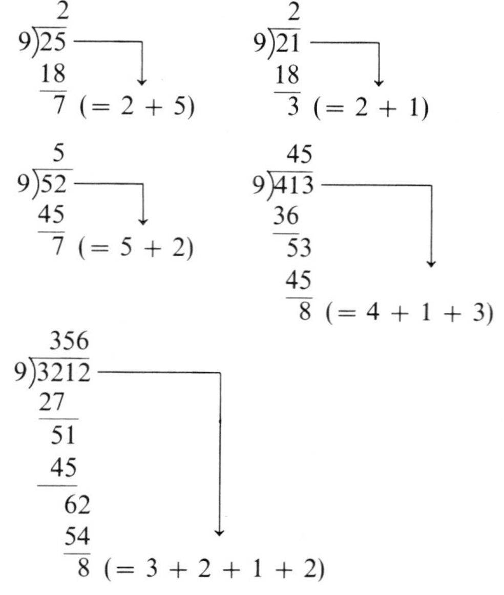

When the sum of the digits of the dividend is greater than 9, as in the examples below, the remainder can be found by adding the digits of the sum of the digits of the dividend.

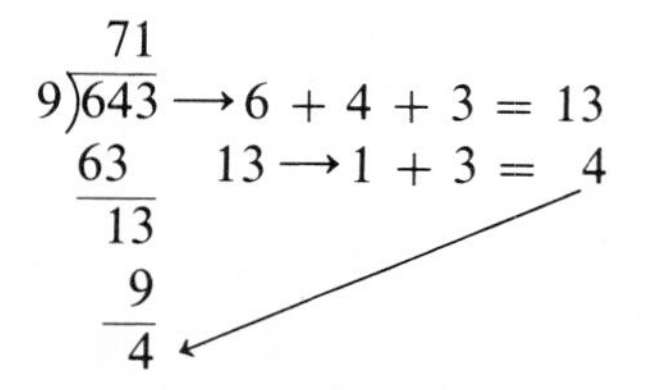

$$
\begin{array}{rl}
95 & \\
9\overline{)856} & \longrightarrow 8 + 5 + 6 = 19 \\
\underline{81} & \quad 19 \longrightarrow 1 + 9 = 10 \\
46 & \quad 10 \longrightarrow 1 + 0 = 1 \\
\underline{45} & \\
1 & \leftarrow
\end{array}
$$

$$
\begin{array}{rl}
48 & \\
9\overline{)432} & \longrightarrow 4 + 3 + 2 = 9 \\
\underline{36} & \quad \text{(excess of nines in 9 is 0)} \\
72 & \\
\underline{72} & \\
0 & \leftarrow
\end{array}
$$

Thus we have a short way for finding the remainder for any number divided by 9. This technique is known as *finding the excess of nines*, or *casting out nines*, and is of great use in checking the results of all four of the basic operations.

(1.3) *To check addition by the excess of nines:*

Step 1. Find the excess of nines for each of the addends.

Step 2. Find the sum of the excesses of Step 1.

Step 3. Find the excess of nines for the sum of Step 2.

Step 4. Find the excess of nines for the sum of the given addition.

Step 5. Compare to see that the excess of Step 3 is the same as the excess of Step 4.

EXAMPLES

1. Check the addition $32 + 41 + 12 = 85$ by the excess of nines.

   Step 1.
   The excess of nines for 32 is $3 + 2$ or 5.
   The excess of nines for 41 is $4 + 1$ or 5.
   The excess of nines for 12 is $1 + 2$ or 3.

   Step 2. The sum of the excesses of the addends is $5 + 5 + 3 = 13$.

   Step 3. The excess of nines for the sum of the excesses (13) is $1 + 3$ or 4.

   Step 4. The excess of nines for the sum of the given addition (85) is $8 + 5 = 13 \longrightarrow 1 + 3$ or 4.

   Step 5. The excess of nines for the sum of the excesses (4) equals the excess of nines for the sum of the given addition (4); so the computation checks.

2. Check the addition below by the excess of nines.

$$
\begin{array}{rl}
 & \overbrace{\hspace{12em}}^{\text{Step 1}} \\
347 & \longrightarrow 3 + 4 + 7 = 14 \longrightarrow 1 + 4 = 5 \\
492 & \longrightarrow 4 + 9 + 2 = 15 \longrightarrow 1 + 5 = 6 \quad \text{Step 2} \\
\underline{363} & \longrightarrow 3 + 6 + 3 = 12 \longrightarrow 1 + 2 = \underline{3} \quad \text{Step 3} \\
1202 & \qquad\qquad\qquad\quad 14 \longrightarrow 1 + 4 = 5 \\
\searrow & \\
 & 1 + 2 + 0 + 2 = 5 \longleftarrow \text{Step 5} \longrightarrow \\
 & \underbrace{\hspace{8em}}_{\text{Step 4}}
\end{array}
$$

3. Check the addition below by the excess of nines.

$$
\begin{array}{rlr}
4365 & \longrightarrow 4 + 3 + 6 + 5 = 18 \longrightarrow 1 + 8 & = 9 \longrightarrow 0 \\
271 & \longrightarrow 2 + 7 + 1 = 10 \longrightarrow 1 + 0 & = 1 \\
8362 & \longrightarrow 8 + 3 + 6 + 2 = 19 \longrightarrow 1 + 9 & = 10 \longrightarrow 1 \\
\underline{21} & \longrightarrow 2 + 1 & = \underline{3} \\
13019 & \longrightarrow 1 + 3 + 0 + 1 + 9 = 14 \longrightarrow 1 + 4 = & 5
\end{array}
$$

With a little experience one can "cast out" nines as he goes along and write the excesses directly. For instance, for Example 3 above:

$$
\begin{array}{rlr}
4365 & \longrightarrow 4 + 3 + 6 = 13 \longrightarrow 4, \quad 4 + 5 & = 9 \longrightarrow 0 \\
271 & \longrightarrow 2 + 7 = 9 \longrightarrow 0, \quad 0 + 1 & = 1 \\
8362 & \longrightarrow 8 + 3 = 11 \longrightarrow 2, \quad 2 + 6 + 2 & = 10 \longrightarrow 1 \\
\underline{21} & \longrightarrow 2 + 1 & = \underline{3} \\
13019 & \longrightarrow 1 + 3 + 0 + 1 + 9 = 14 \longrightarrow 1 + 4 = & 5
\end{array}
$$

It should be noted that this excess-of-nines check is not foolproof. The check will not detect errors of transposition (for instance, if we mistakenly call the sum in the last example 10,319, its excess of nines is also 5). Moreover, erroneous sums which differ from the correct sum by 9 or multiples of 9 will go undetected (for instance, the correct sum 13,019 as well as 13,028 (i.e., 13,019 + 9), 13,037 (i.e., 13,019 + 18), 13,919 (i.e., 13,019 + 900), 13,010 (i.e., 13,019 − 9), etc.—all have 5 for their excess of nines).

(Complete Exercise 1.1)

4. SUBTRACTION

Subtraction is the inverse or reverse operation to addition. To subtract one number (say 8) from a second number (say 12) means to find a third number (in this case 4) which, when added to the first, equals the second.

$$
\begin{array}{r} 12 \\ -8 \\ \hline ? \end{array} \leftrightarrow \begin{array}{r} 8 \\ +? \\ \hline 12 \end{array} \qquad \text{or} \qquad \begin{array}{r} 12 \\ -8 \\ \hline 4 \end{array} \leftrightarrow \begin{array}{r} 8 \\ +4 \\ \hline 12 \end{array}
$$

In effect, subtraction is concerned with finding a missing addend (*difference*) when the other addend (*subtrahend*) and the sum (*minuend*) are known.

In computation, the form for subtracting whole numbers is much like that for adding whole numbers. Numerals for the numbers to be subtracted are arranged so that digits having the same place values fall in the same column; then the result is computed piecemeal, digit by digit, right to left.

EXAMPLE

Subtract: 8365 − 2124.

Thousands	Hundreds	Tens	Ones	
8	3	6	5	← minuend
2	1	2	4	← subtrahend
6	2	4	1	← difference

At times, in any column except the far left, the digit in the subtrahend will have a greater value than that of the corresponding digit in the minuend. When this happens, there are two ways in which to proceed: by *decomposition* or by *equal additions*. We explain both, but encourage the reader to continue using the method which is more familiar to him, and to accept the other as information.

EXAMPLE

The computation:

Hundreds	Tens	Ones
8	2	5
−4	7	3

Decomposition Method

Hundreds	Tens	Ones	
$^{7}\not{8}$	$^{1}2$	5	$825 = 700 + 120 + 5$
−4	7	3	$-473 = 400 + 70 + 3$
3	5	2	$300 \quad 50 + 2$

(a) In the ones column: 3 from 5 is 2. We write 2 as the ones digit of the difference.
(b) In the tens column: we have no whole number for 7 from 2, so we change (decompose) the 8 hundreds in the minuend into 7 hundreds plus 1 hundred. The 1 hundred is then changed to 10 tens and added to the already-present 2 tens, making 12 tens in all. (This is sometimes referred to as "borrowing.") Now it is possible to subtract the 7 tens of the subtrahend from the 12 tens of the minuend. Result: 5 tens. We write 5 as the tens digit of the difference.
(c) In the hundreds column: 4 from 7 is 3. We write 3 as the hundreds digit of the difference.

Equal Additions Method

Hundreds	Tens	Ones	
8	$^{1}2$	5	$825 + 100 = 800 + 120 + 5$
$-^{5}\not{4}$	7	3	$473 + 100 = 500 + 70 + 3$
3	5	2	$300 + 50 + 2$

(a) In the ones column: 3 from 5 is 2. We write 2 as the ones digit of the difference.
(b) In the tens column: we have no whole number for 7 from 2, so we add 1 hundred to *both* the minuend and the subtrahend. (Adding the same number to both minuend and subtrahend does not change the value of the difference.) The 1 hundred is *added to the minuend as 10 tens* (making 12 tens in all) and *added to the subtrahend as 1 hundred* (making 5 hundreds in all). Now it is possible to subtract the 7 tens of the subtrahend from the 12 tens of the minuend. Result: 5 tens. We write 5 as the tens digit of the difference.
(c) In the hundreds column: 5 from 8 is 3. We write 3 as the hundreds digit of the difference.

EXAMPLES

1. Subtract: 373 − 136.

Decomposition:

3	$^{6}\not{7}$	$^{1}3$	Ones:	6 from 13 is 7
−1	3	6	Tens:	3 from 6 is 3
2	3	7	Hundreds:	1 from 3 is 2

Equal Additions:

3	7	$^{1}3$	Ones:	6 from 13 is 7
−1	$^{4}\not{3}$	6	Tens:	4 from 7 is 3
2	3	7	Hundreds:	1 from 3 is 2

2. Subtract: 4276 − 1481.

Decomposition:

$^{3}\not{4}$	$^{11}\not{2}$	$^{1}7$	6	Ones:	1 from 6 is 5
− 1	4	8	1	Tens:	8 from 17 is 9
2	7	9	5	Hundreds:	4 from 11 is 7
				Thousands:	1 from 3 is 2

Equal Additions:

4	$^{1}2$	$^{1}7$	6	Ones:	1 from 6 is 5
$-^{2}\not{1}$	$^{5}\not{4}$	8	1	Tens:	8 from 17 is 9
2	7	9	5	Hundreds:	5 from 12 is 7
				Thousands:	2 from 4 is 2

3. Subtract: 4002 − 1837.

Decomposition:

$^{3}\not{4}$	$^{9}\not{0}$	$^{9}\not{0}$	$^{1}2$	Ones:	7 from 12 is 5
− 1	8	3	7	Tens:	3 from 9 is 6
2	1	6	5	Hundreds:	8 from 9 is 1
				Thousands:	1 from 3 is 2

Equal Additions:

4	$^{1}0$	$^{1}0$	$^{1}2$	Ones:	7 from 12 is 5
$-^{2}\not{1}$	$^{9}\not{8}$	$^{4}\not{3}$	7	Tens:	4 from 10 is 6
2	1	6	5	Hundreds:	9 from 10 is 1
				Thousands:	2 from 4 is 2

5. CHECKS FOR SUBTRACTION

(a) Addition check. The fact that the difference in a subtraction may be thought of as a missing addend is the basis for the most frequent check for subtraction. It is outlined in Program 1.4.

(1.4) *To check subtraction by addition:*
Step 1. Find the sum of the difference and the subtrahend.
Step 2. Compare to see that the sum of Step 1 is the same as the minuend of the given subtraction.

EXAMPLE

Check the subtraction 825 − 473 = 352 by addition.
Step 1. Find the sum of the difference (352) and the subtrahend (473):

$$\begin{array}{r} 352 \\ +473 \\ \hline 825 \end{array}$$

Step 2. The sum of Step 1 (825) is the same as the minuend (825); so the subtraction checks.

(b) Subtraction check. Since both the difference and the subtrahend may be thought of as addends and the minuend as the sum, it follows that if the difference is subtracted from the minuend, the result should equal the subtrahend.

(1.5) *To check subtraction by subtraction:*
Step 1. Subtract the difference from the minuend.
Step 2. Compare to see that the result of Step 1 is the same as the subtrahend of the given subtraction.

EXAMPLE

Check the subtraction 825 − 473 = 352 by subtraction.
Step 1. Subtract the difference (352) from the minuend (825):

$$\begin{array}{r} 825 \\ -352 \\ \hline 473 \end{array}$$

Step 2. The result of Step 1 (473) is the same as the subtrahend of the given subtraction (473); so the subtraction checks.

(c) Excess-of-nines check. By combining the technique of casting out nines with the fact that DIFFERENCE + SUBTRAHEND = MINUEND we have a third method of checking the results of a subtraction.

(1.6) *To check subtraction by the excess of nines:*
Step 1. Find the excess of nines for each term—difference, subtrahend, and minuend.
Step 2. Find the sum of the excesses of the subtrahend and the difference.
Step 3. Find the excess of nines for the sum of Step 2.
Step 4. Compare to see that the excess of Step 3 is the same as the excess of nines for the minuend.

EXAMPLES

1. Check the subtraction below by the excess of nines.

$$\begin{array}{r} 3862 \\ -2731 \\ \hline 1131 \end{array}$$

Step 1. The excess of nines for the difference (1131) is 6. The excess of nines for the subtrahend (2731) is 4. The excess of nines for the minuend (3862) is 1.
Step 2. The sum of the excesses for the subtrahend and difference is 6 + 4 or 10.
Step 3. The excess of nines for the sum of Step 2 (10) is 1.
Step 4. The excess of nines (1) found in Step 3 is the same as the excess of nines for the minuend (1); so the subtraction checks.

2. Check the subtraction below by the excess of nines.

$$\begin{array}{r} 3026 \longrightarrow 2 \\ -1379 \longrightarrow 2 \\ \hline 1647 \longrightarrow 0 \end{array} \quad 2 + 0 = 2$$

3. Check the subtraction below by the excess of nines.

$$\begin{array}{r} 43{,}627 \longrightarrow 4 \\ -986 \longrightarrow 5 \\ \hline 42{,}641 \longrightarrow 8 \end{array} \quad 5 + 8 = 13 \longrightarrow 1 + 3 = 4$$

(Complete Exercise 1.2)

6. MULTIPLICATION

Multiplication is another of the four basic operations. It may be thought of as repeated addition in which all of the addends are the same. For example, the sum 3 + 3 + 3 + 3 + 3 = 15 can be expressed by multiplication as 5 × 3 = 15 (or 5 · 3 = 15). Here the terms 5 and 3 are called *factors* (*multiplier* and *multiplicand*, respectively) and the 15 is called the *product*.

When one factor is expressed by a single-digit numeral, it is convenient in computation to regard it as the multiplier, and to compute the product piecemeal by multiplying it with each digit of the other factor, right to left, carrying where necessary.

EXAMPLE

```
 728    Ones:      6 × 8 = 48; write 8 in the
  ×6               ones place of the product;
----               carry 4 (tens).
4368    Tens:      6 × 2 = 12; add carried 4
                   for 16; write 6 in the tens
                   place of the product; carry
                   1 (hundred).
        Hundreds:  6 × 7 = 42; add carried 1
                   for 43; write 4 and 3 in the
                   thousands and hundreds
                   places of the product, respectively.
```

In expanded notation:

```
 728     700 +  20 +  8
  ×6  =  ×              6
----    ----------------
4368    4200 + 120 + 48
```

When both factors are expressed by multi-digit numerals, it is convenient in computation to regard the factor with the fewer digits as the multiplier. Products (actually *partial products* with respect to the final product) are computed by multiplying each digit of the multiplier with each digit of the multiplicand, making certain that the extreme right digit of each of these partial products is aligned under the multiplying digit. In doing so, we affirm the fact that when the digit in the multiplier represents ones, or tens, or hundreds, the resulting partial product will correspondingly represent ones, or tens, or hundreds. Aligning partial products in this way—sometimes called *indenting*—simplifies the last step: add the partial products for the final product.

EXAMPLES

```
1.   837 ← factor (multiplicand)
     ×26 ← factor (multiplier)
    5022 ← first partial product
   1674  ← second partial product
   21762 ← final product
```

```
2.    3062          3062          3062
      ×403   or     ×403   or     ×403
      9186          9186          9186
     0000         122480        12248
    12248        1233986       1233986
    1233986
```

7. CHECKS FOR MULTIPLICATION

(a) Reverse factor check. One way to check a multiplication computation involving whole numbers is to repeat the multiplication using the *other* factor as the multiplier. Reversing the factors in this way does not affect the value of the product.

EXAMPLE

```
The computation    The check
  384                  53
  ×53                ×384
 1152                 212
1920                 424
20352               159
                    20352
```

(b) Excess-of-nines check. An algebraic proof, which cannot be undertaken here, states that in multiplication the excess of nines in the product is equal to the excess of nines in the product of the excesses of the factors. The following check on multiplication is based upon this fact.

(1.7) *To check multiplication by the excess of nines:*
Step 1. Find the excess of nines for each of the factors.
Step 2. Find the product of the excesses of Step 1.
Step 3. Find the excess of nines for the product of Step 2.
Step 4. Find the excess of nines for the product of the factors.
Step 5. Compare to see that the excess of Step 4 is the same as the excess of Step 3.

EXAMPLES

1. Check the multiplication 32 × 427 = 13,664 by the excess of nines.
   Step 1. The excess of nines for 32 is 5. The excess of nines for 427 is 4.
   Step 2. The product of the excesses of Step 1 is 5 × 4 = 20.
   Step 3. The excess of nines for the product of Step 2 (20) is 2.
   Step 4. The excess of nines for the product of the computation (13,664) is 2.
   Step 5. The excess of nines found in Step 4 (2) is the same as the excess of nines found in Step 3 (2); so the computation checks.
2. Check the multiplication below by the excess of nines.

```
  2437 →  7   (Excess of nines for 2,437)
  ×304 →  7   (Excess of nines for 304)
  9748   49   (Product of the excesses of the factors)
 7311     ↓
740848↘   4   (Excess of nines for product of excesses
       ↘ ↗      of factors)
        4     (Excess of nines for the product of the factors)
```

3. Check the multiplication below by the excess of nines.

```
   5274 → 0
  × 312 → 6
  10548   0
   5274   ↑
 15822    ↓
1645488 → 0
```

(Complete Exercise 1.3)

8. DIVISION

Division may be thought of in two ways:

(a) As a repeated subtraction in which we find the number of times (*quotient*) a given number (*divisor*) can be subtracted from another given number (*dividend*).

EXAMPLE

How many times can 7 be subtracted from 28?
By subtraction:

```
 28
− 7 ← once
 21
− 7 ← twice
 14
− 7 ← three times
  7
− 7 ← four times
  0
```

By division:

```
            4 ← quotient
divisor → 7)28 ← dividend
```

(b) As a way of finding the unknown factor (*quotient*) of a multiplication in which the product (*dividend*) and the other factor (*divisor*) are known.

EXAMPLE

What number when multiplied by 7 yields a product of 56?
As a multiplication:

```
  ? ← unknown factor
× 7 ← known factor
 56 ← product
```

As a division:

```
                  8 ← quotient (unknown factor)
  divisor → 7)56 ← dividend (known product)
(known factor)
```

Division is the most difficult of the whole-number computations. Besides needing to know how to multiply and subtract, the person computing must make guesses at the correct quotient digits. At the left below is the usual procedure for computing the quotient of two whole numbers, explained at the right in terms of repeated subtraction.

The problem: $1504 \div 32 = ?$

First step:

```
     4
32)1504      32)1504
   128        −1280 ← 40 × 32
    22          224
```

Second step:

```
    47
32)1504      32)1504
   128        −1280 ← 40 × 32
    224         224
    224        −224 ←  7 × 32
                       47 × 32
```

In the first step, the 4 in the tens place of the quotient at the left corresponds to the 40 at the right. In the second step, the 7 in the ones place of the quotient at the left corresponds to the 7 of 7 × 32 at the right. The 47 in the quotient at the left means that in 1504 there are 47 of the 32's. Essentially the same thing is shown at the right when 40 of the 32's (or 1280) is subtracted from the dividend (1504), and then 7 more 32's are taken from what is left (224). Thus, in 1504 there are (40 + 7), or 47, increments of 32.

EXAMPLES

```
1.      86
   328)28208   328)  28208
       2624        −26240   80 × 328
        1968         1968
        1968        −1968    6 × 328
                            86 × 328 (= 28208)

2.   352
   18)6336   18)  6336
      54         −5400   300 × 18
       93          936
       90         −900    50 × 18
        36          36
        36         −36     2 × 18
                         352 × 18 (= 6336)
```

9. DIVISION REMAINDERS

There will be many cases in which the divisor is not contained a whole number of times in the dividend. In such cases a final undivided portion of the dividend, called the *remainder*, will result. For instance, in the division $23 \div 7$, the divisor (7) will be contained in the dividend (23) three times (which accounts for 21 of the dividend's 23) with a remainder of 2. Thus

```
                3 ← quotient
divisor → 7)23 ← dividend
            21
             2 ← remainder
```

Depending upon the nature of the problem that gave rise to this division, the remainder may be treated in either of two ways:

(a) As a remaining, undivided portion of whatever the dividend stands for.

EXAMPLE: How many pennies will each of 7 children receive if 23 pennies are available for equal distribution?

ANSWER: $23 \div 7 = 3$ and *remainder* of 2. Hence each child will receive 3 pennies, with 2 pennies left over undistributed.

(b) As the fraction part of the quotient in which the remainder is the numerator and the divisor is the denominator.

EXAMPLE: How many grams of a certain salt will each of 7 chemistry students receive if there were 23 grams available to be distributed equally?

ANSWER: $23 \div 7 = 3\frac{2}{7}$. Hence each student will receive $3\frac{2}{7}$ grams.

The same kind of repeated-subtraction explanation given in Section 8 holds in the case of inexact division, i.e., division involving remainders.

EXAMPLES

1.
```
      82
26)2139        26)2139
   208            2080   80 × 26
    59              59
    52              52    2 × 26
     7               7   82 × 26
```
$2139 \div 26 = 82$ (r 7)

2.
```
     135
64)8671        64)8671
   64             6400   100
   227            2271
   192            1920    30
    351            351
    320            320     5
     31             31   135
```
$8671 \div 64 = 135$ (r 31)

10. CHECKS FOR DIVISION

(a) Multiplication check. Since the quotient in an exact division may be thought of as an unknown factor in a multiplication in which the other factor and the product are known, it follows that the product of the quotient and the divisor should equal the dividend. In cases where there is a remainder, the following is true:

DIVIDEND = (DIVISOR × QUOTIENT) + REMAINDER

EXAMPLES

1.
```
     321          321 ← quotient
23)7383          × 23 ← divisor
   69             963
    48            642
    46           7383 ← dividend
     23
     23
```

2.
```
     173          173 ← quotient
37)6436          × 37 ← divisor
   37            1211
   273           519
   259           6401
    146          + 35 ← remainder
    111          6436 ← dividend
     35
```

3.
```
      142           142 ← quotient
73)10400           × 73 ← divisor
   73               426
   310             994
   292            10366
    180            + 34 ← remainder
    146           10400 ← dividend
     34
```

(b) Excess-of-nines check. By combining the technique of casting out nines with the fact that

DIVIDEND = (DIVISOR × QUOTIENT) + REMAINDER

we get a quick way to check division computations.

(1.8) *To check division by the excess of nines:*

Step 1. *Find the excess of nines for the divisor and quotient.*
Step 2. *Find the product of the excesses of Step 1.*
Step 3. *Add the product of excesses of Step 2 and the excess of nines for the remainder.*
Step 4. *Find the excess of nines for the result of Step 3.*
Step 5. *Find the excess of nines for the dividend.*
Step 6. *Compare to see that the excess of Step 5 is the same as the excess of Step 4.*

EXAMPLES

1. Check the division 6,436 ÷ 37 = 173 (r 35) by the excess of nines.

Step 1. The excess of nines for the divisor (37) is 1. The excess of nines for the quotient (173) is 2.
Step 2. The product of the excesses of Step 1: 1 × 2 = 2.
Step 3. The excess of nines for the remainder (35) is 8; this excess added to the excess of Step 2 is 8 + 2 or 10.
Step 4. The excess of nines for the result of Step 3 (10) is 1.
Step 5. The excess of nines for the dividend (6436) is 1.
Step 6. The excess of nines found in Step 5 is the same as the excess of nines found in Step 4; so the division checks.

2. Check the division below by the excess of nines.

142	Step 1. 73 ⟶ 1; 142 ⟶ 7
73)10400	Step 2. 1 × 7 = 7
73	Step 3 34 ⟶ 7; 7 + 7 = 14
310	Step 4. 14 ⟶ 5
292	Step 5. 10,400 ⟶ 5
180	Step 6. 5 = 5
146	
34	

3. Check the division below by the excess of nines.

421	Step 1. 421 ⟶ 7; 832 ⟶ 4
832)350289	Step 2. 7 × 4 = 28
3328	Step 3. 17 ⟶ 8; 28 + 8 = 36
1748	Step 4. 36 ⟶ 0
1664	Step 5. 350,289 ⟶ 0
849	Step 6. 0 = 0
832	
17	

11. TESTS OF DIVISIBILITY

There is a basic division relationship which can often be used to simplify a division computation: *The value of the quotient remains unchanged when the divisor and dividend are both multiplied or divided by the same non-zero number.* For example:

```
   3
12)36
```

(a) Multiply the divisor (12) and the dividend (36) by 2; the quotient remains unchanged at 3:

```
   3
24)72
```

(b) Multiply divisor and dividend by 5; the quotient remains unchanged at 3:

```
    3
60)180
```

(c) Divide divisor and dividend by 4; the quotient remains unchanged at 3:

```
  3
3)9
```

(d) Divide divisor and dividend by 3; the quotient remains unchanged at 3:

```
   3
4)12
```

and so on.

Thus the quotient in a division such as

```
    11
75)825
   75
    75
    75
```

can also be found after both terms (divisor and dividend) are divided by 25:

75)825 = { 75 ÷ 25 = 3; 825 ÷ 25 = 33 } = 3)33 (quotient 11)

Another illustration of simplifying a division before starting the computation:

3480)36320 = { Divide both terms by 10 and then by 4 } = 348)3632

= 87)908

In determining whether both divisor and dividend are exactly divisible by the same number, the following *Tests of Divisibility* will help. The tests assume that the number is expressed by the standard Hindu-Arabic numeral.

(a) *A number is exactly divisible by 2 if its ones digit is exactly divisible by 2.*

EXAMPLES: 396, 428, 1004 are exactly divisible by 2; 685, 80,081, 397 are not.

(b) *A number is exactly divisible by 3 if the sum of its digits is exactly divisible by 3.*

EXAMPLES: 462 ⟶ 4 + 6 + 2 = 12, 12 is exactly divisible by 3, therefore 462 is exactly divisible by 3; 863 ⟶ 8 + 6 + 3 = 17, 17 is not exactly divisible by 3, so 863 is not exactly divisible by 3.

(c) *A number is exactly divisible by 4 if the number expressed by its tens and ones digits is exactly divisible by 4.*

EXAMPLES: 532 is exactly divisible by 4 because 32 is exactly divisible by 4; so are 904, 796, and 144; however, 805, 766, and 530 are not.

(d) *A number is exactly divisible by 5 if its ones digit is 0 or 5.*

EXAMPLES: 645, 3070, 4000, 385 are exactly divisible by 5; any number whose ones digit is other than 0 or 5 is not exactly divisible by 5.

(e) *A number is exactly divisible by 6 if it is exactly divisible both by 2 and by 3.* [Use tests (a) and (b) above.]

EXAMPLES: 834 is exactly divisible by 2 since its ones digit is exactly divisible by 2; also 834 is exactly divisible by 3 since the sum of its digits (15) is exactly divisible by 3; hence 834 is exactly divisible by 6. So are 7422 and 10,134.

(f) *A number is exactly divisible by 7 if the difference between twice its ones digit and the number expressed by its remaining digits is exactly divisible by 7.*

EXAMPLES: 294 is exactly divisible by 7 because the difference between twice its ones digit ($2 \times 4 = 8$) and the number expressed by its other digits (29) is 21 ($29 - 8 = 21$), which is exactly divisible by 7. On the other hand, 436 is not exactly divisible by 7, since $436 \longrightarrow 43 - (2 \times 6) = 43 - 12 = 31$, and 31 is not exactly divisible by 7. For larger numbers, the test may be repeated until a sufficiently small test number can be had: Is 30,275 exactly divisible by 7? $30{,}275 \longrightarrow 3027 - (2 \times 5) = 3017$; $3017 \longrightarrow 301 - (2 \times 7) = 287$; $287 \longrightarrow 28 - (2 \times 7) = 14$. Since 14 is exactly divisible by 7, then 30,275 is exactly divisible by 7.

(g) *A number is exactly divisible by 8 if the number expressed by its hundreds, tens, and ones digits is exactly divisible by 8.*

EXAMPLES: 7432, 970,816, 864, 47,712 are exactly divisible by 8 since 432, 816, 864, and 712, respectively, are exactly divisible by 8.

(h) *A number is exactly divisible by 9 if its excess of nines is 0* (in other words, if the sum of its digits is exactly divisible by 9).

EXAMPLES: 46,323, 378, 50,049, 368,427,654 are each exactly divisible by 9 since the sum of each number's digits is exactly divisible by 9; 735, 9847, 6,876,824 are not exactly divisible by 9 because "casting out nines" in each case yields a remainder, or excess, other than 0.

(i) *A number is exactly divisible by 10 if its ones digit is 0.*

EXAMPLES: 460, 3840, 4000 are exactly divisible by 10.

(Complete Exercise 1.4)

Name.. Date............................ **EXERCISE 1.1**

Write the standard Hindu-Arabic numeral for the following.

1. Nine thousand, five hundred twenty-seven 9,527

2. Sixteen thousand, four hundred twenty 16,420

3. Seventy-six thousand, three hundred seven 76,307

4. Six hundred thousand, three hundred fifty 600,350

5. Twenty-two million, nine hundred eighty-six thousand, forty-five 22,986,045

Express the following numerals in words.

6. 6,842,375 ....

7. 42,320,750 ....

8. 60,004,326 ....

9. 700,000,000 ....

10. 6,300,000,007 ....

Give the excess of nines for each.

11. 17 .... **12.** 92 .... **13.** 387 .... **14.** 639 .... **15.** 8326 ....

16. 2468 .... **17.** 4682 .... **18.** 6248 .... **19.** 4,273,968 ....

Add, and check by the reverse addend check. When adding downward, place the sum below the line; when adding upward place the sum in the box at the top.

904	2868	4830	18989	155609
20. 364	**21.** 523	**22.** 3842	**23.** 8543	**24.** 49,675
172	946	26	762	7 248
+368	871	900	47	4 267
904	+528	+62	+9637	37
	2868	4830	18989	+94,382
				155609

Add, and check by the withheld addend check at the right of each problem.

25.	**26.**	**27.**	**28.**
572	4384	87,639	4 436
438	1267	4 275	98,721
+675	9583	8 862	52,763
	+4262	476	4 854
		+2 728	+12,588

Add, and to the right of each problem check by the accountant's check.

29.
4675
8932
+ 7521

30.
35,638
12,929
79,584
+ 31,006

31.
523,637
644,395
38,669
+ 249,072

Add, and check by the excess of nines.

32.
412
331
402
+ 142

33.
513
635
919
+ 542

34.
5632
797
6217
428
+ 351

35.
6375
6943
2107
5333
+ 7853

36.
6379
524
8676
321
+ 5842

37.
7463
27,957
8362
57,617
9036
+ 27

38.
126
4971
86,382
47
4362
+ 10,013

39.
4,683,792
8,630,427
5,639,175
4,260,871
+ 6,385,246

40. Below are the three-game scores for a bowling team of four members.

Al	153	201	164
Bob	175	194	162
Cy	183	176	189
Don	206	183	167

(a) Best 3-game set, name and score: ....

(b) Best team game (score): ....

(c) Total of 3 team games: ....

(d) Total of members' total scores: ....

41. Turnstile count at a football game was 6795, which included 426 free admissions, but did not include 135 band members, 18 cheerleaders, and 26 ushers. What was the actual attendance?

....

42. Road distance between Baltimore and Washington is 38 miles, between Baltimore and Philadelphia 99 miles, between Philadelphia and New York 88 miles, between New York and Boston 222 miles. How far is it (a) from Boston to Baltimore via New York and Philadelphia? (b) from Washington to New York via Baltimore and Philadelphia?

(a) .... (b) ....

Name.. Date........................ **EXERCISE 1.2**

Subtract. Check by addition.

1. 63 − 31

2. 65 − 42

3. 84 − 14

4. 59 − 34

5. 89 − 76

6. 79 − 32

7. 68 − 63

8. 53 − 24

9. 32 − 27

10. 41 − 29

11. 82 − 13

12. 65 − 38

13. 70 − 46

14. 78 − 19

15. 268 − 74

16. 369 − 99

17. 808 − 195

18. 734 − 484

19. 365 − 209

20. 308 − 149

21. 609 − 497

22. 302 − 164

23. 830 − 257

24. 700 − 138

25. 525 − 149

26. 666 − 389

27. 842 − 397

28. 422 − 388

29. 836 − 257

30. 8746 − 3524

31. 6372 − 4957

32. 67,942 − 34,985

33. 76,372 − 5796

34. 6,030,381 − 5,182,472

Subtract, and check by subtracting the difference from the minuend.

35. 5743 − 3268

36. 50,003 − 7 621

37. 80,004 − 15,926

Subtract, and check by the excess of nines.

38. 5635 − 4251

39. 86,000 − 47,025

40. 470,007 − 63,729

41. 500,463 − 9 007

42. What number is 867 less than 900? ....

43. How much must be added to 479 to equal 1037? ....

44. How many years are there in the period between 1695 and 1972? ....

45. Al's salary for the year was $9609, of which $2748 was withheld (taxes, social security, etc.). How much spendable income did Al have for the year? ....

46. At the end of a 482-mile automobile trip the mileage indicator (odometer) read 12,567. What must it have read at the start of the trip? ....

47. A company's net worth is the difference between what it "owns" and what it "owes." What is the net worth of a company that owns $240,000 in assets but owes $38,642 in debts? ....

48. This year so far the Titans played before 697,523 fans. How many more are needed to match last year's total of 842,061? ....

49. How much larger is California (156,740 sq mi) than Montana (145,878 sq mi)? ....

50. At the beginning of the month there was $347 in a checking account. During the month checks were drawn in the amounts of $37, $112, $450, and $72; deposits during the month were $142, $74, $116, and $20. What was the balance at the end of the month? ....

51. Readings from an electric meter were:

Jan. 1:	47,368 kilowatt hours
Feb. 1:	51,627
Mar. 1:	54,822
Apr. 1:	57,901
May 1:	60,004

How many kilowatt hours were used between (a) Jan. 1 and May 1? (b) Mar. 1 and May 1? (c) Feb. 1 and Apr. 1?

(a).............. (b).............. (c)..............

Name.. Date............................ **EXERCISE 1.3**

Multiply.

1. 21 × 4

2. 33 × 3

3. 36 × 4

4. 58 × 6

5. 37 × 5

6. 72 × 4

7. 18 × 9

8. 63 × 8

9. 52 × 7

10. 90 × 8

11. 403 × 8

12. 625 × 7

13. 347 × 9

14. 853 × 4

15. 2061 × 6

Multiply, and check by the reverse factor check.

16. 672 × 48

17. 436 × 271

18. 8632 × 56

Multiply, and check by the excess of nines.

19. 562 × 37

20. 526 × 431

21. 477 × 635

22. 5625 × 406

23. 10,907 × 5 043

24. 12,467 × 11,005

25. 3549 × 3737

26. 74,305 × 20,503

27. On a map, a length of 1 inch represents 225 miles. How many miles will 7 inches represent?

....

28. Sound travels at a speed of 332 meters per second. How far will it travel in 1 minute?

....

29. Light travels at a speed of 186,324 miles per second. How many miles will it travel in 3 minutes?

....

30. How far can an automobile run on a tank of gas if the tank holds 19 gallons, and the auto gets 17 miles to the gallon?

....

31. At an average ground speed of 578 miles per hour, how far will an aircraft travel in 7 hours?

....

32. A used-car costs $2742 cash, or may be paid for in 36 monthly payments of $93 each. How much does the car cost on the monthly pay plan?

....

33. A clothier placed an order for 2 dozen shirts at $5.95 per shirt, and 16 dozen pairs of socks at 97¢ a pair. What is the total bill?

....

34. How many minutes are there in one week?

....

35. If a certain type of engine uses 4 gallons of fuel oil per day for each horsepower, how much will it take to run a 375 horsepower engine constantly for 7 days?

....

36. A fruit grower has two fields of fruit trees. In field A there are 46 rows of 17 trees each. In field B there are 38 rows of 19 trees each. Expected yield for field A is 27 bushels per tree; in field B, 29 bushels per tree. Which field is expected to yield the more fruit, and by how many bushels?

....

Name.. Date.............................. **EXERCISE 1.4**

Divide, and express the remainder in parentheses beside the quotient. Check by multiplication.

1.

135
36) 4 9 6 8
36
136
118
188
180
8

2.

390) 2 6 7, 1 6 8

3.

503) 7 2 1, 1 1 7

Divide, and express the remainder in parentheses beside the quotient. Check by the excess of nines.

4.

42) 6 7 8 3

5.

73) 6 7, 2 3 3

6.

97) 2 3, 8 8 6

7.

72) 1 1, 8 0 8

8.

941) 4 2, 6 5 1

9.

345) 1, 6 2 7, 8 9 9

10.

827) 5, 0 0 0, 6 0 0

11.

5004) 2, 7 4 5, 6 1 2

12.

27,597) 1, 4 7 3, 8 6 4

In answering the following, use only the Test of Divisibility.

13. Circle to show the numbers that are exactly divisible by 2:

64 307 512 700 1401 627 5325 8,426,816

14. Circle to show the numbers that are exactly divisible by 3:

87 421 337 8472 403 6,745,212 8,345,101 621,621,126

15. Circle to show the numbers that are exactly divisible by 4:

3642 724 8384 81,000 42,367,300 85,426 8,245,101 14 billion

16. Circle to show the numbers that are exactly divisible by 5:

627 545 38,420 7770 5634 10,101 20,004 84,625

17. Circle to show the numbers that are exactly divisible by 6:

4326 5328 462 8127 1292 4634 4,268,754 62,010,030

18. Circle to show the numbers that are exactly divisible by 7:

441 574 365 832 1904 2532 3535 294,224

19. Circle to show the numbers that are exactly divisible by 8:

820 4320 6746 36,848 81,648 5764 93,808 10,096

20. Circle to show the numbers that are exactly divisible by 9:

4637 8424 79,362 4005 62,597 79,199 627,498 777,777,777

21. What number must be multiplied by 62 to yield a product of 59,706? 963 ....

22. What number divided by 87 yields a quotient of 327 and a remainder of 23? 28472 ....

23. How often will a slot machine pay off in 544 tries if it is geared to pay off three times in every 17 tries? 96 ....

24. If one machine can turn out 9615 parts in 15 days and another can turn out 3924 parts in 12 days, how long will it take both machines running together to turn out 16,456 parts? 17 ....

Write in standard Hindu-Arabic notation.

1. Seventy-two thousand, four hundred eight.

2. One hundred fifteen thousand, three hundred eighty-six.

3. One million, two hundred seven thousand, six hundred fifteen.

4. Seven million, eight.

5. Nine billion, twenty-three million.

Write the excess of nines for these.

6. 15 **7.** 86 **8.** 33

9. 104 **10.** 99 **11.** 763

12. 2471 **13.** 3885 **14.** 2071

Add, and check.

15. 23 +54 **16.** 76 +19 **17.** 67 +49

18. 8 3 +7 **19.** 24 16 +83 **20.** 3 7 19 +12

21. 35 84 96 +42 **22.** 6 7 9 2 +4 **23.** 33 80 43 65 +82

24. 389 +549 **25.** 3281 +2739 **26.** 76,057 +34,697

27. 625 398 +867 **28.** 7389 4678 +5793 **29.** 34,625 17,952 +67,241

30. 346 974 268 +493 **31.** 5246 3271 4002 +3617 **32.** 83,246 72,061 93,777 +82,467

33. 3627 4987 2431 5628 +9130 **34.** 26,384 62,157 389 64,271 + 6385 **35.** 34,271 87 6352 40,138 + 1462

36. Add: 2632; 571; 4679; 3001.

37. Add: 46,278; 3264; 17,829; 52.

Subtract, and check.

38. 68 − 6 **39.** 53 − 7 **40.** 69 −45

41. 95 −67 **42.** 44 −14 **43.** 57 −52

44. 90 −66 **45.** 547 −216 **46.** 387 −139

47. 521 −372 **48.** 683 −479 **49.** 600 −492

50. 803 −675 **51.** 364 − 57 **52.** 6325 −1314

53. 3647 −1239 **54.** 7634 −2817 **55.** 4678 −2099

56. 4000 −3642 **57.** 8361 − 71 **58.** 36,470 −12,352

59. 34,062 −13,741 **60.** 76,342 −17,530 **61.** 82,657 −17,882

62. 42,635 −17,896 **63.** 34,627 − 5773 **64.** 64,205 − 1387

65. Subtract: 36,423 − 17,881

66. Subtract: 39,426 from 40,301

67. Find the difference between 4,326,715 and 3,841,737.

Multiply, and check.

68. 42 × 3 **69.** 63 × 2 **70.** 31 × 4

71. 57 × 6 **72.** 312 × 2 **73.** 413 × 4

74. 865 × 6 **75.** 3241 × 5 **76.** 3265 × 8

77. 50 × 7 **78.** 500 × 4 **79.** 603 × 2

80. 3204×9 **81.** 4002×8 **82.** $13,067 \times 3$

83. 64×32 **84.** 87×56 **85.** 4×24

86. 501×32 **87.** 635×27 **88.** 841×63

89. 423×214 **90.** 267×169 **91.** 308×415

92. 367×428 **93.** 705×412 **94.** 368×205

95. 3800×426 **96.** 5031×207 **97.** 4162×2009

98. Multiply: 62 × 41 × 37.

99. Find the product: 324 × 106 × 81.

100. Multiply: 32 × 63 × 27 × 59 × 0.

Divide, and check.

101. 3)69 **102.** 4)64 **103.** 2)246

104. 5)615 **105.** 7)364 **106.** 4)327

107. 8)452 **108.** 7)367 **109.** 9)450

110. 7)634 **111.** 8)4624 **112.** 4)32,615

113. 23)69 **114.** 47)97 **115.** 38)143

116. 12)324 **117.** 13)456 **118.** 34)671

119. 45)895 **120.** 63)4972 **121.** 38)5641

122. 625)3791 **123.** 343)8264 **124.** 104)30,007

125. Divide: 34,627 ÷ 871.

126. Divide 64,379 by 5021.

127. Divide 428,369 by 3047.

Use the Tests of Divisibility to determine which of the given numbers are divisible . . .

128. by 2: 326; 521; 60,741; 752,017.

129. by 3: 142; 375; 6429; 30,461.

130. by 4: 265; 3426; 47,136; 42,754.

131. by 5: 204; 553; 6245; 874,130.

132. by 6: 2715; 2432; 12,342; 463,014.

133. by 7: 3374; 4263; 3462; 346,381.

134. by 8: 1120; 7464; 2483; 98,308.

135. by 9: 3465; 2795; 46,295; 38,880.

136. A tournament golfer has scores of 72, 67, 69, and 84. What is his total for the tournament?

137. What is the total enrollment for a college in which there are 4326 freshmen, 3462 sophomores, 2794 juniors, and 2867 seniors enrolled?

138. What is the land area of New England if the six states that make up the region have the following individual areas: Connecticut, 4899 sq mi; Vermont, 9278 sq mi; New Hampshire, 9017 sq mi; Maine, 31,040 sq mi; Massachusetts, 7867 sq mi; and Rhode Island, 1058 sq mi?

139. In an election, Alf received 426 votes and Ted received 387 votes. Who won, and by how many votes?

140. How much in cash will a new car cost if the price is $3046 and the trade-in is worth $689?

141. A salesman put 20,046 miles on his car in one year, 8627 miles for pleasure and the rest for business. How much of the mileage was for business?

142. For an 8-month job, an employee earned $6000. Withheld from his pay were $748 for income tax and $316 for other deductions. How much remained of his earnings?

143. Texas has a land area of 263,513 sq mi. How much larger is Texas than all of New England? (See Exercise 138.)

144. A fast reader averages 675 words per minute. At that rate how many words can he read in 8 minutes?

145. How many oranges in 137 dozen oranges?

146. Each month a man earns $1053 and pays $236 of it on his mortgage. What is his annual pay, and how much does he pay on his mortgage each year?

147. Seattle to Los Angeles is 1127 miles. At an average speed of 49 mph, how long will it take to drive between the two cities?

148. On a map 1 in. represents 125 actual miles. What distance on the map would represent 1000 miles?

149. At 568 names per page, how many pages will it take to list 132,500 names?

150. At 7 letters per word, 9 words a line, 53 lines a page, how many words would there be in a total of 268 pages?

2

FRACTIONS

1. DEFINITIONS

Fractions are also numerals. In the same way that the Hindu-Arabic numeral, 50, names the number of states in the United States, the numerals, $\frac{1}{2}, \frac{2}{4}, \frac{3}{6}, \frac{10}{20}$, are four equivalent names for the number that is exactly midway in value between 0 and 1. Although our language is often loose on this point, a fraction is actually a three-part symbol consisting of two numerals and a bar or mark between them. The numeral above the bar is called the *numerator*, and the numeral below the bar is called the *denominator*. Thus $\frac{1}{2}$, $\frac{3}{4}$, $\frac{5}{2}$ are fractions, as are $\pi/3$, x/y, and $2/\sqrt{5}$. In arithmetic we are concerned primarily with numbers whose fraction names can be expressed by whole-number numerators and whole-number denominators. Such numbers belong to a set of numbers called the *rationals*.

Since a number may have many different fraction names, we need some way by which to test whether two fractions name the same or different numbers—that is, whether or not two fractions are equivalent; and if they are not equivalent, which fraction names the larger (or smaller) number.

(2.1) *To test whether two fractions name the same or different numbers; and if different, which names the larger (or smaller) number:*

*Step 1. Multiply the numerator of the first fraction and the denominator of the second fraction.**

Step 2. Multiply the denominator of the first fraction and the numerator of the second fraction.

Step 3.

(a) *If the products of Steps 1 and 2 are equal, then the fractions name the same number.*

(b) *If the product of Step 1 is greater than the product of Step 2, then the first fraction names the larger number.*

(c) *If the product of Step 1 is less than the product of Step 2, then the first fraction names the smaller number.*

EXAMPLES

1. Do $\frac{3}{4}$ and $\frac{6}{8}$ name the same number?
   Step 1. Consider $\frac{3}{4}$ the first and $\frac{6}{8}$ the second fraction. Numerator of first (3) × denominator of second (8) = 24.
   Step 2. Denominator of first (4) × numerator of second (6) = 24.
   Step 3. The products of Steps 1 and 2 are equal: 24 = 24. So $\frac{3}{4}$ and $\frac{6}{8}$ name the same number.
2. Do $\frac{3}{5}$ and $\frac{1}{2}$ name the same number?
   Step 1. $3 \times 2 = 6$
   Step 2. $5 \times 1 = 5$
   Step 3. $6 > 5$ (read: 6 is greater than 5). So $\frac{3}{5}$ names a number that is greater than the number named by $\frac{1}{2}$, i.e., $\frac{3}{5} > \frac{1}{2}$.
3. Is $\frac{3}{4}$ equal to, greater than, or less than $\frac{7}{9}$?
   Step 1. $\frac{3}{4} \searrow \frac{7}{9} \rightarrow 27$
   Step 2. $\frac{3}{4} \nearrow \frac{7}{9}$, $7 \rightarrow 28$
   Step 3. $27 < 28$ (read: 27 is less than 28). So $\frac{3}{4} < \frac{7}{9}$.
4. $\frac{9}{4} ? \frac{15}{7}$
   Steps 1 and 2. $\frac{9}{4} \times \frac{15}{7}$: $15 \rightarrow 60$, $7 \rightarrow 63$
   Step 3. $63 > 60$; so $\frac{9}{4} > \frac{15}{7}$.

* Strictly speaking, we should say: Multiply *the number named by* the numerator of the first fraction and *the number named by* the denominator of the second fraction. However, when the distinction between a number and its symbol is not fundamentally important to the discussion, it is permissible to use the simpler though less precise language.

When the numerator of a fraction expresses a number that is less than that expressed by the denominator, the fraction is called a *proper* fraction. When the reverse is true, when the numerator expresses a number that is greater than or equal to that expressed by the denominator, the fraction is called an *improper* fraction. For the fraction $\frac{n}{d}$, then:

$n < d$: proper (e.g., $\frac{2}{3}, \frac{7}{10}, \frac{93}{94}, \frac{100}{101}$)
$n \geq d$; improper (e.g., $\frac{8}{3}, \frac{14}{11}, \frac{6}{6}, \frac{17}{17}, \frac{9}{2}$)

($\geq$ means "is greater than or equal to"; $\leq$ means "is less than or equal to.")

Numbers named by improper fractions may also be named by numerals said to be in *mixed form*, i.e., by numerals that combine a whole-number numeral and a fraction. For example: $\frac{3}{2} = 1\frac{1}{2}$; $\frac{5}{3} = 1\frac{2}{3}$; $\frac{18}{5} = 3\frac{3}{5}$. The expressions $1\frac{1}{2}$, $1\frac{2}{3}$, $3\frac{3}{5}$ are numerals in mixed form.

A fraction takes on different meanings in different situations. The most common is the part-whole interpretation of the proper fraction. In this the denominator tells how many equal parts there are to the whole, and the numerator tells how many of those parts there are under consideration. Thus $\frac{2}{3}$ may mean "two (numerator) of three (denominator) equal parts of the whole." The whole, expressed by a fraction, would be $\frac{3}{3}$. Other examples:

$\frac{3}{5}$ means 3 of 5 equal parts:

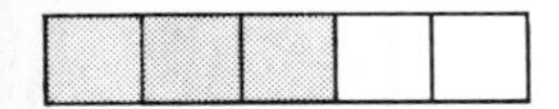

$\frac{1}{4}$ means 1 of 4 equal parts:

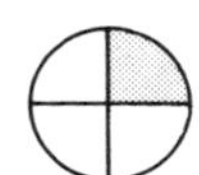

A second interpretation applies equally well to both proper and improper fractions. In this, the fraction is considered to be the quotient of a division in which the numerator is the dividend and the denominator is the divisor. For example:

$\frac{3}{4} = 3 \div 4 = 4\overline{)3}$ $\qquad$ $\frac{7}{2} = 7 \div 2 = 2\overline{)7}$

Still another interpretation is that of *ratio*—a comparison of one number to another. For example, if a boy is 4 ft tall and his mother is 5 ft tall, the boy's height compares to his mother's height in the ratio of 4 to 5 (sometimes written 4 : 5). By associating the "compared" number with the numerator of a fraction and the "compared to" number with the denominator, the ratio 4:5 may be written as $\frac{4}{5}$. It may be said that the boy's height is $\frac{4}{5}$ that of his mother. Conversely, the mother's height compares to her son's height in the ratio of 5 : 4, or $\frac{5}{4}$; it may be said that the mother's height is $\frac{5}{4}$ that of her son.

2. RENAMING FRACTIONS

In computation it is sometimes convenient, sometimes necessary, to rename or re-express a number by an equivalent numeral. Among these renaming procedures are the following.

(a) To write a fraction in lowest terms equivalent to a given fraction. Here the division interpretation of fractions is useful. Since dividing the dividend and the divisor by the same nonzero number does not affect the quotient (Section 11, Chapter 1), it follows that dividing the numerator and the denominator of a fraction by the same number will not affect the value of the resulting fraction.

(2.2) *To write a fraction in lowest terms equivalent to a given fraction, divide the numerator and denominator of the given fraction by all the common factors which exist between the two terms.*

(The Tests of Divisibility developed in Section 11 of Chapter 1 can be useful here.)

EXAMPLES

1. Rename $\frac{6}{9}$ as a fraction in lowest terms.
Divide both 6 and 9 by 3. Result: $\frac{2}{3}$.

2. Express $\frac{36}{60}$ in lowest terms.
Divide both 36 and 60 by 12. Result: $\frac{3}{5}$.
[Note: If it is not recognized immediately that 12 is the greatest common divisor for 36 and 60, the renaming or reducing can be done in stages. For example, should it be recognized at first that 2 will divide both 36 and 60, the fraction can be reduced to $\frac{18}{30}$, and then further reduced by dividing both terms by 2 and then by 3, or by 6:

$$\frac{36}{60} = \frac{\overset{18}{\cancel{36}}}{\underset{30}{\cancel{60}}} = \frac{\overset{9}{\cancel{18}}}{\underset{15}{\cancel{30}}} = \frac{\overset{3}{\cancel{9}}}{\underset{5}{\cancel{15}}} = \frac{3}{5}]$$

(b) To write a fraction in higher terms equivalent to a given fraction. This is the reverse of (a) above and, as we shall see, the necessary first step in adding and subtracting with fractions whose denominators are unequal. Basic is the fact that multiplying the dividend and divisor (equivalently, numerator and denominator) by the same nonzero number does not affect the value of the quotient (equivalently, the resulting fraction). Consequently, it is always possible to produce a fraction equivalent to a given fraction by some appropriate multiplication, so that the new fraction has a desired denominator or desired numerator.*

(2.3) *To write a fraction in higher terms equivalent to a given fraction:*

Step 1. Divide the desired numerator (or denominator) by the numerator (or denominator) of the given fraction.
Step 2. Multiply both numerator and denominator of the given fraction by the quotient obtained in Step 1.

EXAMPLES

1. Write a fraction equivalent to $\frac{1}{4}$ whose denominator is 12.
   Step 1. $12 \div 4 = 3$
   Step 2. $\frac{1}{4} = \frac{3 \times 1}{3 \times 4} = \frac{3}{12}$
2. Write a fraction equivalent to $\frac{2}{3}$ whose numerator is 10.
   Step 1. $10 \div 2 = 5$
   Step 2. $\frac{2}{3} = \frac{5 \times 2}{5 \times 3} = \frac{10}{15}$
3. Express $\frac{8}{7}$ as an equivalent fraction whose denominator is 63.
   Step 1. $63 \div 7 = 9$
   Step 2. $\frac{8}{7} = \frac{9 \times 8}{9 \times 7} = \frac{72}{63}$

Renaming fractions in this way can be a useful alternative to Program 2.1 in comparing the relative values of fractions. When the denominators are equal, the fraction representing the larger number will have the greater numerator. When the numerators are equal, the fraction representing the larger number will have the smaller denominator. When two fractions differ in both numerator and denominator, they can be renamed by Program 2.3 so that their equivalents agree in either numerator or denominator; then a comparison of the other terms will establish which fraction represents the greater number.

EXAMPLES

1. Which is greater: $\frac{3}{5}$ or $\frac{5}{8}$?
   (a) Express both fractions as equivalent fractions with equal denominators, e.g., fortieths:
   $\frac{3}{5} = \frac{?}{40} = \frac{24}{40}$, $\frac{5}{8} = \frac{?}{40} = \frac{25}{40}$
   Since $\frac{25}{40} > \frac{24}{40}$, then $\frac{5}{8} > \frac{3}{5}$.
   (b) Express both fractions as equivalent fractions with equal numerators, e.g., fifteens:
   $\frac{3}{5} = \frac{15}{?} = \frac{15}{25}$, $\frac{5}{8} = \frac{15}{?} = \frac{15}{24}$
   Since $\frac{1}{24} > \frac{1}{25}$, $\frac{15}{24} > \frac{15}{25}$; then $\frac{5}{8} > \frac{3}{5}$.
2. Which is larger: $\frac{7}{8}$ or $\frac{17}{20}$?
   (a) Making denominators equal:
   $\frac{7}{8} = \frac{35}{40}$, $\frac{17}{20} = \frac{34}{40}$
   Since $\frac{35}{40} > \frac{34}{40}$, then $\frac{7}{8} > \frac{17}{20}$.
   (b) Making numerators equal:
   $\frac{7}{8} = \frac{17 \times 7}{17 \times 8} = \frac{119}{136}$
   $\frac{17}{20} = \frac{7 \times 17}{7 \times 20} = \frac{119}{140}$
   Since $\frac{119}{136} > \frac{119}{140}$, then $\frac{7}{8} > \frac{17}{20}$.

(c) To write a numeral in mixed form equivalent to a given improper fraction. A fraction, we have noted, may be thought of as the quotient of numerator divided by denominator; consequently:

(2.4) *To write a numeral in mixed form equivalent to a given improper fraction, divide the numerator by the denominator of the improper fraction, expressing the remainder as the numerator of a fraction whose denominator is the original denominator.*

EXAMPLES

1. Write a numeral in mixed form equivalent to $\frac{19}{12}$.
   $$12\overline{)19} = 1,\quad 19 - 12 = 7$$
   Thus $\frac{19}{12} = 1\frac{7}{12}$.
2. Express $\frac{37}{5}$ equivalently in mixed form.
   $$5\overline{)37} = 7,\quad 37 - 35 = 2$$
   Thus $\frac{37}{5} = 7\frac{2}{5}$.
3. Express $\frac{52}{8}$ as an equivalent numeral in mixed form.
   $$8\overline{)52} = 6,\quad 52 - 48 = 4$$
   Thus $\frac{52}{8} = 6\frac{4}{8} = 6\frac{1}{2}$.

(d) To write an improper fraction equivalent to a given numeral in mixed form. Any number expressed in mixed form (e.g., $4\frac{7}{9}$) may be thought to represent the sum of two addends, one named by the whole-number part (in this case, 4) and the other

* In some cases the result may be a *complex fraction*, which is defined and discussed in Section 10 of this chapter.

named by the fraction part (in this case, $\frac{7}{9}$). Thus:

$$4\tfrac{7}{9} = 4 + \tfrac{7}{9}$$

If we express the whole-number part (4) as a fraction having the same denominator as that of the fraction part (in this case, 9), we have:

$$4\tfrac{7}{9} = 4 + \tfrac{7}{9} = \tfrac{36}{9} + \tfrac{7}{9}$$

By Program 2.6 of the next section, the sum of $\frac{36}{9}$ and $\frac{7}{9}$ may be expressed by the improper fraction $\frac{43}{9}$:

$$4\tfrac{7}{9} = 4 + \tfrac{7}{9} = \tfrac{36}{9} + \tfrac{7}{9} = \tfrac{36+7}{9} = \tfrac{43}{9}$$

All of this may be condensed into the following program.

(2.5) *To write an improper fraction equivalent to a given numeral in mixed form:*
Step 1. Multiply the whole-number part of the numeral in mixed form by the denominator of the fraction part.
Step 2. Add the numerator of the fraction part to the product of Step 1.
Step 3. Express the sum of Step 2 as the numerator of a fraction whose denominator is the denominator of the fraction part.

EXAMPLES

1. Write an improper fraction equivalent to $4\frac{7}{9}$.
   Step 1. Multiply the whole-number part, 4, by the denominator of the fraction part, 9, for a product of 36.
   Step 2. To 36 add 7, the numerator of the fraction part, for 43.
   Step 3. Express the 43 of Step 2 as the numerator of a fraction whose denominator is 9; result $\frac{43}{9}$, an improper fraction equivalent to $4\frac{7}{9}$.
2. Rename $5\frac{2}{3}$ as an equivalent improper fraction.
   Step 1. $3 \times 5 = 15$
   Step 2. $15 + 2 = 17$
   Step 3. $\frac{17}{3}$ $(= 5\frac{2}{3})$
3. Write an improper fraction equivalent to $6\frac{3}{4}$.

$$6\tfrac{3}{4} = 6\ \frac{3}{4} \longrightarrow \frac{(4 \times 6) + 3}{4} = \frac{27}{4}$$

(Complete Exercise 2.1)

3. ADDITION AND SUBTRACTION WITH FRACTIONS

Two fractions are said to be alike if they have equal denominators. Thus $\frac{3}{4}$ and $\frac{1}{4}$ are alike, $\frac{5}{7}$ and $\frac{3}{7}$ are alike, but $\frac{3}{4}$ and $\frac{3}{5}$ are *not* alike, and $\frac{5}{8}$ and $\frac{4}{7}$ are also *not* alike. Procedures for computing sums and differences of numbers expressed as fractions fall into one of two categories—when the fractions are alike (Program 2.6), and when the fractions are not alike (Program 2.7).

(2.6) *To compute the sum (difference) of numbers expressed as fractions having equal denominators:*
Step 1. Add (subtract) the numerators.
Step 2. Express the result of Step 1 as the numerator of a fraction whose denominator is the common or equal denominator.

EXAMPLES

1. Add: $\frac{3}{7} + \frac{2}{7}$.
   Step 1. $3 + 2 = 5$
   Step 2. $\frac{5}{7}$ $(= \frac{3}{7} + \frac{2}{7})$
2. Add: $\frac{8}{21} + \frac{4}{21}$.
   $$\frac{8}{21} + \frac{4}{21} = \frac{8+4}{21} = \frac{12}{21} = \frac{4}{7}$$
3. Subtract: $\frac{9}{11} - \frac{2}{11}$.
   Step 1. $9 - 2 = 7$
   Step 2. $\frac{7}{11}$ $(= \frac{9}{11} - \frac{2}{11})$
4. Simplify: $\frac{5}{24} + \frac{11}{24} - \frac{1}{24}$.
   $$\frac{5}{24} + \frac{11}{24} - \frac{1}{24} = \frac{5 + 11 - 1}{24} = \frac{15}{24} = \frac{5}{8}$$

When the denominators are *not* equal, renaming some or all of the fractions is necessary to make the fractions alike. Once the fractions have been renamed to equivalent fractions having a common denominator, Program 2.6 is applied to find the required sum or difference.

(2.7) *To compute the sum (difference) of numbers expressed as fractions having unequal denominators:*
Step 1. Rename the fractions as necessary so that all have equal denominators.
Step 2. Add (subtract) with the fractions of Step 1 according to Program 2.6.

EXAMPLES

1. Add: $\frac{3}{4} + \frac{1}{12}$.
   Step 1. $\frac{3}{4} = \frac{9}{12}$
   $\frac{1}{12} = \frac{1}{12}$
   Step 2. $\frac{9}{12} + \frac{1}{12} = \frac{10}{12} = \frac{5}{6}$
2. Add: $\frac{3}{5} + \frac{1}{6}$.
   Step 1. $\frac{3}{5} = \frac{18}{30}$
   $\frac{1}{6} = \frac{5}{30}$
   Step 2. $\frac{18}{30} + \frac{5}{30} = \frac{23}{30}$

3. Add: $\frac{3}{4} + \frac{5}{8} + \frac{1}{6}$.

$$\frac{3}{4} + \frac{5}{8} + \frac{1}{6}$$
$$\downarrow \quad \downarrow \quad \downarrow$$
$$\frac{18}{24} + \frac{15}{24} + \frac{4}{24} = \frac{37}{24}$$

4. Subtract: $\frac{3}{4} - \frac{1}{5}$.

Step 1. $\frac{3}{4} = \frac{15}{20}$

$\frac{1}{5} = \frac{4}{20}$

Step 2. $\frac{15}{20} - \frac{4}{20} = \frac{11}{20}$

5. Simplify: $\frac{2}{5} - \frac{1}{4} + \frac{3}{8}$.

$$\frac{2}{5} - \frac{1}{4} + \frac{3}{8}$$
$$\downarrow \quad \downarrow \quad \downarrow$$
$$\frac{16}{40} - \frac{10}{40} + \frac{15}{40}$$
$$= \frac{16 - 10 + 15}{40} = \frac{21}{40}$$

4. LEAST COMMON DENOMINATOR

When the denominators of the several fractions in an addition or subtraction computation are such that a common denominator cannot be readily found by inspection, determining the *least common denominator* (LCD) by either of the two methods given below will usually save considerable work.

(2.8) *To compute the least common denominator by the prime-factor method:*

*Step 1. Express each denominator as a product of prime numbers.**

Step 2. Form a product from the prime-number factors of Step 1 such that each different factor appears in it the number of times it appears in the denominator which contains it most. This product is the least common denominator.

EXAMPLES

1. Compute the least common denominator for $\frac{1}{6}$, $\frac{9}{20}$, $\frac{3}{8}$, $\frac{11}{75}$ by the prime-factor method.

Step 1.

6 in prime factor form is 2×3

20 in prime factor form is $2 \times 2 \times 5$

8 in prime factor form is $2 \times 2 \times 2$

75 in prime factor form is $3 \times 5 \times 5$

Step 2. Factors of the LCD would be: three 2's (as in the 8), one 3 (as in 6 and 75), two 5's (as in 75). Hence, the desired LCD is $2 \times 2 \times 2 \times 3 \times 5 \times 5$, or 600.

2. Find the LCD for $\frac{7}{36}$, $\frac{1}{8}$, $\frac{13}{45}$, $\frac{17}{20}$.

In prime factor form:

36: $2 \times 2 \times 3 \times 3$

8: $2 \times 2 \times 2$

45: $3 \times 3 \times 5$

20: $2 \times 2 \times 5$

LCD: $2 \times 2 \times 2 \times 3 \times 3 \times 5 = 360$

(2.9) *To compute the least common denominator by the continued-division method:*

Step 1. Write the denominators in a horizontal row.

Step 2. Divide the numbers expressed in this row by the smallest whole number greater than one that will divide two or more of the numbers exactly; bring down to a new row all quotients and numbers not exactly divisible.

Step 3. Repeat Step 2 until there are no two numbers expressed in the resulting row which are exactly divisible by any whole number except one.

Step 4. Form a product of the divisors and the remaining terms in the last row for the LCD.

EXAMPLES

1. Compute the least common denominator for $\frac{1}{6}$, $\frac{9}{20}$, $\frac{3}{8}$, $\frac{11}{75}$ by the continued-division method.

Step 2 → 2)6 20 8 75 ← *Step 1*

Step 3 → 2)3 10 4 75

3)3 5 2 75

5)1 5 2 25

1 1 2 5

Step 4. LCD $= 2 \times 2 \times 3 \times 5 \times 2 \times 5$

$= 600$

2. Find the LCD for $\frac{7}{36}$, $\frac{1}{8}$, $\frac{13}{45}$, $\frac{17}{20}$.

2)36 8 45 20

2)18 4 45 10

3) 9 2 45 5

3) 3 2 15 5

5) 1 2 5 5

1 2 1 1

LCD $= 2 \times 2 \times 3 \times 3 \times 5 \times 2 = 360$

5. ADDITION AND SUBTRACTION WITH NUMERALS IN MIXED FORM

As is true of most computational procedures, that for computing sums and differences of numbers expressed in mixed form is a piecemeal operation.

(2.10) *To compute the sum of numbers expressed in mixed form:*

Step 1. Add the whole-number parts of the addends.

* A *prime number* is a whole number greater than one which has for exact whole-number divisors *only* itself and one. The prime numbers less than 100 are: 2, 3, 5, 7, 11, 13, 17, 19, 23, 29, 31, 37, 41, 43, 47, 53, 59, 61, 67, 71, 73, 79, 83, 89, 97.

Step 2. Add the fraction parts of the addends. If this results in an improper fraction, express it in mixed form (Program 2.4).
Step 3. Add the results of Steps 1 and 2.

EXAMPLES

1. Add: $3\frac{3}{8} + 4\frac{1}{8}$.
   Step 1. $3 + 4 = 7$
   Step 2. $\frac{3}{8} + \frac{1}{8} = \frac{4}{8} = \frac{1}{2}$
   Step 3. $7 + \frac{1}{2} = 7\frac{1}{2}$
   [Note: It is possible, and in some cases advantageous, to express the mixed-form addends as improper fractions and add. Thus:
   $3\frac{3}{8} + 4\frac{1}{8} = \frac{27}{8} + \frac{33}{8} = \frac{60}{8} = \frac{15}{2} = 7\frac{1}{2}$.]
2. Add: $15\frac{3}{5} + 12\frac{2}{3}$.
   Step 1. $15 + 12 = 27$
   Step 2. $\frac{3}{5} + \frac{2}{3} = \frac{9}{15} + \frac{10}{15} = \frac{19}{15} = 1\frac{4}{15}$
   Step 3. $27 + 1\frac{4}{15} = 28\frac{4}{15}$
3. Add: $16\frac{3}{8} + 12\frac{5}{6} + 37\frac{2}{9}$.
   Step 1. $16 + 12 + 37 = 65$
   Step 2. $\frac{3}{8} + \frac{5}{6} + \frac{2}{9}$ (LCD = 72) $= \frac{27}{72} + \frac{60}{72} + \frac{16}{72} = \frac{103}{72} = 1\frac{31}{72}$
   Step 3. $65 + 1\frac{31}{72} = 66\frac{31}{72}$
   Often it is more convenient to arrange mixed-form addends in a column:
   $$\begin{aligned} 16\tfrac{3}{8} &= 16\tfrac{27}{72} \\ 12\tfrac{5}{6} &= 12\tfrac{60}{72} \\ +37\tfrac{2}{9} &= 37\tfrac{16}{72} \\ & \overline{65\tfrac{103}{72}} = 65 + \tfrac{103}{72} \\ &= 65 + 1\tfrac{31}{72} = 66\tfrac{31}{72} \end{aligned}$$

(2.11) *To compute the difference of numbers expressed in mixed form:*
Step 1. Replace the fraction parts of the minuend and subtrahend by equivalent fractions having equal denominators.
Step 2. Subtract the fraction part of the subtrahend from the fraction part of the minuend. If the fraction part of the minuend is less than the fraction part of the subtrahend, reduce the whole-number part of the minuend by one and add it as a fraction to the fraction part of the minuend.
Step 3. Subtract the whole-number parts.
Step 4. Add the results of Step 2 and Step 3.

EXAMPLES

1. Subtract: $18\frac{3}{8} - 12\frac{1}{8}$.
   Step 2. $\frac{3}{8} - \frac{1}{8} = \frac{2}{8} = \frac{1}{4}$
   Step 3. $18 - 12 = 6$
   Step 4. $6 + \frac{1}{4} = 6\frac{1}{4}$
   or
   $$\begin{array}{r} 18\tfrac{3}{8} \\ -12\tfrac{1}{8} \\ \hline 6\tfrac{2}{8} = 6\tfrac{1}{4} \end{array}$$
2. Subtract: $36\frac{1}{8} - 14\frac{5}{8}$
   $$\begin{aligned} 36\tfrac{1}{8} &= 36 + \tfrac{1}{8} = 35 + 1 + \tfrac{1}{8} \\ &= 35 + \tfrac{8}{8} + \tfrac{1}{8} = 35 + \tfrac{9}{8} = 35\tfrac{9}{8} \\ -14\tfrac{5}{8} &= \qquad -14\tfrac{5}{8} \\ & \qquad \overline{21\tfrac{4}{8}} = 21\tfrac{1}{2} \end{aligned}$$
3. Subtract: $173\frac{1}{3} - 76\frac{1}{2}$.
   $$\begin{aligned} 173\tfrac{1}{3} &= 173\tfrac{2}{6} = 172\tfrac{8}{6} \\ -76\tfrac{1}{2} &= -76\tfrac{3}{6} = -76\tfrac{3}{6} \\ & \qquad\qquad \overline{96\tfrac{5}{6}} \end{aligned}$$
4. Subtract: $425\frac{1}{6} - 321\frac{7}{10}$.
   $$\begin{aligned} 425\tfrac{1}{6} &= 425\tfrac{5}{30} = 424\tfrac{35}{30} \\ -321\tfrac{7}{10} &= -321\tfrac{21}{30} = -321\tfrac{21}{30} \\ & \qquad\qquad \overline{103\tfrac{14}{30}} = 103\tfrac{7}{15} \end{aligned}$$
5. Subtract: $43 - 26\frac{3}{4}$.
   $$\begin{aligned} 43 &= 42\tfrac{4}{4} \\ -26\tfrac{3}{4} &= -26\tfrac{3}{4} \\ & \quad \overline{16\tfrac{1}{4}} \end{aligned}$$

(Complete Exercise 2.2)

6. MULTIPLICATION WITH FRACTIONS

Computing products of numbers expressed as fractions is simple and straightforward.

(2.12) *To compute the product of numbers expressed as fractions:*
Step 1. Multiply the numerators for the numerator of the product.
Step 2. Multiply the denominators for the denominator of the product.
Step 3. (Optional, but usual.) Express the resulting product fraction in lowest terms.

EXAMPLES

1. Compute the product: $\frac{3}{5} \times \frac{7}{11} \times \frac{1}{2}$.
   Step 1. $3 \times 7 \times 1 = 21$
   Step 2. $5 \times 11 \times 2 = 110$
   Step 3. $\frac{21}{110}$ (already in lowest terms)
2. Compute the product: $\frac{3}{4} \times \frac{2}{9} \times \frac{1}{5}$.
   Step 1. $3 \times 2 \times 1 = 6$
   Step 2. $4 \times 9 \times 5 = 180$
   Step 3. $\frac{6}{180} = \frac{1}{30}$
3. Compute the product: $\frac{3}{7} \times \frac{5}{6} \times \frac{4}{15}$.
   $$\frac{3}{7} \times \frac{5}{6} \times \frac{4}{15} = \frac{3 \times 5 \times 4}{7 \times 6 \times 15} = \frac{60}{630} = \frac{2}{21}$$
4. Compute the product: $\frac{3}{4} \times 8 \times \frac{2}{3} \times 5$.
   $$\frac{3}{4} \times 8 \times \frac{2}{3} \times 5 = \frac{3}{4} \times \frac{8}{1} \times \frac{2}{3} \times \frac{5}{1} = \frac{3 \times 8 \times 2 \times 5}{4 \times 1 \times 3 \times 1} = \frac{240}{12} = 20$$

7. REDUCING PRODUCTS

In multiplication we have a technique which anticipates the final step of Program 2.12 (Express the resulting product in lowest terms). By dividing out all common factors that exist between numerators and denominators, the ultimate product is found more quickly and with simpler computations. For instance, in Example 2 of Section 6 above, the product could have been computed more easily had we divided out a 2 and 3 *before* performing Steps 1 and 2, instead of dividing out the 6 in Step 3. Thus:

$$\frac{\overset{1}{\cancel{3}}}{\underset{2}{\cancel{4}}} \times \frac{\overset{1}{\cancel{2}}}{\underset{3}{\cancel{9}}} \times \frac{1}{5} = \frac{1 \times 1 \times 1}{2 \times 3 \times 5} = \frac{1}{30}$$

In Example 3 the final product, $\frac{2}{21}$, could also have been computed more easily by dividing out common factors before multiplying. Actually if *all* common whole-number factors are divided out before the multiplication takes place, the final step will be unnecessary because the final product *must* be in lowest terms, since all common factors between numerator and denominator will have been eliminated. Thus:

$$\frac{\overset{1}{\cancel{3}}}{7} \times \frac{\overset{1}{\cancel{5}}}{\underset{\underset{1}{\cancel{2}}}{\cancel{6}}} \times \frac{\overset{2}{\cancel{4}}}{\underset{3}{\cancel{15}}} = \frac{1 \times 1 \times 2}{7 \times 1 \times 3} = \frac{2}{21}$$

Similarly for Example 4:

$$\frac{\overset{1}{\cancel{3}}}{\underset{1}{\cancel{4}}} \times \overset{2}{\cancel{8}} \times \frac{2}{\underset{1}{\cancel{3}}} \times 5 = \frac{1 \times 2 \times 2 \times 5}{1 \times 1} = \frac{20}{1} = 20$$

8. DIVISION WITH FRACTIONS

Division of numbers expressed as fractions is handled most conveniently by transforming the computation into an equivalent one of another type. There are two possibilities:

(a) Replace the division by an equivalent division of two whole numbers. This is known as the *common-denominator method.*

(b) Replace the division by an equivalent multiplication with two fractions. This is known as the *inverted-divisor method.*

Although we shall explain both, the reader is advised to continue to compute by the method more familiar to him and to accept the other as information.

(a) Common-Denominator Method. Multiplying the divisor and dividend of a division by the same nonzero number does not affect the value of the quotient (Section 11, Chapter 1). When the divisor and/or dividend are expressed as fractions, multiplying both terms by their LCD results in an equivalent division computation (i.e., same quotient), but one in which the divisor and dividend are whole numbers.

(2.13) *To compute the quotient of two numbers expressed as fractions by the common-denominator method:*

Step 1. *Change the divisor and dividend to whole numbers by multiplying each by the* LCD *of the given divisor and dividend.*

Step 2. *Divide as with whole numbers for the desired quotient.*

EXAMPLES

1. Divide $\frac{3}{8}$ by $\frac{1}{2}$ (or $\frac{3}{8} \div \frac{1}{2}$).
Step 1. LCD of $\frac{3}{8}$ and $\frac{1}{2} = 8$:
$$\left(\cancel{8} \times \frac{3}{\cancel{8}}\right) \div \left(\overset{4}{\cancel{8}} \times \frac{1}{\cancel{2}}\right)$$
Step 2. $\quad 3 \div 4 = \frac{3}{4}$

2. Divide: $\frac{3}{7} \div \frac{5}{8}$
Step 1. LCD $= 56$:
$$\left(\overset{8}{\cancel{56}} \times \frac{3}{\cancel{7}}\right) \div \left(\overset{7}{\cancel{56}} \times \frac{5}{\cancel{8}}\right)$$
Step 2. $\quad 24 \div 35 = \frac{24}{35}$

3. Divide: $\frac{6}{11} \div \frac{15}{22}$.
LCD $= 22$:
$$\left(\overset{2}{\cancel{22}} \times \frac{6}{\cancel{11}}\right) \div \left(\cancel{22} \times \frac{15}{\cancel{22}}\right) = 12 \div 15 = \frac{12}{15} = \frac{4}{5}$$

4. Divide: $6 \div \frac{3}{4}$.
LCD $= 4$:
$$(4 \times 6) \div \left(\cancel{4} \times \frac{3}{\cancel{4}}\right) = 24 \div 3 = 8$$

5. Divide: $\frac{3}{8} \div 12$.
LCD $= 8$:
$$\left(\cancel{8} \times \frac{3}{\cancel{8}}\right) \div (8 \times 12) = 3 \div 96 = \frac{3}{96} = \frac{1}{32}$$

(b) Inverted-Divisor Method. Here, too, we use the fact that multiplying divisor and dividend by the same nonzero number does not affect the value of the quotient. But this time we multiply both terms by the *reciprocal of the divisor.** In this way the new dividend becomes the product of the original dividend and the reciprocal of the original divisor; the new divisor becomes the product of the original divisor and the reciprocal of the original divisor, or *one*. Since division by 1 yields a quotient equal to the dividend, the original division of two fractions has been essentially transformed into an equivalent multiplication of two fractions (the original dividend and the reciprocal of the original divisor), the product of which is the desired quotient.

To compute the quotient for $\frac{3}{8} \div \frac{2}{3}$:
The reciprocal of the divisor, $\frac{2}{3}$, is $\frac{3}{2}$.
Multiply both terms of the division by the reciprocal of the divisor.

$$\left(\frac{3}{2} \times \frac{3}{8}\right) \div \left(\frac{\not{3}}{\not{2}} \times \frac{\not{2}}{\not{3}}\right)$$

$$\downarrow \qquad \downarrow$$

$$\left(\frac{3}{2} \times \frac{3}{8}\right) \div 1 = \left(\frac{3}{2} \times \frac{3}{8}\right) = \frac{9}{16}$$

This illustrates the following program.

(2.14) *To compute the quotient of two numbers expressed as fractions by the inverted-divisor method:*
Step 1. Invert the divisor. (Find its reciprocal.)
Step 2. Multiply the dividend and the inverted divisor for the desired quotient.

EXAMPLES

1. Divide: $\frac{3}{8} \div \frac{1}{2}$.
   Step 1. Reciprocal of $\frac{1}{2}$ is 2.
   Step 2. $\frac{3}{8} \times 2 = \frac{6}{8} = \frac{3}{4}$
2. Divide: $\frac{3}{7} \div \frac{5}{8}$.
   Step 1. Reciprocal of $\frac{5}{8}$ is $\frac{8}{5}$.
   Step 2. $\frac{3}{7} \times \frac{8}{5} = \frac{24}{35}$
3. Divide: $\frac{6}{11} \div \frac{15}{22}$.
   Step 1. Reciprocal of $\frac{15}{22}$ is $\frac{22}{15}$.
   Step 2. $\frac{\overset{2}{\not{6}}}{\not{11}} \times \frac{\overset{2}{\not{22}}}{\underset{5}{\not{15}}} = \frac{4}{5}$
4. Divide: $6 \div \frac{3}{4}$.
   Step 1. Reciprocal of $\frac{3}{4}$ is $\frac{4}{3}$.
   Step 2. $\overset{2}{\not{6}} \times \frac{4}{\not{3}} = 8$
5. Divide: $\frac{3}{8} \div 12$.
   Step 1. Reciprocal of 12 is $\frac{1}{12}$.
   Step 2. $\frac{\not{3}}{8} \times \frac{1}{\underset{4}{\not{12}}} = \frac{1}{32}$

Note: There are characteristic patterns to simple multiplication and division with fractions, which may be contrasted as follows.

Multiplication	Division
$\frac{3}{4} \times \frac{1}{5} = \frac{3}{20}$	$\frac{3}{4} \div \frac{1}{5} = \frac{15}{4} = 3\frac{3}{4}$
$\frac{2}{3} \times \frac{3}{4} = \frac{6}{12} = \frac{1}{2}$	$\frac{2}{3} \div \frac{3}{4} = \frac{8}{9}$
$\frac{4}{5} \times \frac{7}{3} = \frac{28}{15} = 1\frac{13}{15}$	$\frac{4}{5} \div \frac{7}{3} = \frac{12}{35}$

9. MULTIPLICATION AND DIVISION WITH NUMERALS IN MIXED FORM

The usual procedure for computing products and quotients of numbers expressed in mixed form involves a combination of Programs 2.5, 2.12, and 2.14.

(2.15) *To compute the product (quotient) of numbers expressed in mixed form by the improper-fraction method:*
Step 1. Express the numerals in mixed form as equivalent improper fractions.
Step 2. Multiply (divide) as with fractions.

EXAMPLES

1. Multiply: $3\frac{2}{5} \times 8\frac{1}{3}$.
   Step 1. $3\frac{2}{5} = \frac{17}{5}$; $8\frac{1}{3} = \frac{25}{3}$
   Step 2. $\frac{17}{\not{5}} \times \frac{\overset{5}{\not{25}}}{3} = \frac{85}{3} = 28\frac{1}{3}$
2. Multiply: $36\frac{2}{3} \times 18\frac{3}{4}$.
   $\frac{\overset{55}{\not{110}}}{\underset{1}{\not{3}}} \times \frac{\overset{25}{\not{75}}}{\underset{2}{\not{4}}} = \frac{55 \times 25}{1 \times 2} = \frac{1375}{2} = 687\frac{1}{2}$
3. Divide: $6\frac{3}{4} \div 4\frac{1}{2}$.
   Step 1. $\frac{27}{4} \div \frac{9}{2}$

* The *reciprocal* of a given number is that number which when multiplied by the given number yields a product of one. Consequently, the reciprocal of $\frac{2}{3}$ is $\frac{3}{2}$, since $\frac{2}{3} \times \frac{3}{2} = 1$; the reciprocal of 8 is $\frac{1}{8}$; the reciprocal of $\frac{1}{3}$ is 3. Often it is said that the reciprocal of a fraction is produced by "inverting" the fraction—i.e., interchanging numerator and denominator.

$$\text{Step 2. } \frac{\overset{3}{\cancel{27}}}{\underset{2}{\cancel{4}}} \times \frac{\cancel{2}}{\cancel{9}} = \frac{3}{2} = 1\frac{1}{2}$$

4. Divide: $18\frac{2}{3} \div 3\frac{8}{9}$.

$$\frac{56}{3} \div \frac{35}{9} = \frac{\overset{8}{\cancel{56}}}{\cancel{3}} \times \frac{\overset{3}{\cancel{9}}}{\underset{5}{\cancel{35}}} = \frac{24}{5} = 4\frac{4}{5}$$

There is also a four-step procedure for multiplying pairs of numbers expressed in mixed form.

(2.16) *To compute the product of two numbers expressed in mixed form:*

Step 1. Multiply the whole-number parts together.
Step 2. Multiply the fraction parts of each factor with the whole-number part of the other factor.
Step 3. Multiply the fraction parts together.
Step 4. Add the results of Steps 1, 2, and 3.

EXAMPLES

1 Multiply: $3\frac{2}{5} \times 8\frac{1}{3}$.

$$\begin{array}{rll} & 8\frac{1}{3} & \\ \times & 3\frac{2}{5} & \\ \hline & 24 & (3 \times 8) \leftarrow \text{Step 1} \\ & 3\frac{1}{5} & (\frac{2}{5} \times 8) \\ & 1 & (3 \times \frac{1}{3}) \quad \} \text{ Step 2} \\ & \frac{2}{15} & (\frac{2}{5} \times \frac{1}{3}) \leftarrow \text{Step 3} \\ \hline & 28\frac{5}{15} = 28\frac{1}{3} & \leftarrow \text{Step 4} \end{array}$$

2. Multiply: $32\frac{3}{5} \times 75\frac{5}{8}$.

$$\begin{array}{rll} & 75\frac{5}{8} & \\ \times & 32\frac{3}{5} & \\ \hline & 150 & \\ & 225 & \} \ (32 \times 75) \\ & 45 & \} \ (\frac{3}{5} \times 75) \\ & 20 & \} \ (32 \times \frac{5}{8}) \\ & \frac{3}{8} & \} \ (\frac{3}{5} \times \frac{5}{8}) \\ \hline & 2465\frac{3}{8} & \end{array}$$

10. COMPLEX AND COMPOUND FRACTIONS

A *complex fraction* is one in which the numerator or denominator, or both, contain fractions, e.g.,

$$\frac{\frac{2}{3}}{8}, \quad \frac{7}{\frac{1}{2}}, \quad \frac{\frac{3}{4}}{\frac{5}{8}}, \quad \frac{3\frac{1}{2}}{\frac{2}{3}}, \quad \frac{9}{6\frac{3}{4}}$$

Fractions of this type often cause difficulty, though they need not if one remembers that a fraction may be thought of as the quotient of numerator divided by denominator. Consequently, any complex fraction can be simplified by transforming it into an equivalent division.

EXAMPLES

1. $\frac{\frac{2}{3}}{8} = \frac{2}{3} \div 8 = \frac{2}{3} \times \frac{1}{8} = \frac{1}{12}$

2. $\frac{7}{\frac{1}{2}} = 7 \div \frac{1}{2} = 7 \times \frac{2}{1} = 14$

3. $\frac{\frac{3}{4}}{\frac{5}{8}} = \frac{3}{4} \div \frac{5}{8} = \frac{3}{4} \times \frac{8}{5} = \frac{6}{5} = 1\frac{1}{5}$

4. $\frac{3\frac{1}{2}}{\frac{2}{3}} = 3\frac{1}{2} \div \frac{2}{3} = \frac{7}{2} \times \frac{3}{2} = \frac{21}{4} = 5\frac{1}{4}$

5. $\frac{9}{6\frac{3}{4}} = 9 \div 6\frac{3}{4} = 9 \times \frac{4}{27} = \frac{4}{3} = 1\frac{1}{3}$

A *compound fraction* is one in which the numerator or denominator, or both, involve symbols of operation, e.g.,

$$\frac{3+2}{7}, \quad \frac{6(2+4)}{8\frac{1}{2}}, \quad \frac{4}{6-2}, \quad \frac{\frac{1}{2}}{\frac{1}{4} \div \frac{3}{5}}, \quad \frac{6 \times \frac{1}{4}}{3 - \frac{1}{2}}$$

Compound fractions may be simplified by first carrying out the symbolized operations in the numerator and/or denominator, then, if necessary, further simplified by one or more of the many renaming procedures.

EXAMPLES

1. $\frac{3+2}{7} = \frac{5}{7}$

2. $\frac{6(2+4)}{8\frac{1}{2}} = \frac{6(6)}{8\frac{1}{2}} = \frac{36}{8\frac{1}{2}} = \frac{36}{\frac{17}{2}}$
$= 36 \div \frac{17}{2} = 36 \times \frac{2}{17} = \frac{72}{17} = 4\frac{4}{17}$

3. $\frac{4}{6-2} = \frac{4}{4} = 1$

4. $\frac{\frac{1}{2}}{\frac{1}{4} \div \frac{3}{5}} = \frac{\frac{1}{2}}{\frac{1}{4} \times \frac{5}{3}} = \frac{\frac{1}{2}}{\frac{5}{12}}$
$= \frac{1}{2} \div \frac{5}{12} = \frac{1}{2} \times \frac{12}{5} = \frac{6}{5} = 1\frac{1}{5}$

5. $\frac{6 \times \frac{1}{4}}{3 - \frac{1}{2}} = \frac{\frac{3}{2}}{2\frac{1}{2}} = \frac{\frac{3}{2}}{\frac{5}{2}} = \frac{2 \times \frac{3}{2}}{2 \times \frac{5}{2}} = \frac{3}{5}$

6. $\frac{2-1}{1 + \frac{3}{2 - \frac{1}{2}}} = \frac{1}{1 + \frac{3}{\frac{3}{2}}}$
$= \frac{1}{1 + (3 \times \frac{2}{3})} = \frac{1}{1+2} = \frac{1}{3}$

(Complete Exercise 2.3)

Name.. Date.............................. **EXERCISE 2.1**

Insert the appropriate symbol for *is less than* (<), *is greater than* (>), *equals* (=) between each pair of fractions.

1. $\frac{3}{5} = \frac{9}{15}$ **2.** $\frac{5}{8} < \frac{7}{8}$ **3.** $\frac{3}{5} > \frac{6}{11}$ **4.** $\frac{1}{2} < \frac{17}{33}$

5. $\frac{14}{35} > \frac{4}{10}$ **6.** $\frac{7}{10} > \frac{7}{11}$ **7.** $\frac{7}{8} < \frac{9}{10}$ **8.** $\frac{5}{3} = \frac{20}{12}$

9. $\frac{7}{4} > \frac{8}{5}$ **10.** $\frac{16}{17} < \frac{14}{15}$ **11.** $\frac{17}{16} > \frac{15}{14}$ **12.** $\frac{35}{17} > \frac{37}{19}$

Write equivalent fractions in lowest terms.

13. $\frac{24}{36} = \frac{2}{3}$ **14.** $\frac{6}{8} = \frac{3}{4}$ **15.** $\frac{36}{60} = \frac{2}{5}$ **16.** $\frac{28}{56} = \frac{3}{5}$

17. $\frac{49}{14} = \frac{7}{2}$ **18.** $\frac{120}{600} = \frac{2}{15}$ **19.** $\frac{84}{70} = \frac{6}{5}$ **20.** $\frac{63}{81} = \frac{7}{9}$

21. $\frac{750}{2400} = \frac{5}{16}$ **22.** $\frac{720}{320} = \frac{9}{4}$ **23.** $\frac{186}{378} = \frac{31}{63}$ **24.** $\frac{680}{2520} =$

Write equivalent fractions having denominators or numerators as indicated.

25. $\frac{2}{3} = \frac{4}{6}$ **26.** $\frac{3}{5} = \frac{60}{100}$ **27.** $\frac{7}{8} = \frac{\quad}{240}$ **28.** $\frac{3}{31} = \frac{\quad}{93}$

29. $\frac{5}{4} = \frac{20}{16}$ **30.** $\frac{11}{7} = \frac{132}{84}$ **31.** $\frac{15}{40} = \frac{9}{24}$ **32.** $\frac{6}{39} = \frac{\quad}{26}$

33. $\frac{2}{3} = \frac{6}{\quad}$ **34.** $\frac{3}{7} = \frac{21}{\quad}$ **35.** $\frac{2}{5} = \frac{14}{\quad}$ **36.** $\frac{4}{9} = \frac{96}{\quad}$

37. $\frac{3}{2} = \frac{75}{50}$ **38.** $\frac{18}{71} = \frac{54}{\quad}$ **39.** $\frac{4}{14} = \frac{10}{35}$ **40.** $\frac{7}{21} = \frac{4}{12}$

Write equivalent numerals in mixed form.

41. $\frac{17}{5} =$ **42.** $\frac{28}{9} =$ **43.** $\frac{32}{7} =$ **44.** $\frac{47}{5} =$

45. $\frac{301}{150} =$ **46.** $\frac{86}{22} =$ **47.** $\frac{408}{125} =$ **48.** $\frac{1132}{75} =$

Write equivalent improper fractions in lowest terms.

49. $6\frac{2}{3} =$ **50.** $7\frac{5}{8} =$ **51.** $4\frac{2}{7} =$ **52.** $12\frac{1}{4} =$

53. $37\frac{2}{3} =$ **54.** $81\frac{2}{9} =$ **55.** $40\frac{3}{7} =$ **56.** $302\frac{4}{5} =$

Arrange each of the following sets of fractions in order, from least to greatest, left to right.

57. $\frac{5}{16}, \frac{1}{4}, \frac{3}{8}, \frac{3}{16}$ ...., ..., ..., ...

58. $\frac{5}{6}, \frac{2}{3}, \frac{1}{2}, \frac{1}{6}$ ...., ..., ..., ...

59. $\frac{11}{12}, \frac{5}{6}, \frac{7}{8}, \frac{2}{3}$ ...., ..., ..., ...

60. $\frac{31}{7}, \frac{9}{2}, \frac{22}{5}, \frac{17}{4}$ ...., ..., ..., ...

61. $\frac{2}{9}, \frac{3}{8}, \frac{3}{10}, \frac{2}{7}$ ...., ..., ..., ...

62. $\frac{9}{2}, \frac{21}{5}, \frac{13}{3}, \frac{25}{6}$ ...., ..., ..., ...

63. Write a fraction in lowest terms that is:

(a) twice the size of $\frac{6}{7}$ ....

(b) two larger than $\frac{6}{7}$ ....

(c) half the size of $\frac{3}{8}$ ....

(d) twice the size of $\frac{3}{8}$ ....

Name.. Date.............................. **EXERCISE 2.2**

Find the least common denominator for each set of fractions by the prime-factor method.

1. $\frac{7}{20}, \frac{5}{16}, \frac{11}{30}, \frac{1}{24}$ ....

2. $\frac{1}{30}, \frac{17}{45}, \frac{4}{75}, \frac{8}{81}$ ....

Find the least common denominator for each set of fractions by the continued-division method.

3. $\frac{3}{20}, \frac{1}{16}, \frac{7}{30}, \frac{5}{24}$ ....

4. $\frac{3}{32}, \frac{5}{42}, \frac{19}{48}, \frac{1}{52}$ ....

Add, and express sums in lowest terms.

5. $\frac{3}{8} + \frac{1}{3} =$

6. $\frac{1}{2} + \frac{3}{8} =$

7. $\frac{7}{8} + \frac{1}{6} =$

8. $\frac{1}{3} + \frac{1}{4} + \frac{1}{6} =$

9. $\frac{5}{8} + \frac{1}{6} + \frac{7}{12} =$

10. $\frac{5}{6} + \frac{3}{10} + \frac{1}{2} =$

11. $\frac{9}{35} + \frac{3}{28} + \frac{7}{45} + \frac{5}{36} =$

Subtract, and express differences in lowest terms.

12. $\frac{3}{5} - \frac{1}{4} =$

13. $\frac{3}{4} - \frac{1}{8} =$

14. $\frac{13}{80} - \frac{1}{16} =$

15. $\frac{17}{18} - \frac{20}{27} =$

16. $\frac{3}{11} - \frac{4}{19} =$

17. $\frac{19}{35} - \frac{17}{42} =$

Add, and express sums in simplest terms.

18. $\begin{array}{r} 8\frac{1}{4} \\ 17\frac{3}{8} \\ +\ 9\frac{5}{12} \\ \hline \end{array}$

19. $\begin{array}{r} 14\frac{1}{3} \\ 17\frac{5}{6} \\ +21\frac{1}{12} \\ \hline \end{array}$

20. $\begin{array}{r} 62\frac{1}{8} \\ 27\frac{13}{16} \\ +29\frac{3}{4} \\ \hline \end{array}$

21. $\begin{array}{r} 47\frac{7}{10} \\ 86\frac{19}{100} \\ +37\frac{3}{4} \\ \hline \end{array}$

22. $\begin{array}{r} 19\frac{17}{45} \\ 64 \\ +37\frac{8}{27} \\ \hline \end{array}$

23. $\begin{array}{r} 44\frac{15}{44} \\ 36\frac{11}{36} \\ 18\frac{7}{8} \\ +\ 2\frac{1}{2} \\ \hline \end{array}$

Subtract, and express differences in simplest terms.

24. $\begin{array}{r} 18\frac{2}{3} \\ -12\frac{1}{2} \\ \hline \end{array}$

25. $\begin{array}{r} 17\frac{3}{4} \\ -12\frac{1}{3} \\ \hline \end{array}$

26. $\begin{array}{r} 639\frac{3}{5} \\ -428\frac{3}{8} \\ \hline \end{array}$

27. $\begin{array}{r} 74\frac{3}{8} \\ -42\frac{3}{4} \\ \hline \end{array}$

28. $\begin{array}{r} 40\frac{1}{5} \\ -17\frac{19}{20} \\ \hline \end{array}$

29. $\begin{array}{r} 600\frac{3}{7} \\ -199\frac{5}{8} \\ \hline \end{array}$

30. $\begin{array}{r} 36 \\ -14\frac{13}{16} \\ \hline \end{array}$

31. $\begin{array}{r} 5000 \\ -\ 267\frac{17}{113} \\ \hline \end{array}$

Compute, and express results in simplest terms.

32. $\frac{3}{5} + 6\frac{3}{7} - \frac{18}{35} =$

33. $\frac{3}{4} - \frac{5}{8} + \frac{1}{2} =$

34. $\frac{9}{4} - \frac{3}{5} - \frac{11}{20} =$

35. $3\frac{2}{5} - 1\frac{2}{3} + \frac{13}{15} =$

36. $7\frac{1}{4} - 2\frac{1}{3} - 4\frac{3}{8} =$

37. What must be added to $\frac{3}{8}$ to make $\frac{2}{3}$?

....

38. How much larger is $6\frac{2}{3}$ than $4\frac{3}{4}$?

....

39. An instructor marks $\frac{1}{3}$ of a set of test papers before lunch and $\frac{3}{5}$ of them after lunch. What part of them still remains to be corrected?

....

40. Four boards of thicknesses $\frac{5}{8}$ in., $\frac{15}{32}$ in., $\frac{3}{4}$ in., and $\frac{3}{8}$ in. are to be glued together. Allowing $\frac{1}{16}$ in. for glue between each two boards, how thick will the final product be?

....

Name.. Date.............................. **EXERCISE 2.3**

Multiply, and express products in simplest terms.

1. $\frac{3}{4} \times \frac{2}{9} =$

2. $\frac{3}{5} \times \frac{1}{3} =$

3. $\frac{5}{7} \times 2 =$

4. $\frac{3}{25} \times \frac{2}{3} \times 8 =$

5. $\frac{3}{4} \times 2 \times \frac{1}{2} =$

6. $\frac{12}{35} \times \frac{3}{4} \times \frac{7}{10} =$

7. $\frac{3}{4} \times \frac{2}{3} \times 8 =$

8. $\frac{3}{2} \times \frac{5}{7} \times \frac{11}{8} =$

9. $2\frac{1}{3} \times 3\frac{1}{5} =$

10. $8 \times 4\frac{3}{16} =$

11. $2 \times 4\frac{3}{4} \times 0 =$

12. $2\frac{1}{4} \times \frac{3}{4} \times 2\frac{8}{9} =$

Divide, and express quotients in simplest terms.

13. $\frac{3}{4} \div \frac{2}{5} =$ $\frac{15}{8}$ or $1\frac{7}{8}$

14. $\frac{3}{5} \div \frac{4}{5} =$

15. $\frac{3}{4} \div \frac{1}{3} =$

16. $\frac{7}{22} \div \frac{2}{11} =$

17. $\frac{2}{3} \div 6 =$

18. $8 \div \frac{3}{4} =$

19. $\frac{7}{36} \div \frac{14}{27} =$

20. $\frac{24}{7} \div \frac{2}{3} =$

21. $17 \div \frac{34}{35} =$

22. $1\frac{2}{3} \div 6\frac{2}{3} =$

23. $6\frac{3}{4} \div 9\frac{1}{3} =$ $\frac{81}{112}$

24. $4\frac{8}{19} \div 5\frac{1}{4} =$

25. $9\frac{1}{3} \div 4\frac{2}{3} =$

26. $6\frac{3}{5} \div 11 =$

27. $51 \div 3\frac{2}{5} =$

Multiply, using the four-step method—Program 2.16.

28. $\begin{array}{r} 24\frac{2}{7} \\ \times 14\frac{1}{5} \\ \hline \end{array}$

29. $\begin{array}{r} 32\frac{3}{5} \\ \times 15\frac{3}{4} \\ \hline \end{array}$

30. $\begin{array}{r} 320\frac{3}{7} \\ \times \ \ 35\frac{1}{4} \\ \hline \end{array}$

31. $\begin{array}{r} 62\frac{3}{5} \\ \times 24\frac{2}{3} \\ \hline \end{array}$

Simplify.

32. $\dfrac{\frac{3}{4}}{7} =$

33. $\dfrac{6}{1\frac{1}{2}} =$

34. $\dfrac{\frac{2}{3}}{\frac{3}{4}} =$

35. $\dfrac{2\frac{1}{5}}{4\frac{2}{5}} =$

36. $\dfrac{11\frac{1}{4}}{3\frac{3}{8}} =$

37. $\dfrac{\frac{1}{\frac{1}{2}}}{4} =$

38. $\dfrac{16 \div (8 \times 3)}{4 + 17 - 6} =$

39. $\dfrac{(24 \div 5) \times 15}{\frac{3}{4}} =$

40. $\dfrac{25 + (10 \times 2\frac{1}{4})}{12\frac{1}{2}} =$

41. $\dfrac{2 - \frac{1}{3}}{3 + \dfrac{1}{1 + \frac{1}{2}}} =$

42. How many $3\frac{1}{4}$'s are there in $5\frac{1}{5}$?

....

43. How many hamburgers, each $\frac{2}{9}$ of a pound, can be made from $2\frac{2}{3}$ lb of ground beef?

12

....

44. If vegetables, when dried, lose $\frac{9}{14}$ of their fresh weight, how much will 126 kg of fresh vegetables weigh when dried?

....

45. What must the dividend be if the divisor is $16\frac{2}{3}$, the quotient $5\frac{1}{4}$, and the remainder $2\frac{1}{2}$?

....

SUPPLEMENTARY EXERCISES

Replace the □ with the symbol for *is less than* (<), *is greater than* (>), or *is equal to* (=).

1. $\frac{1}{2} \square \frac{2}{3}$ **2.** $\frac{3}{5} \square \frac{2}{3}$ **3.** $\frac{5}{6} \square \frac{3}{4}$

4. $\frac{5}{8} \square \frac{7}{11}$ **5.** $\frac{2}{7} \square \frac{4}{15}$ **6.** $\frac{4}{6} \square \frac{14}{21}$

7. $\frac{6}{5} \square \frac{9}{10}$ **8.** $\frac{9}{4} \square \frac{8}{3}$ **9.** $\frac{12}{7} \square \frac{18}{11}$

Change to equivalent fractions in lowest terms.

10. $\frac{55}{77}$ **11.** $\frac{44}{110}$ **12.** $\frac{45}{60}$

13. $\frac{54}{63}$ **14.** $\frac{56}{77}$ **15.** $\frac{21}{98}$

16. $\frac{52}{91}$ **17.** $\frac{138}{174}$ **18.** $\frac{147}{189}$

Change to equivalent fractions with numerator or denominator as indicated.

19. $\frac{2}{3} = \frac{\quad}{12}$ **20.** $\frac{2}{5} = \frac{\quad}{15}$ **21.** $\frac{1}{3} = \frac{8}{\quad}$

22. $\frac{1}{4} = \frac{\quad}{64}$ **23.** $\frac{9}{24} = \frac{3}{\quad}$ **24.** $\frac{7}{10} = \frac{21}{\quad}$

25. $\frac{7}{8} = \frac{\quad}{32}$ **26.** $\frac{10}{14} = \frac{25}{\quad}$ **27.** $\frac{4}{24} = \frac{\quad}{18}$

Change to equivalent numerals in mixed form.

28. $\frac{9}{2}$ **29.** $\frac{10}{7}$ **30.** $\frac{17}{12}$

31. $\frac{27}{6}$ **32.** $\frac{11}{2}$ **33.** $\frac{12}{9}$

34. $\frac{34}{7}$ **35.** $\frac{25}{20}$ **36.** $\frac{24}{8}$

Change to equivalent improper fractions.

37. $2\frac{1}{5}$ **38.** $1\frac{3}{4}$ **39.** $2\frac{3}{7}$

40. $4\frac{1}{3}$ **41.** $3\frac{1}{2}$ **42.** $6\frac{1}{8}$

43. $5\frac{2}{11}$ **44.** $13\frac{1}{4}$ **45.** $82\frac{1}{3}$

Arrange in order, from least to greatest.

46. $\frac{1}{2}, \frac{3}{10}, \frac{7}{10}, \frac{2}{5}$

47. $\frac{2}{3}, \frac{5}{6}, \frac{11}{12}, \frac{1}{2}$

48. $\frac{1}{2}, \frac{3}{4}, \frac{7}{12}, \frac{9}{16}$

49. $\frac{7}{8}, \frac{4}{5}, \frac{2}{3}, \frac{5}{6}$

Add. Express sums in simplest terms.

50. $\frac{5}{6} + \frac{5}{12}$

51. $\frac{3}{4} + \frac{1}{16}$

52. $\frac{3}{8} + \frac{4}{8} + \frac{1}{8}$

53. $\frac{2}{3} + \frac{5}{6} + \frac{3}{4}$

54. $\frac{3}{16} + \frac{2}{8} + \frac{1}{4}$

55. $\frac{7}{24} + \frac{5}{12} + \frac{1}{16}$

56. $\frac{5}{6} + \frac{3}{10} + \frac{5}{16}$

57. $\frac{1}{3} + \frac{3}{8} + \frac{3}{4} + \frac{3}{8}$

58. $\frac{5}{8} + \frac{7}{12} + \frac{3}{16} + \frac{4}{9}$

Subtract. Express differences in simplest terms.

59. $\frac{3}{4} - \frac{2}{5}$ **60.** $\frac{7}{8} - \frac{3}{16}$

61. $\frac{3}{7} - \frac{1}{3}$ **62.** $\frac{2}{3} - \frac{1}{8}$

63. $\frac{2}{3} - \frac{1}{6}$ **64.** $\frac{11}{12} - \frac{3}{4}$

65. $\frac{5}{8} - \frac{9}{20}$ **66.** $\frac{15}{16} - \frac{2}{3}$

67. $\frac{11}{14} - \frac{1}{2}$ **68.** $\frac{11}{12} - \frac{5}{8}$

69. $\frac{3}{4} - \frac{7}{10}$ **70.** $\frac{7}{13} - \frac{1}{3}$

Compute as indicated.

71. $\begin{array}{r} 4\frac{1}{3} \\ +3\frac{1}{2} \\ \hline \end{array}$ **72.** $\begin{array}{r} 6\frac{2}{3} \\ +5\frac{1}{5} \\ \hline \end{array}$ **73.** $\begin{array}{r} 9\frac{3}{5} \\ +4\frac{2}{7} \\ \hline \end{array}$

74. $\begin{array}{r} 5\frac{6}{7} \\ +\ \frac{3}{4} \\ \hline \end{array}$ **75.** $\begin{array}{r} 7\frac{2}{3} \\ +5\frac{3}{4} \\ \hline \end{array}$ **76.** $\begin{array}{r} 7\frac{9}{10} \\ +12\frac{4}{5} \\ \hline \end{array}$

77. $\begin{array}{r} 7\frac{2}{3} \\ -4\frac{1}{3} \\ \hline \end{array}$ **78.** $\begin{array}{r} 9\frac{4}{5} \\ -2\frac{1}{3} \\ \hline \end{array}$ **79.** $\begin{array}{r} 9\frac{2}{7} \\ -\ \frac{3}{4} \\ \hline \end{array}$

80. $\begin{array}{r} 8\frac{2}{5} \\ -1\frac{3}{4} \\ \hline \end{array}$ **81.** $\begin{array}{r} 8 \\ -1\frac{3}{4} \\ \hline \end{array}$ **82.** $\begin{array}{r} 16\frac{1}{5} \\ -\ 2\frac{5}{8} \\ \hline \end{array}$

83. $\begin{array}{r} 3\frac{1}{2} \\ 4\frac{1}{3} \\ +7\frac{1}{2} \\ \hline \end{array}$ **84.** $\begin{array}{r} 1\frac{3}{4} \\ 14\frac{7}{10} \\ +\ 3\frac{2}{5} \\ \hline \end{array}$ **85.** $\begin{array}{r} 4\frac{3}{7} \\ 6\frac{2}{3} \\ +2\frac{1}{2} \\ \hline \end{array}$

86. $7\frac{3}{7} + 6\frac{2}{3} + 4\frac{1}{3}$

87. $14\frac{1}{3} + 1\frac{6}{7} - 1\frac{3}{7}$

88. $7\frac{1}{2} - 1\frac{3}{4} + 5\frac{1}{5} - 2\frac{1}{4}$

89. $5\frac{3}{4} + 4\frac{2}{5} - 3\frac{1}{8} + 14$

90. $19\frac{1}{2} - 3\frac{1}{4} - 2\frac{4}{5} - 3\frac{5}{8}$

Multiply. Simplify products.

91. $\frac{3}{7} \times \frac{1}{5}$ **92.** $\frac{8}{9} \times \frac{3}{4}$

93. $\frac{5}{8} \times \frac{8}{5}$ **94.** $\frac{3}{7} \times \frac{28}{33}$

95. $\frac{21}{22} \times \frac{6}{35}$ **96.** $\frac{3}{5} \times \frac{10}{21}$

97. $\frac{15}{42} \times \frac{14}{125}$ **98.** $\frac{12}{13} \times \frac{34}{30}$

99. $\frac{9}{10} \times \frac{15}{16} \times \frac{2}{3}$ **100.** $\frac{24}{32} \times \frac{24}{81} \times \frac{8}{9}$

101. $2\frac{3}{4} \times \frac{1}{8} \times \frac{5}{6}$ **102.** $1\frac{1}{2} \times \frac{4}{5} \times 3\frac{1}{6}$

103. $2\frac{1}{5} \times 1\frac{1}{4} \times 2\frac{1}{3}$ **104.** $3\frac{1}{7} \times 1\frac{3}{4} \times 1\frac{3}{5}$

105. $6\frac{5}{8} \times 7\frac{1}{3} \times 0$ **106.** $\frac{9}{16} \times 2\frac{5}{8} \times 4\frac{1}{7}$

107. $\begin{array}{r} 34\frac{1}{2} \\ \times\ 2\frac{1}{4} \\ \hline \end{array}$ **108.** $\begin{array}{r} 26\frac{3}{8} \\ \times 16\frac{3}{13} \\ \hline \end{array}$ **109.** $\begin{array}{r} 142\frac{3}{5} \\ \times\ 25\frac{1}{7} \\ \hline \end{array}$

110. $\begin{array}{r} 214\frac{3}{5} \\ \times\ 26\frac{1}{5} \\ \hline \end{array}$ **111.** $\begin{array}{r} 426\frac{3}{4} \\ \times\ 12\frac{3}{4} \\ \hline \end{array}$ **112.** $\begin{array}{r} 16\frac{5}{8} \\ \times 16\frac{5}{8} \\ \hline \end{array}$

Divide. Simplify quotients.

113. $\frac{3}{4} \div \frac{3}{5}$ **114.** $\frac{2}{5} \div \frac{4}{5}$ **115.** $\frac{3}{10} \div \frac{3}{8}$

116. $\frac{11}{12} \div \frac{4}{3}$ **117.** $\frac{7}{8} \div \frac{1}{4}$ **118.** $\frac{2}{5} \div \frac{1}{8}$

119. $\frac{9}{16} \div \frac{3}{4}$ **120.** $\frac{15}{28} \div \frac{5}{14}$ **121.** $\frac{36}{55} \div \frac{42}{75}$

122. $\frac{11}{13} \div \frac{5}{5}$ **123.** $\frac{15}{32} \div \frac{8}{5}$ **124.** $\frac{25}{24} \div \frac{10}{9}$

125. $7 \div \frac{2}{3}$ **126.** $3 \div \frac{4}{9}$ **127.** $30 \div \frac{3}{7}$

128. $5\frac{6}{7} \div 8\frac{1}{5}$ **129.** $2\frac{1}{4} \div \frac{9}{10}$ **130.** $7\frac{1}{4} \div 2\frac{1}{3}$

131. $\frac{8}{9} \div 4$ **132.** $5\frac{1}{9} \div 1\frac{2}{3}$ **133.** $3\frac{1}{16} \div 2\frac{1}{4}$

134. $4\frac{3}{8} \div 7$ **135.** $26 \div 2\frac{1}{6}$ **136.** $24 \div 3\frac{1}{3}$

Simplify.

137. $\dfrac{\frac{2}{9}}{\frac{2}{3}}$ **138.** $\dfrac{\frac{11}{14}}{\frac{5}{21}}$ **139.** $\dfrac{\frac{28}{75}}{\frac{18}{35}}$

140. $\dfrac{5\frac{1}{3}}{2\frac{5}{6}}$ **141.** $\dfrac{24\frac{3}{4}}{17\frac{3}{8}}$ **142.** $\dfrac{100}{7\frac{1}{7}}$

143. $\dfrac{2 - \frac{1}{4}}{3 + \frac{5}{8}}$ **144.** $\dfrac{6 \times \frac{5}{8}}{9 - 1\frac{1}{2}}$

145. A two-piece dress requires $2\frac{1}{4}$ yards of material for one piece and $3\frac{2}{3}$ yards for the other. How much material is needed to complete the dress?

146. How thick a piece of wood will result if you glue three pieces together that are $\frac{1}{4}$ in., $\frac{3}{8}$ in., and $\frac{7}{10}$ in. thick, allowing $\frac{1}{16}$ in. for each layer of glue?

147. If you drive a nail that is $3\frac{1}{2}$ in. long straight into a piece of lumber $2\frac{5}{8}$ in. thick, how much of the nail will stick out the back side?

148. A customer bought $26\frac{3}{8}$ yards of wire from a new roll that was 81 yards long. How much wire is left on the roll?

149. What is the inside diameter of a piece of tubing if the outside diameter is $3\frac{1}{4}$ in. and the metal is $\frac{3}{16}$ in. thick?

150. A man says he budgets $\frac{3}{10}$ of his income for shelter, $\frac{1}{4}$ for food, $\frac{1}{8}$ for clothing, and the rest he spends foolishly. What part of his income does he spend foolishly?

151. How much is a piece of meat that weighs $3\frac{5}{16}$ lb. if the price is $1.28 a pound?

152. At a cruising speed of $16\frac{1}{2}$ mph, how far will a ship travel in $6\frac{2}{5}$ hours?

153. The top speed of a motor boat is $19\frac{1}{2}$ mph. What is its most economical cruising speed if it is $\frac{3}{5}$ top speed?

154. A distant heir is to receive $\frac{1}{2}$ of $\frac{2}{3}$ of $\frac{3}{4}$ of an estate. What part of the estate is he entitled to?

155. How many pieces of rope, each $\frac{3}{4}$ yard long, can be cut from a length of rope 27 yards long?

156. If $3\frac{1}{5}$ gallons of fuel costs $1.76, how much is it a gallon?

157. If the scale of a drawing is $\frac{3}{8}$ in. = 1 ft, how long is a wall that is $3\frac{1}{4}$ in. on the drawing?

158. A driver intends to travel $87\frac{1}{2}$ mi in $2\frac{1}{2}$ hr. What should be his average speed?

159. If 13 floor tiles in a row extend to a length of $107\frac{1}{4}$ in., how wide must each tile be?

160. The volume of a tank is $126\frac{3}{4}$ cu ft. How many gallons will it hold if 1 gal takes up $7\frac{1}{2}$ cu ft?

REVIEW

Add.

1. $\begin{array}{r} 362 \\ 435 \\ 246 \\ +189 \\ \hline \end{array}$ **2.** $\begin{array}{r} 3248 \\ 4639 \\ 5278 \\ +1970 \\ \hline \end{array}$ **3.** $\begin{array}{r} 36,421 \\ 1793 \\ 462 \\ +\ 5834 \\ \hline \end{array}$

4. $\begin{array}{r} 826 \\ -317 \\ \hline \end{array}$ **5.** $\begin{array}{r} 4263 \\ -3147 \\ \hline \end{array}$ **6.** $\begin{array}{r} 8625 \\ -7849 \\ \hline \end{array}$

7. $\begin{array}{r} 4000 \\ -\ 298 \\ \hline \end{array}$ **8.** $\begin{array}{r} 34,622 \\ -33,852 \\ \hline \end{array}$ **9.** $\begin{array}{r} 60,083 \\ -42,199 \\ \hline \end{array}$

Multiply.

10. $\begin{array}{r} 383 \\ \times\ 25 \\ \hline \end{array}$ **11.** $\begin{array}{r} 462 \\ \times\ 68 \\ \hline \end{array}$ **12.** $\begin{array}{r} 430 \\ \times 291 \\ \hline \end{array}$

13. $\begin{array}{r} 5026 \\ \times\ 103 \\ \hline \end{array}$ **14.** $\begin{array}{r} 6247 \\ \times\ 438 \\ \hline \end{array}$ **15.** $\begin{array}{r} 3261 \\ \times\ 597 \\ \hline \end{array}$

Divide.

16. $8\overline{)464}$ **17.** $71\overline{)3657}$ **18.** $42\overline{)8491}$

19. $30\overline{)6040}$ **20.** $77\overline{)8326}$ **21.** $124\overline{)80,465}$

Replace the □ with $<$ (is less than), $>$ (is greater than), or $=$ (is equal to).

22. 436 + 279 + 842 □ 593 + 648 + 316

23. 314 × 21 □ 2465 + 4128

24. 83 × 305 □ 53,640 ÷ 2

25. 394 + 2567 − 287 □ 14 × 14 × 14

26. 29,444 ÷ 34 □ 9477 ÷ 13

3

DECIMALS AND PERCENT

1. DECIMAL NOTATION

Decimals, like fractions, are a class of numerals—names for numbers. Unlike fractions, however, decimals are a direct extension of the system of notation that we use to express whole numbers. The extension is to the right, with the ones placed at the center of the system.

←	Millions	Hundred-thousands	Ten-thousands	Thousands	Hundreds	Tens	**Ones**	Tenths	Hundredths	Thousandths	Ten-thousandths	Hundred-thousandths	→
		3	4	2 ,	7	6	7 .	2	1	4	5		

The *decimal point* (.) in the numeral locates the ones digit, which is immediately to its left. Once we know which digit is the ones digit, the values of the other digits, as well as that of the complete numeral, become known. The decimal in the display above is read: Three hundred forty-two thousand, seven hundred sixty-seven *and* two thousand one hundred forty-five ten-thousandths. The word "and" signifies the location of the decimal point.

Fractions in which the denominators are powers of ten (i.e., 10, 100, 1000, etc.) are readily expressed in equivalent decimal notation. For example:

$\frac{7}{10} = .7$ (both read: "seven *tenths*")
$\frac{7}{100} = .07$ (both read: "seven *hundredths*")
$\frac{7}{1000} = .007$ (both read: "seven *thousandths*")
$\frac{7}{10000} = .0007$ (both read: "seven *ten-thousandths*")

Note that there are as many zeros in the denominator of such fractions as there are digits to the right of the decimal point in its equivalent decimal expression. Thus:

$\frac{37}{100} = .37$
$\frac{429}{1000} = .429$
$\frac{41}{1000} = .041$
$\frac{8}{10000} = .0008$

This also holds true for numbers expressed in mixed form:

$3\frac{63}{100} = \frac{363}{100} = 3.63$
$5\frac{3}{4} = 5\frac{75}{100} = \frac{575}{100} = 5.75$
$63\frac{3}{5} = 63\frac{6}{10} = \frac{636}{10} = 63.6$

2. ADDITION AND SUBTRACTION WITH DECIMALS

Computational procedures with decimals are essentially the same as those for whole numbers. In fact, the major advantage to using decimals is that no fundamentally different rules are needed for computing sums, differences, products, and quotients, as is true when the numbers are expressed as fractions.

To compute sums and differences using decimals, the numerals are arranged so that digits having the same place value fall in the same column. A simple way to promote such an alignment is to see that the decimal points align vertically.

(3.1) *To compute the sum (difference) of numbers expressed as decimals:*

Step 1. *Arrange the decimals so that their decimal points align vertically.*

Step 2. *Add (subtract) as though the numerals represented whole numbers.*

Step 3. *Place a decimal point in the result so that it aligns with the decimal points of the other terms.*

EXAMPLES

1. Add: 4.62 + 3.97 + 14.83 + .62.

$$\begin{array}{r} 4.62 \\ 3.97 \\ 14.83 \\ +\ .62 \\ \hline 24.04 \end{array}$$

2. Add: 83.86 + .003 + 9 + 4.625.

$$\begin{array}{r} 83.86 \\ .003 \\ 9. \\ +\ 4.625 \\ \hline 97.488 \end{array} \qquad \begin{array}{r} 83.860 \\ .003 \\ 9.000 \\ +\ 4.625 \\ \hline 97.488 \end{array}$$

Terminal zeros may be annexed to those digits to the right of the decimal point without changing values.

3. Subtract: 8.637 − 4.971.

$$\begin{array}{r} 8.637 \\ -4.971 \\ \hline 3.666 \end{array}$$

4. Subtract: 8.062 − 4.3.

$$\begin{array}{r} 8.062 \\ -4.3 \\ \hline 3.762 \end{array} \quad \text{or} \quad \begin{array}{r} 8.062 \\ -4.300 \\ \hline 3.762 \end{array}$$

5. Subtract: .0057 − .004.

$$\begin{array}{r} .0057 \\ -.004 \\ \hline .0017 \end{array} \quad \text{or} \quad \begin{array}{r} .0057 \\ -.0040 \\ \hline .0017 \end{array}$$

6. Subtract: 8 − .036.

$$\begin{array}{r} 8. \\ -\ .036 \\ \hline 7.964 \end{array} \quad \text{or} \quad \begin{array}{r} 8.000 \\ -\ .036 \\ \hline 7.964 \end{array}$$

3. MULTIPLICATION WITH DECIMALS

Consider the multiplication examples, 3.4 × .16 and 1.62 × .857, carried out using fraction equivalents for each term. In the first instance:

$$3.4 = \frac{34}{10}; \qquad .16 = \frac{16}{100}$$

$$3.4 \times .16 = \frac{34}{10} \times \frac{16}{100}$$

$$= \frac{34 \times 16}{10 \times 100} = \frac{544}{1000} = .544$$

Similarly, for the second example:

$$1.62 \times .857 = \frac{162}{100} \times \frac{857}{1000}$$

$$= \frac{138{,}834}{100{,}000} = 1.38834$$

Had we ignored the decimal points in these two examples and multiplied as though the numerals represented whole numbers, we would have arrived at products having the same sequence of digits as those found in the respective products above:

$$\begin{array}{r} 3.4 \longrightarrow 34 \\ .16 \longrightarrow 16 \\ \hline 204 \\ 34 \\ \hline 544 \\ \uparrow\ \ \end{array} \qquad \begin{array}{r} .857 \longrightarrow 857 \\ 1.62 \longrightarrow 162 \\ \hline 1714 \\ 5142 \\ 857 \\ \hline 138834 \\ \uparrow\ \ \ \ \ \ \end{array}$$

From the products obtained using fraction notation we know where the decimal points *should* go in the product-numerals (indicated by ↑). By comparing these two attacks on the problems, it appears—and can be substantiated by other arguments and by additional examples—that the position of the decimal point in the product has as many digits to its right as there are *collectively* digits to the right of the decimal points in the factors.

(3.2) *To compute the product of numbers expressed as decimals:*

Step 1. *Ignore the decimal points in the factors and multiply as though the numerals represented whole numbers.*

Step 2. *Locate the decimal point in the product so that there are as many digits to its right as there are collectively digits to the right of the decimal points in the factors.*

EXAMPLES

1. Multiply: .453 × 78.2.

$$\begin{array}{rl} 78.2 & (\ \ 1 \text{ digit right of decimal point}) \\ \times .453 & (+3 \text{ digits right of decimal point}) \\ \hline 2346 & (\ \ 4 \text{ digits collectively}) \\ 3910 & \\ 3128 & \\ \hline 35.\underbrace{4246}_{(4 \text{ digits})} & \end{array}$$

2. Multiply: .12 × .372

$$\begin{array}{rl} .372 & (\ \ 3 \text{ digits right of decimal point}) \\ \times .12 & (+2 \text{ digits right of decimal point}) \\ \hline 744 & (\ \ 5 \text{ digits collectively}) \\ 372 & \\ \hline .\underbrace{04464}_{(5 \text{ digits})} & \end{array}$$

[Note: If there are insufficient digits in the product-numeral, place as many zeros as necessary between the decimal point and the left digit.]

3. Multiply: .0032 × .065.

$$\begin{array}{rl} .065 & (\ \ 3 \text{ digits right of decimal point}) \\ \times .0032 & (+4 \text{ digits right of decimal point}) \\ \hline 130 & (\ \ 7 \text{ digits collectively}) \\ 195 & \\ \hline .0002080 & (\ \ 7 \text{ digits to the right of the decimal point}) \end{array}$$

4. Compute the product: $.3 \times .2 \times 1.2 \times .01$.
   $3 \times 2 \times 12 \times 1 = 72$
   Digits to the right of the decimal point, collectively, in the factors: .3 (one), .2 (one), 1.2 (one), .01 (two); total: five. So
   $.3 \times .2 \times 1.2 \times .01 = \underbrace{.00072}_{5\text{ digits}}$

(Complete Exercise 3.1)

4. DIVISION WITH DECIMALS

When one or both of the terms of a division computation involve decimals, the procedure for finding their quotient rests upon two facts:

(a) The quotient is unaffected when dividend and divisor are multiplied by the same number.

(b) Decimal points of the numerals for the quotient and dividend align vertically in the usual division form when the divisor is a whole number.

By (a), any division computation with a decimal divisor can be converted to an equivalent division with a whole-number divisor by multiplying the original divisor and dividend by an appropriate power of ten. The net effect of such a multiplication is an *apparent movement* of the decimal point to the right in the numerals for each term. For instance, since

$$100 \times 3.62 = 362. = 362$$

and

$$100 \times 29.684 = 2968.4$$

then $3.62\overline{)29.684}$ may be written equivalently as $362\overline{)2968.4}$.

The "before" and "after" state of the respective numerals may be interpreted as the decimal point *moving* the same number of places to the right in both terms.

Before: $3.62\overline{)29.684}$
After: $362\overline{)2968.4}$
$\Big\}$ $3.62_\wedge\overline{)29.68_\wedge4}$

Now that the original division has been replaced by an equivalent division involving a whole number divisor (362) and a new dividend (2968.4), according to (b) above the decimal point in the quotient will line up vertically with the decimal point in the new dividend—and by (a) this equivalent division will have the same quotient as that of the original division. However, instead of rewriting the terms to show the equivalent division, or to show two sets of decimal points in the one computation, a frequent tactic is to insert a caret ($\wedge$) to mark the new location of the decimal point in the dividend.

(3.3) *To compute the quotient of two numbers expressed as decimals:*

Step 1. Count the number of digits or places to the right of the decimal point in the divisor.

Step 2. Count off to the right of the decimal point in the dividend the same number of digits as in Step 1 and set a caret ($\wedge$) to the right of the last digit counted. (Annex zeros if necessary.)

Step 3. Divide as though the numerals represented whole numbers, i.e., ignoring decimal points and carets.

Step 4. Locate the decimal point in the quotient of Step 3 so that it aligns vertically with the caret mark in the dividend.

EXAMPLES

1. Divide: $.23\overline{)3.611}$.
   Step 1. .23 (two digits to the right of the decimal point in the divisor).
   Step 2. In the dividend, 3.611, we place a caret two places to the right of the decimal point: $3.61_\wedge1$.
   Step 3.
```
      157
  23)3611
     23
     131
     115
      161
      161
```
   Step 4. $.23\overline{)3.61_\wedge1}$ with quotient 15.7 (quotient)

2. Divide: $2.627 \div 3.7$.
```
       .71
  3.7)2.6^27
      2 5 9
         37
         37
```
   Since there is one digit to the right of the decimal point in the divisor, the caret in the dividend is located one place to the right of the decimal point.

3. Divide: $8.4 \div .35$.
```
        24.
  .35)8.40^
      7 0
      1 40
      1 40
```
   Since there are two digits to the right of the decimal point in the divisor, the caret in the dividend is located two places to the right of the decimal point—which requires the annexing of one zero.

4. Divide: $29{,}664 \div .36$.
```
        824 00.
  .36)29664.00^
      288
       86
       72
       144
       144
```
   Since there are two digits to the right of the decimal point in the divisor, the caret in the dividend is located two places to the right of the decimal point—which requires the annexing of two zeros.

5. Divide: .005832 ÷ 3.6.

```
      .00162
3.6).0‸05832
      36
      223
      216
        72
        72
```

In this computation the first quotient digit (1) occurs over the 8 of the dividend; zeros must be inserted in the empty places between the decimal point and the first quotient digit.

5. ROUNDING OFF DECIMALS

A decimal is *rounded off* by retaining the digits for a certain number of places, and discarding the rest. The value of the resulting numeral is an approximation of the original. The following program serves as a useful guide.

(3.4) *To round off a decimal:*

Step 1. Discard all digits beyond the last to be retained when they are to the right of the decimal point; replace them with zeros when they are to the left of the decimal point.

Step 2. When the digit immediately to the right of the last retained digit is a 5, 6, 7, 8, or 9, increase the last retained digit by 1.

Step 3. When the digit immediately to the right of the last retained digit is a 0, 1, 2, 3, or 4, the last retained digit is left unchanged.

EXAMPLES

1. Rounded off to *thousands*:
   463,827 becomes 464,000
   52,326.4 becomes 52,000
   1279 becomes 1000
   55,500 becomes 56,000
   499 becomes 0

2. Rounded off to *tens*:
   364 becomes 360
   2365 becomes 2370
   6.35 becomes 10
   34.99 becomes 30
   3.7 becomes 0

3. Rounded off to *tenths*:
   .3499 becomes .3
   37.06 becomes 37.1
   427.02 becomes 427.0
   3.0514 becomes 3.1
   .036 becomes 0

4. Rounded off to *hundredths*:
   .3499 becomes .35
   3.0514 becomes 3.05
   7.645 becomes 7.65
   48.002 becomes 48.00
   .004 becomes 0

5. 839.46072 = 839.4607 to nearest *ten-thousandth*
   = 839.461 to nearest *thousandth*
   = 839.46 to nearest *hundredth*
   = 839.5 to nearest *tenth*
   = 839 to nearest *one*
   = 840 to nearest *ten*
   = 800 to nearest *hundred*
   = 1000 to nearest *thousand*
   = 0 to nearest *ten-thousand*

Frequently a division computation fails to yield an exact quotient. When this is so, the usual practice is to carry the computation to one place beyond that which is desired, so that the quotient-numeral can be properly rounded off.

EXAMPLES

1. Divide 3.652 by 3.4, and express the quotient to the nearest tenth.

```
      1.07 ⟶ 1.1 (quotient)
3.4)3.6‸52
    3 4
      252
      238
       14
```

2. Divide .4637 by .42, and express the quotient to the nearest hundredth.

```
      1.104 ⟶ 1.10 (quotient)
.42).46‸37
     42
      43
      42
      170
      168
        2
```

3. Divide 837.2 by .024, and express the quotient to the nearest hundred.

```
        34 88x. ⟶ 34,900 (quotient)
.024)837.200‸
     72
     117
      96
      21 2
      19 2
       2 00
       1 92
          8
```

6. EQUIVALENT DECIMALS AND FRACTIONS

Particularly in problems of percent, it is often necessary to replace a decimal with an equivalent fraction, and vice versa. Following are two procedures by which these equivalents may be written.

(3.5) *To write a fraction equivalent to a given decimal:*

Step 1. Count the number of digits to the right of the decimal point in the decimal.

Step 2. Write a fraction whose numerator is the decimal numeral without its decimal point, and whose denominator is 1 followed by as many zeros as there are digits in the count of Step 1.

Step 3 (Optional). Express the fraction of Step 2 in lowest terms.

EXAMPLES

1. Write a fraction in lowest terms equivalent to .375.

   Step 1. .375 (three digits to the right of the decimal point)

   Step 2. $\frac{375}{1000}$ ← (three zeros)

   Step 3. $\frac{\overset{3}{\cancel{375}}}{\underset{8}{\cancel{1000}}} = \frac{3}{8}$

2. Express .042 as an equivalent fraction in lowest terms.

   Step 1. .042 (three digits to the right of the decimal point)

   Step 2. $\frac{42}{1000}$

   Step 3. $\frac{\overset{21}{\cancel{42}}}{\underset{500}{\cancel{1000}}} = \frac{21}{500}$

3. Express 3.67 as an equivalent fraction in lowest terms.

   Step 1. 3.67 (two digits to the right of the decimal point)

   Step 2. $\frac{367}{100}$

   Step 3. $\frac{367}{100}$ (or, in mixed form, $3\frac{67}{100}$)

(3.6) *To write a decimal equivalent to a given fraction, divide the numerator by the denominator, using decimal notation.*

EXAMPLES

1. Write a decimal equivalent to $\frac{3}{4}$.

```
   .75 (answer)
 4)3.00
   2 8
     20
     20
```

2. Express $\frac{9}{40}$ as an equivalent decimal.

```
    .225 (answer)
 40)9.000
    8 0
    1 00
      80
     200
     200
```

3. Write a decimal equivalent to $3\frac{1}{4}$.

   One approach:

   Express $3\frac{1}{4}$ as an improper fraction: $\frac{13}{4}$.

   Divide numerator (13) by denominator (4) (Program 3.6):

```
    3.25 (= 3¼)
 4)13.00
   12
    1 0
      8
     20
     20
```

   Another approach:

   Think of $3\frac{1}{4}$ as $3 + \frac{1}{4}$.

   Express $\frac{1}{4}$ decimally by dividing numerator (1) by denominator (4) (Program 3.6):

```
   .25 (= ¼)
 4)1.00
    8
    20
    20
```

   Add the whole-number part (3) to the fraction equivalent (.25):

   $3 + .25 = 3.25$ $(= 3\frac{1}{4})$

4. Express $\frac{2}{3}$ as an equivalent decimal.

```
   .666 ⟶ .67 (answer)
 3)2.000
   1 8
     20
     18
      20
      18
       2
```

[Note: Whenever a fraction *in lowest terms* contains a denominator whose only prime factors* are 2's and/or 5's, the equivalent decimal will eventually terminate—i.e., an exact decimal equivalent can be obtained. If, when the fraction is reduced to lowest terms, the denominator contains any other prime factors besides 2's and/or 5's, the decimal equivalent will never terminate and a rounded-off decimal is the best that can be obtained in that notation.]

(Complete Exercise 3.2)

7. PERCENT

Another important type of numeral is the percent. It expresses ratios of numbers compared to 100. The term percent comes from the Latin, *per cento*, meaning "by the hundred." The symbol for percent is %. Thus:

50% means 50 compared to 100
27% means 27 compared to 100
4% means 4 compared to 100
.3% means .3 compared to 100
121% means 121 compared to 100
$\frac{1}{2}$% means $\frac{1}{2}$ compared to 100

Since fractions express ratios also, and since fractions have decimal equivalents, percent numerals have their equivalents in fraction form as well as in decimal form. Thus:

Percent	*Fraction*	*Decimal*
50%	$= \frac{50}{100} = \frac{1}{2}$	$= .50$
27%	$= \frac{27}{100}$	$= .27$
4%	$= \frac{4}{100} = \frac{1}{25}$	$= .04$
.3%	$= \frac{.3}{100} = \frac{3}{1000}$	$= .003$
121%	$= \frac{121}{100} = 1\frac{21}{100}$	$= 1.21$
$\frac{1}{2}$%	$= \frac{\frac{1}{2}}{100} = \frac{1}{200}$	$= .005$

Problems involving percent arise frequently. Normally in solving them we do not compute with percent expressions, but replace them with equivalent decimals or fractions. Also there will be occasions to replace a decimal or fraction with an equivalent percent. The following procedures may be used to translate numerals in one form to equivalent numerals in another form.

The first two of these procedures (Programs 3.7 and 3.8) make use of the fact that the percent numeral is essentially the same as its equivalent decimal, except that the decimal point has been shifted two places to the right and the percent symbol has been affixed.

For example:

.5 = 50.% (or simply 50%)
.75 = 75.% (or simply 75%)
.625 = 62.5%
1.35 = 135%

(3.7) *To write a percent equivalent to a given decimal:*
Step 1. *Move the decimal point in the given decimal two places to the right (annexing zeros if necessary).*
Step 2. *Affix the percent symbol (%) to the numeral developed in Step 1.*

EXAMPLES

1. Write a percent equivalent to .32.
   Step 1. $.32 \rightarrow 32.$
   Step 2. 32.% (or simply 32%)
2. Write a percent equivalent to .7.
   Step 1. $.7 \rightarrow 70.$ (annex a zero)
   Step 2. 70.% (or simply 70%)
3. Express .06 as a percent.
   $.06 \rightarrow 06. \rightarrow 6\%$
4. Express .4268 as a percent.
   $.4268 \rightarrow 42.68 \rightarrow 42.68\%$
5. Express 43.2 as a percent.
   $43.2 \rightarrow 4320. \rightarrow 4{,}320\%$
6. Express .00037 as a percent.
   $.00037 \rightarrow 00.037 \rightarrow .037\%$

(3.8) *To write a percent equivalent to a given fraction:*
Step 1. *Express the given fraction as a decimal (Program 3.6)*
Step 2. *Express the decimal of Step 1 as a percent (Program 3.7).*

EXAMPLES

1. Write a percent equivalent to $\frac{2}{5}$.
   Step 1. $\frac{2}{5} = 5\overline{)2.0} = .4$ (quotient .4)
   Step 2. $.4 \rightarrow 40. \rightarrow 40\%$ (i.e., $\frac{2}{5} = 40\%$)
2. Write a percent equivalent to $\frac{3}{8}$.
   Step 1. $\frac{3}{8} = 8\overline{)3.000} = .375$ (quotient .375)
   Step 2. $.375 \rightarrow 37.5 \rightarrow 37.5\%$ (i.e., $\frac{3}{8} = 37.5\%$)
3. Express $\frac{9}{160}$ as a percent.
   Step 1. $\frac{9}{160} = 160\overline{)9.00000} = .05625$ (quotient .05625)
   Step 2. $.05625 \rightarrow 05.625 \rightarrow 5.625\%$ (or $5\frac{5}{8}\%$)

* See footnote, page 27.

4. Express $\frac{7}{4}$ as a percent.

Step 1. $\frac{7}{4} = 4\overline{)7.00}^{\,1.75} = 1.75$
Step 2. $1.75 \rightarrow 175. \rightarrow 175\%$

5. Express $\frac{1}{3}$ as a percent.

Step 1. $\frac{1}{3} = 3\overline{)1.000}^{\,.333\ldots}$
$= .333\ldots (= .33\frac{1}{3})$
Step 2. $.333\ldots \rightarrow 33.3\ldots \rightarrow 33.3\ldots\%$
(exactly $33\frac{1}{3}\%$; or less precisely 33.3%; or even less precisely 33%).

In producing a fraction or decimal that is equivalent to a percent, it is well to keep in mind that the percent is essentially a decimal expression with its decimal point located two places to the right of the ones place (i.e., immediately to the right of the hundredths digit).

(3.9) *To write a decimal equivalent to a given percent, drop the percent symbol in the numeral for the given percent and move the decimal point two places to the left (prefixing zeros if necessary).*

EXAMPLES

1. Write a decimal equivalent to 37%.
$37\% \rightarrow 37. \rightarrow .37$ (i.e., $37\% = .37$)

2. Write a decimal equivalent to 2.5%.
$2.5\% \rightarrow 2.5 \rightarrow .025$ (i.e., $2.5\% = .025$)

3. Express .05% as a decimal.
$.05\% \rightarrow .05 \rightarrow .0005$ (i.e., $.05\% = .0005$)

4. Express 326% as a decimal.
$326\% \rightarrow 326. \rightarrow 3.26$ (i.e., $326\% = 3.26$)

5. Express $\frac{1}{2}\%$ as a decimal.
$\frac{1}{2}\% = .00\frac{1}{2} = .005$ (or $\frac{1}{2}\% = .5\% = .005$)

(3.10) *To write a fraction equivalent to a given percent:*
Step 1. Express the given percent as a decimal (Program 3.9).
Step 2. Express the decimal of Step 1 as a common fraction (Program 3.5).

EXAMPLES

1. Write a fraction in lowest terms equivalent to 28%.
Step 1. $28\% = .28$
Step 2. $.28 = \frac{28}{100} = \frac{7}{25}$ (i.e., $28\% = \frac{7}{25}$)

2. Write a fraction in lowest terms equivalent to 20.5%.

Step 1. $20.5\% = .205$
Step 2. $.205 = \frac{205}{1000} = \frac{41}{200}$ (i.e., $20.5\% = \frac{41}{200}$)

3. Express .07% as a fraction.
$.07\% = .0007 = \frac{7}{10000}$

4. Express $\frac{1}{2}\%$ as a fraction.
$\frac{1}{2}\% = .00\frac{1}{2} = .005 = \frac{5}{1000} = \frac{1}{200}$

[Note: An alternative approach proves to be more convenient in some cases. Replace the % symbol with a denominator of 100 (i.e., $\frac{}{100}$) and reduce to lowest terms. Thus:

$28\% = \frac{28}{100} = \frac{7}{25}$
$\frac{1}{2}\% = \frac{\frac{1}{2}}{100} = \frac{1}{200}$
$37\% = \frac{37}{100}$]

(Complete Exercise 3.3)

8. BASIC PERCENTAGE PROBLEMS

The greatest number of percentage problems involve some kind of comparison of a part to the whole. Such comparisons may be expressed by a ratio in the form of 3:4, for example, or in equivalent fraction form, $\frac{3}{4}$, where the numerator 3 represents the part and the denominator 4 represents the whole. (Review Section 1, Chapter 2.) Comparison problems become percentage problems when the ratio fraction (e.g., $\frac{3}{4}$) is expressed as a percent (75%). Thus, if there are 31 male students in a class of 50, the 31 represents a specified part of the whole class and the males may be compared to the whole class either by a ratio, 31:50, or by a fraction, $\frac{31}{50}$ ($\frac{31}{50}$ of the class is male), or by a percent, 62% (62% of the class is male).

(3.11) *To find the percent one number is of another number:*
Step 1. Express the number representing the part as the numerator of a fraction and the number representing the whole as the denominator of that fraction.
Step 2. Express the fraction developed in Step 1 as a percent (Program 3.8).

EXAMPLES

1. 28 is what percent of 32?

Step 1. $\frac{28 \text{ (part)}}{32 \text{ (whole)}}$

Step 2. $\frac{28}{32} = 32\overline{)28.000}^{\,.875} = 87.5\%$

2. What percent of 16 is 4?

Step 1. $\frac{4 \text{ (part)}}{16 \text{ (whole)}}$

Step 2. $\frac{4}{16} = 16\overline{)4.00}^{\,.25} = 25\%$

3. What percent of 2.4 is .3?

Step 1. $\frac{.3 \text{ (part)}}{2.4 \text{ (whole)}}$

Step 2. $\frac{.3}{2.4} = 2.4\overline{).3_\wedge000}$ $= 12.5\%$ (quotient .125)

4. $\frac{3}{4}$ is what percent of $1\frac{7}{8}$?

Step 1. $\frac{\frac{3}{4}\text{(part)}}{1\frac{7}{8}\text{(whole)}}$

Step 2. $\frac{\frac{3}{4}}{1\frac{7}{8}} = \frac{\frac{3}{4}}{\frac{15}{8}} = \frac{3}{4} \div \frac{15}{8} = \frac{3}{4} \times \frac{8}{15} = \frac{2}{5}$

$\frac{2}{5} = .40 = 40\%$

A frequent percentage problem is one in which both the ratio of the part to the whole (expressed as a percent) and the value of the whole are known, but the value of the part is unknown. By multiplying the number associated with the whole by either the decimal or common fraction equivalent of the percent (remember the percent is not a computational form) we get the number associated with the unknown part.

(3.12) *To find a percent of a given number:*

Step 1. Express the percent either as an equivalent decimal or fraction (Program 3.9 or 3.10).

Step 2. Multiply the given number by either of the equivalent expressions of Step 1.

EXAMPLES

1. What is 62% of 80?
   Step 1. Write 62% as a decimal: .62
   Step 2. Multiply: $.62 \times 80 = 49.6$
2. What is 40% of 25?
   Step 1. $40\% = \frac{40}{100} = \frac{2}{5}$
   Step 2. $\frac{2}{5} \times 25 = 10$
3. What is .4% of 3.2?
   Step 1. $.4\% = .004$
   Step 2. $.004 \times 3.2 = .0128$
4. What is $87\frac{1}{2}\%$ of $\frac{5}{21}$?
   Step 1. $87\frac{1}{2}\% = \frac{7}{8}$
   Step 2. $\frac{7}{8} \times \frac{5}{21} = \frac{5}{24}$

The third type of basic percentage problem is the one in which both the ratio of the part to the whole (again expressed as a percent) and the number associated with the part are known, but the number associated with the whole is unknown. Problems of this type are often expressed in language of this sort:

12% of what number is 24?

If we remember that the whole (the unknown in this instance) is 100% of itself, we can easily calculate the number associated with the whole by means of this two-step approach:

(1) Find 1%
If 12% of the unknown number is 24...
Then 1% of the unknown number must be one-twelfth of 24, or 2:

$24 \div 12 = 2$

(2) Find 100%
Since 1% of the unknown number is 2...
Then 100% of the unknown number must be 100×2, or 200.

Note that the known part (24) was first divided by the percent number (12), and then the resulting quotient was multiplied by 100:

$(24 \div 12) \times 100 = 200$

This same result* could have been obtained by the single step of dividing the known part (24) by .12, the decimal equivalent of the known percent ($12\% = .12$):

$24 \div .12 = 200$

(3.13) *To find a number when a percent of it is known:*

Step 1. Express the percent as a decimal (Program 3.9).

Step 2. Divide the number associated with the known part by the decimal of Step 1.

EXAMPLES

1. 30% of what number is 45?
   Step 1. Express 30% as a decimal: .3.
   Step 2. Divide 45 by the decimal of Step 1.
   $.3\overline{)45.0_\wedge}$ = 150. (answer)
2. 22% of what number is 3.08?
   Step 1. $22\% = .22$
   Step 2. $.22\overline{)3.08_\wedge}$ = 14. (answer)
3. 1.926 is 6% of what number?
   Step 1. $6\% = .06$
   Step 2. $.06\overline{)1.92_\wedge6}$ = 32.1 (answer)
4. What is the number if $\frac{1}{2}\%$ of it is 2?
   Step 1. $\frac{1}{2}\% = .005$
   Step 2. $.005\overline{)2.000_\wedge}$ = 400. (answer)

---

* Note the equivalences:

$$(24 \div 12) \times 100 = \frac{24}{12} \times 100 = 24 \times \frac{100}{12}$$
$$= 24 \div \frac{12}{100} = 24 \div .12$$

5. 36% of what number is $\frac{9}{13}$?
[Note: Here it is more convenient to work with the fraction equivalent of the percent than with the decimal equivalent.]

Step 1. $36\% = \frac{36}{100}$
Step 2. $\frac{9}{13} \div \frac{36}{100} = \frac{9}{13} \times \frac{100}{36}$
$= \frac{25}{13} = 1\frac{12}{13}$ (answer)

(Complete Exercise 3.4)

9. PERCENT GREATER THAN 100%

At times we meet percent expressions that are greater than 100%. If we try to interpret such expressions in terms of the part-to-whole comparison used in the previous section, we are in the awkward position of suggesting that the part can be greater than the whole.

However, if we consider all percentage problems in terms of a simple comparison of one number to another, we have an approach that is equally applicable to part-to-whole situations as well as those that involve percents greater than 100%. Essentially this is a ratio approach, one that draws upon the interpretation given in Section 1, Chapter 2, in which the number to be compared is associated with the numerator of a fraction, and the number that is the basis for the comparison is associated with the denominator. Only here the ratio fraction is expressed by an equivalent percent—sometimes called the *rate.*

Thus, when the "number compared" is greater than the "base number" or "number compared to," the resulting percent will exceed 100%. Notice in the following examples (all involving comparisons expressed in percents greater than 100%) that once the comparison has been stated in fraction form, the equivalent greater-than-100% expression can be found by simply carrying out the steps of Program 3.11: Divide numerator by denominator decimally and express the quotient as a percent.

EXAMPLES

1. There are 36 students in Section 1 of a certain course and 24 students in Section 2. Compare the enrollment of Section 1 to that of Section 2.

Step 1. $\frac{36}{24}$ $\frac{\text{(Section 1 – number compared)}}{\text{(Section 2 – basis for comparison)}}$

Step 2. $36 \div 24 = 1.5 = 150\%$
Enrollment of Section 1 is 150% of the enrollment of Section 2.

2. Compare the cost of bread today at 48¢ a loaf with the cost 7 years ago, when it sold for 36¢.

Step 1. $\frac{48¢}{36¢}$ $\frac{\text{(number compared)}}{\text{(number compared to)}}$

Step 2. $48 \div 36 = 1.33+ = 133\%$ (approx.)
Bread today costs 133% (approx.) of what it cost 7 years ago.

3. How does 33 compare to 8 in ratio?

Step 1. $\frac{33}{8}$ $\frac{\text{(number compared)}}{\text{(basis for comparison)}}$

Step 2. $33 \div 8 = 4.125 = 412\frac{1}{2}\%$
33 is $412\frac{1}{2}\%$ of 8

4. 72 is what percent of 60?

Step 1. $\frac{72}{60}$ $\frac{\text{(number compared)}}{\text{(basis for comparison)}}$

Step 2. $72 \div 60 = 1.2 = 120\%$
72 is 120% of 60

5. What percent of 16 is 40?

Step 1. $\frac{40}{16}$ $\frac{\text{(compared)}}{\text{(compared to)}}$

Step 2. $40 \div 16 = 2.5 = 250\%$
40 is 250% of 16

An added word of caution is necessary here. Notice how the percent-greater-than-100% was stated in the examples above:

- Enrollment of Section 1 is 150% *of* that of Section 2.
- Bread today costs 133% *of* what it cost 7 years ago.
- 40 is 250% *of* 16.

Note that we *did not* say:

× Enrollment of Section 1 is 150% *larger* than Section 2.
× Bread today costs 133% *more* than it cost 7 years ago.
× 40 is 250% *greater* than 16.

This last group of statements represents a frequent misinterpretation of percents-greater-than-100%. The mistake is easily avoided if we remember that 100% represents the whole; hence 150% = 100% + 50%, or the whole (100%) plus another half of the whole (50%). When something is *150% greater*, it means that the *excess* is equal to another whole (100%) and a half (50%) again. Thus:

That number which is *150% of 8* is 12:

100% of 8 = 8; 50% of 8 = 4;
150% of 8 = 8 + 4 = 12

That number which is *150% greater than 8* is 20:

8 is 100%; 150% of 8 = 12; hence
a number 150% greater than 8
is 8 + 12 or 20

Note that "*150% greater than* 8" is equivalent to "*250% of* 8."

Another illustration:
That number which is *120% of* 6 is 7.2:

100% of 6 = 6; 20% of 6 = 1.2;
120% of 6 = 6 + 1.2 = 7.2

That number which is *120% greater than* 6 is 13.2:

6 is 100%; 120% of 6 = 7.2;
that number which is 120% greater
than 6 is 6 + 7.2 = 13.2

Note that "*120% larger than* 6" is equivalent to "*220% of* 6."

In problems such as these, instead of thinking "150% of 8 is 100% of 8, plus 50% of 8" and "120% of 6 is 100% of 6, plus 20% of 6," we can arrive at the same result by using Program 3.12: *To find the percent of a given number, multiply the given number by the decimal* (*or fraction*) *equivalent of the percent.* Therefore:

150% of 8 = 1.5 × 8 = 12
120% of 6 = 1.2 × 6 = 7.2

EXAMPLES

1. What is 135% of 44?
   135% of 44 = 1.35 × 44 = 59.4
2. What number is 250% of 16?
   250% of 16 = 2.5 × 16 = 40
3. What number is 250% greater than 16?
   250% greater than 16 = 350% of 16
   = 3.5 × 16 = 56
4. What number is 112% greater than 82?
   112% greater than 82 = 212% of 82
   = 2.12 × 82 = 173.84
5. What number is 65% more than 18?
   65% more than 18 = 165% of 18
   = 1.65 × 18 = 29.7
6. What number is 100% more than 37.5?
   100% more than 37.5 = 200% of 37.5
   = 2. × 37.5 = 75
7. What number is $2\frac{1}{4}$ times the size of 15?
   $2\frac{1}{4}$ times the size of 15 = 225% of 15
   = 2.25 × 15 = 33.75

By the same reasoning, problems such as "What number is 40% *less* than 8?" can be equivalently stated as "What number is 60% *of* 8?" since *40% less than the whole* (100%) is 60% (100% − 40%) *of* the whole. Similarly:

"62% less than" = "38% of"
"25% less than" = "75% of"
"92% less than" = "8% of"
"100% less than" = "0% of"

Obviously expressions such as "120% less than" and "200% less than" are meaningless.

Cases of the type covered by Program 3.13 (basis for comparison unknown) involving percents-greater-than-100% are not so numerous in daily life. One of the most frequent, however, is the kind which involves "cost, tax included." For example, suppose a person was charged $33.18 for an article, including 5% sales tax. This means that the $33.18 includes not only the price of the article (100% of the price) but also the tax (5% of the price). Consequently the $33.18 represents 105% (100% price + 5% tax) of the basic price of the article. The unknown basic price is found, as in any case where the percent of a number is known (Program 3.13), by dividing the "part number," $33.18 by the decimal equivalent of 105%:

$33.18 ÷ 105% = $33.18 ÷ 1.05
= $31.60 (price of article)

To verify, knowing now the price of the article ($31.60), we calculate the tax (5% of $31.60 = .05 × $31.60 = $1.58 tax) and add it to the price of the article ($31.60 + $1.58 tax = $33.18).

Before citing other examples, it may be useful to note a frequent mistake made in problems of this type. Its "misreasoning" goes this way:

× Cost of article, tax included: $33.18.
5% of $33.18 = $1.659, or $1.66 tax
Cost of article, tax included ($33.18) less tax ($1.66)
= price of article before tax ($31.52) (WRONG)

The fact that the answer found by this invalid reasoning comes within several cents of the correct answer is due to the relatively small difference which exists between multiplying by the equivalent of 95% (which in effect was done in the wrong approach: .95 × $33.18 = $31.52) and dividing by the equivalent of 105% (which was done in the correct approach: $33.18 ÷ 1.05 = $31.60). Had the tax been of a higher rate, say 30%, the difference between the correct and incorrect approaches would have been more noticeable:

$33.18 ÷ 130% = $25.52 (correct)
70% × $33.18 = $23.23 (incorrect)

The error comes about through a failure to recognize in the problem the datum which is the basis for the comparison, that which corresponds to the 100% —in this case, the price of the article before tax.

EXAMPLES

1. 160% of what number is 144?
   144 ÷ 160% = 144 ÷ 1.6 = 90

2. 216 is 108% of what number?
   $216 \div 108\% = 216 \div 1.08 = 200$
3. 99 is $12\frac{1}{2}\%$ greater than what number? (This is equivalent to saying "99 is $112\frac{1}{2}\%$ *of* what number?")
   $99 \div 112\frac{1}{2}\% = 99 \div 1.125 = 88$
4. 105 is 250% greater than what number? (This is equivalent to saying "105 is 350% *of* what number?")
   $105 \div 350\% = 105 \div 3.5 = 30$
5. What is the price before tax of an article for which a man paid $17.94, including 4% tax?
   $17.94 = 104% of the price before tax
   $17.94 ÷ 1.04 = $17.25 (price before tax)
6. Tickets for a show were $1.65, including 10% amusement tax. How much is the government entitled to from each ticket?
   $1.65 = 100% admission charge + 10% tax, therefore
   $1.65 = 110% of the admission charge
   $1.65 ÷ 110% = $1.65 ÷ 1.1
   = $1.50 admission charge
   This can be verified by taking 10% of the admission charge (.10 × $1.50 = $.15) and adding it to the admission charge ($1.50 + $.15 = $1.65).

(Complete Exercise 3.5)

Name.. Date.............................. **EXERCISE 3.1**

Insert the appropriate symbol, $>$ for *is greater than* or $<$ for *is less than*, between the pairs of decimals.

1. .036 .03 **2.** .374 .41 **3.** 2.03 .97 **4.** 62.037 62.05

5. .4 .089 **6.** .0003 .001 **7.** .042 .04 **8.** 3.86 3.9

Arrange the following decimals in order, left to right, from the least to greatest value.

9. 3.62, 3.528, 3, .36, .3

..........., ..., ..., ..., ...

10. 3.04, .304, .340, .431, 3.25

..........., ..., ..., ..., ...

11. 6.2, 6.19, 6.01, 6.03, 6.214

..........., ..., ..., ..., ...

12. .0003, .002, .000102, .0006, .00099

..........., ..., ..., ..., ...

Add, and check by the excess of nines.

13. 43.625 + 17.91 + 8.65 + 4.31 **14.** 39.6 + 24.37 + 80.003 + .427 **15.** 39.06 + 42.17 + 387 + 4.002

16. .6379 + 41.76 + .304 + .006 **17.** 4632.7 + 390 + 6.2007 + .307 **18.** .0062 + 3.04 + .007 + .00005

Subtract, and check by addition.

19. 46.37 − 18.263 **20.** 364.07 − 42.68 **21.** 397.14 − 43.875

22. 67.46 − .003 **23.** .826 − .0003 **24.** .00402 − .003907

Multiply, and check by the excess of nines.

25. 46.83 × 2.07

26. 53.86 × 5.07

27. .4963 × 3.82

28. .636 × 5.21

29. 374 × .0062

30. 4.007 × .3004

31. .0062 × .0075

32. .8 × 1.3 × 4.9 × .006

Solve; assume all data are *exact*.

33. Six free-style swimmers were clocked at the following times for the 100-yd dash: 53.2, 54.1, 57.0, 56.3, 53.1, and 52.9 seconds. What would be the expected time for the best 4-man relay team in a 400-yd event?

....

34. The difference between two numbers is 86.395. If 43.287 is added to the larger of the two numbers, what would be the new difference?

....

35. A tract of land contains 86.04 acres. If it is subdivided into four parcels and three of these parcels contain 16.07, 22.35, and 4.23 acres, what is the size of the remaining parcel?

....

36. An object traveling at 93.16 ft/sec overtakes another object traveling at 78.24 ft/sec. How far apart were they 2.3 sec before they met?

....

37. Assume the diagonal of a square is 1.414 times the length of its edge. If the distance from home plate to first base on a regulation baseball field is 90 ft, how far apart are first and third bases?

....

38. Cork weighs .24 times the weight of water, which weighs 62.5 lb/cu ft. How much would 6.7 cu ft of cork weigh?

....

Name.. Date.............................. **EXERCISE 3.2**

Round off to the place values indicated.

1. 5463.247 rounded off to:

(hundredths)..............................

(tenths)..............................

(ones)..............................

(tens)..............................

(hundreds)..............................

2. 397.998 rounded off to:

(hundredths)..............................

(tenths)..............................

(ones)..............................

(tens)..............................

(hundreds)..............................

3. 37.502 rounded off to:

(hundredths)..............................

(tenths)..............................

(ones)..............................

(tens)..............................

(hundreds)..............................

Divide, and express the quotients to the nearest hundredth.

4. $47.28 \div 3.7$

5. $6.387 \div .42$

6. $.83216 \div .301$

7. $.79003 \div 52.1$

8. $807.9 \div .502$

9. $.04638 \div 27.1$

Write the fraction equivalent in simplest terms for each decimal.

10. $.46 =$

11. $.16 =$

12. $.32 =$

13. $.875 =$

14. $.37 =$

15. $.07 =$

16. $1.4 =$

17. $36.625 =$

18. $1.75 =$

19. $1.750 =$

20. $.035 =$

21. $.0035 =$

22. $.625 =$

23. $.62\frac{1}{2} =$
($62\frac{1}{2}$ hundredths)

24. $.6\frac{1}{4} =$
($6\frac{1}{4}$ tenths)

Express as decimal equivalents. If the decimal equivalent is unending, round it off to thousandths.

25. $\frac{16}{55} =$ **26.** $\frac{3}{8} =$ **27.** $\frac{17}{50} =$ **28.** $\frac{3}{25} =$

29. $\frac{17}{125} =$ **30.** $\frac{19}{5} =$ **31.** $\frac{16}{64} =$ **32.** $\frac{3}{13} =$

33. $\frac{30}{7} =$ **34.** $\frac{2}{7} =$ **35.** $\frac{5}{4} =$ **36.** $\frac{147}{3125} =$

Solve; assume all data are *exact.*

37. If an automobile traveled 156.6 miles in 3.6 hours, what was its average speed?

....

38. Gasoline costs 68.9¢ a gallon and an auto averages 15.5 miles to the gallon. How much will the gas cost for a trip of 902 miles?

....

39. What must .863 be multiplied by to yield a product of 34.6063?

....

40. If the maximum capacity of a dump truck is 6.7 tons, how many trips will be necessary to transport (a) 91.3 tons, (b) 81.5 tons?

(a) ....

(b) ....

Name.. Date.............................. **EXERCISE 3.3**

Express as equivalent percents.

1.	.35 =	.87 =	.43 =	.52 =
2.	.9 =	.3 =	.6 =	.1 =
3.	.09 =	.04 =	.036 =	1.87 =
4.	.0032 =	1 =	4 =	1.06 =
5.	.1005 =	6.274 =	1.3 =	.0006 =
6.	.0024 =	1.0024 =	3.65 =	.02 =

Express as equivalent percents. If necessary, round off to the nearest *hundredth of a percent*.

7.	$\frac{3}{5} =$	$\frac{5}{8} =$	$\frac{3}{4} =$
8.	$\frac{5}{4} =$	$\frac{3}{2} =$	$\frac{9}{8} =$
9.	$\frac{2}{7} =$	$\frac{17}{25} =$	$\frac{4}{11} =$
10.	$\frac{9}{40} =$	$\frac{11}{200} =$	$\frac{5}{12} =$

Express as decimal equivalents.

11.	9% =	.6% =	35% =	18% =
12.	220% =	100% =	.04% =	300% =
13.	165% =	16.5% =	1.65% =	.165% =
14.	$\frac{1}{2}$% =	$\frac{3}{4}$% =	.1% =	10.3% =

Express as fraction equivalents in lowest terms.

15.	26% =	32% =	1.6% =
16.	125% =	12.5% =	1.25% =
17.	$82\frac{1}{2}$% =	27% =	5% =
18.	37.5% =	.05% =	4.25% =

Supply the missing equivalents. If necessary, round off decimals to nearest ten-thousandth, percents to nearest hundredth of a percent. In the Fraction column, numerals in mixed form may be used, but all fractions are to be reduced to lowest terms.

	Fraction	Decimal	Percent
19.	$\frac{2}{5}$		
20.	$\frac{1}{8}$		
21.		.06	
22.		.875	
23.			48%
24.			$16\frac{2}{3}$%
25.	$\frac{4}{15}$		
26.	$1\frac{3}{4}$		
27.		.0005	
28.		1.08	
29.			$\frac{1}{5}$%
30.			160%
31.	$2\frac{1}{2}$		
32.		.25	
33.			$2\frac{1}{2}$%
34.	$\frac{2}{9}$		
35.	$\frac{5}{16}$		
36.		.0735	
37.		1.28	
38.			.004%
39.			$\frac{1}{8}$%
40.	$2\frac{5}{16}$		
41.		.0525	
42.			.03%
43.	$\frac{11}{160}$		
44.		.0022	
45.			$.3\frac{1}{3}$%
46.			$\frac{9}{7}$%

Name.. Date.............................. **EXERCISE 3.4**

Complete the following. If data contain fractions, express answers with fractions; otherwise use decimals. Round off all decimals to the nearest hundredth and all percents to the nearest tenth of a percent.

1. 18 is ....% of 72.

2. ....% of 25 is 15.

3. 14% of 800 is ....

4. 25% of .... is 8.

5. 8 is ....% of 64.

6. 3.92 is 4% of ....

7. .... is 36% of 17.

8. ....% of 55 is 33.

9. .... is 23% of 46.

10. 6.3 is .9% of ....

11. ....% of 3.2 is .4.

12. .3% of 40 is ....

13. 18.36 is 90% of ....

14. 30 is ....% of 45.

15. 1.06% of 210 is ....

16. .006 is 30% of ....

17. $\frac{1}{2}$ is ....% of $\frac{5}{8}$.

18. .... is 20% of .035.

19. $\frac{3}{5}$ is 21% of ....

20. ....% of .005 is .0005.

21. 9.24 is 7% of ....

22. .05% of 16 is ....

23. .... is 80% of $\frac{7}{16}$.

24. 15 is ....% of 19.

25. 4.02% of 167.9 is ....

26. .06 is $\frac{1}{2}$% of ....

27. $\frac{1}{2}$% of $\frac{3}{16}$ is ....

28. $\frac{2}{3}$ is ....% of $3\frac{1}{3}$.

29. 47 is $\frac{1}{2}$% of ....

30. .... is .9% of .07.

31. At a fire sale all prices are reported to be reduced 25%. What should be the sale price on an article previously priced at $17.60?

....

32. A used car is advertised at $2,125 cash or $318.75 down, balance in monthly payments. What is the percent of down payment?

....

33. How much must a salesman, working on a 15% commission, sell per month to earn $840 a month?

....

34. A broker's commission on house sales is 6%. How much will a homeowner receive if a broker sold the house for $46,500?

....

35. An electric bill of $16.50 may be paid in full within ten days for $15.29. What percent saving is involved?

....

36. An employee earning $200 a week was given a 10% increase. A year later, business difficulties necessitated a 10% decrease to all employees. What was this employee's salary after the decrease?

....

37. Membership in a consumers' union entitles a member to a 15% discount on all goods. What will a television set, listed at $374.85, cost a member if he lives in a state where there is a 5% sales tax?

....

38. How many pounds of $4\frac{1}{2}\%$ butterfat milk will be necessary to give $3\frac{3}{8}$ lb of butterfat?

....

39. A coat is advertised for $65, which represents a 20% reduction from its former price. What was the former price?

....

40. A baseball player has made 5 hits in 20 times at bat ($\frac{5}{20} = 25\% = .250$ batting average). How many hits must he make in the next 30 times at bat to bring his average up to .340?

....

Name.. Date.............................. **EXERCISE 3.5**

Supply the missing equivalents.

	Fraction	Decimal	Percent
1.	$\frac{5}{4}$		
2.	$1\frac{3}{8}$		
3.		1.6	
4.		1.02	
5.			300%
6.			225%
7.	$\frac{31}{2}$		
8.		16	
9.			$333\frac{1}{3}\%$
10.	$1\frac{5}{7}$		
11.		1.52	

Complete the following.

12. ....% of 80 is 96.

13. 125% of 16 is ....

14. 200% of .... is 18.

15. 42 is ....% of 7.

16. .... is 212% of 50.

17. 150% of .... is $\frac{3}{8}$.

18. 120% more than 15 is ....

19. 16.32 is ....% of 16.

20. 16.32 is ....% more than 16.

21. 130% of $\frac{1}{26}$ is ....

22. 175% of .... is 1.

23. $17\frac{1}{2}$ is ....% more than 14.

24. The goal for a church benefit is set at 135% of earnings of the previous year. If $1860 was cleared the previous year, how much money must be cleared this year?

....

25. Six years ago a market basket of certain groceries cost $12. Today that same basket costs $15.60. State the comparison of today's cost against that of six years ago as a percent.

....

26. An article, bought in a state having a 3% sales tax, cost a buyer $701.43, tax included. What was the cost of the article before tax?

....

27. In five years the school population in a district jumped from 9600 to 12,800. Express in percent form the comparison of the former school population against that of the latter.

....

28. In a certain location the mean summer temperature is 120% greater than the winter mean of 35°. What is the summer mean?

....

29. A live chicken weighs 140% of its dressed weight. How many pounds of undressed chickens should be bought to realize 65 lb of dressed chicken?

....

30. If today medical expenses are 187% of those eight years ago, what would a $563.65 medical bill today have been eight years ago?

....

31. Improvement of a manufactured item boosted sales from $5500 yearly to $11,635 yearly. What percent increase in sales can be attributed to the improvement?

....

Replace the □ with $<$ (is less than), $>$ (is greater than), or $=$ (is equal to).

1. .042 □ .41 **2.** 3.4 □ 3.05

3. .3 □ .214 **4.** 2.06 □ 3.05

5. .002 □ .0019 **6.** 0.021 □ .021

Add.

7. $\begin{array}{r} 9.16 \\ 3.24 \\ +1.79 \\ \hline \end{array}$ **8.** $\begin{array}{r} 5.82 \\ .63 \\ +7.4 \\ \hline \end{array}$ **9.** $\begin{array}{r} 82.63 \\ 1.9 \\ +\ 3.004 \\ \hline \end{array}$

10. $6.34 + 40.5 + .39$

11. $29 + 16.325 + 1.7 + 24$

12. $.042 + .0671 + .1934 + 2$

13. $.007 + 5.8 + 30.42 + 7.2657$

14. $226 + .39 + 10.425 + 2.63 + 5$

15. $27.046 + 3.047 + 321.5 + .1621$

Subtract.

16. $\begin{array}{r} 59.35 \\ -40.25 \\ \hline \end{array}$ **17.** $\begin{array}{r} 91.703 \\ -54.037 \\ \hline \end{array}$ **18.** $\begin{array}{r} 20.041 \\ -\ 1.637 \\ \hline \end{array}$

19. $86.41 - 3.25$ **20.** $250.6 - 31.08$

21. $38.46 - 19.3$ **22.** $52.615 - 37.897$

23. $250 - .69$ **24.** $34.627 - 33.697$

Multiply.

25. $\begin{array}{r} 52 \\ \times .46 \\ \hline \end{array}$ **26.** $\begin{array}{r} .39 \\ \times 63 \\ \hline \end{array}$ **27.** $\begin{array}{r} 8.24 \\ \times 1.6 \\ \hline \end{array}$

28. $.05 \times 4.8$ **29.** $.003 \times 675$

30. 1.24×1.03 **31.** $.078 \times 49.3$

32. 3.24×16.58 **33.** 4.07×3.295

Round off as indicated.

34. 402.16 to (a) tenths; (b) hundreds.

35. 34.64 to (a) ones; (b) tens.

36. 43,256.3 to (a) thousands; (b) hundreds.

37. 46.2795 to (a) hundredths; (b) thousandths.

38. 2463.047 to (a) thousands; (b) hundredths.

39. .36453 to (a) thousandths; (b) ones.

Divide. Express quotients to nearest hundredth.

40. $7\overline{)5.12}$ **41.** $24\overline{)6.271}$

42. $.4\overline{)826}$ **43.** $.8\overline{)74.62}$

44. $.07\overline{)4.963}$ **45.** $2.1\overline{)96.471}$

46. $4.8\overline{)39.641}$ **47.** $.24\overline{).6751}$

48. $78.5 \div .24$ **49.** $110.9 \div 5.6$

50. $4.289 \div 1.7$ **51.** $5631.9 \div .42$

Supply the missing equivalents.

	Fraction	Decimal	Percent
52.	$\frac{7}{100}$	....	....
53.	$\frac{5}{8}$	....	....
54.	$\frac{27}{50}$	....	....
55.	....	.64	....
56.	....	.36	....
57.	....	.97	....
58.	....	....	8%
59.	....	....	.6%
60.	....	....	$12\frac{1}{2}\%$
61.	$\frac{7}{250}$	....	....
62.	....	1.42	....
63.	....	....	180%
64.	$\frac{9}{4}$	....	....
65.	....	24.0	....
66.	....	....	736%

Complete the following.

67. 18% of 31 is ....

68. 4 is ....% of 5.

69. 15% of .... is 45.

70. 12 is ....% of 48.

71. 16% of 180 is ....

72. .... is 68% of 40.

73. 5.6 is .8% of ....

74. 90% of 17.4 is ....

75. .007 is 40% of ....

76. $\frac{1}{2}$ is ....% of $\frac{12}{16}$.

77. .... is 65% of 112.

78. 31 is $27\frac{1}{2}\%$ of ....

79. $3\frac{1}{4}\%$ of 1.6 is ....

80. .75 is ....% of 7.5.

81. 125% of 24 is ....

82. .... is 212% of 72.

83. 68.4 is ....% of 60.

84. 135% more than 9 is ....

85. A mark of 80% on a test of 60 items means that how many items were correct?

86. There are 19 boys and 21 girls in a class. What percent of the class is girls?

87. If you pay $16 for an article that has been reduced 20%, how much was it originally?

88. In a shipment of 300 bottles, 96% came through undamaged. How many bottles in the shipment were undamaged?

89. On a test of 35 items, how many may a student miss and still get a grade of 75% or better?

90. A student had 38 problems right and 7 problems wrong. If all problems were of equal weight, what percent grade should he get?

91. A house worth $43,000 is insured for 80% of its value. How much is that?

92. A lawyer charged $78 to collect a bill of $500. What percent of the bill was the lawyer's collection fee?

93. A bank charges 2.3% to process a mortgage. How much will that come to on a mortgage of $28,560?

94. A manufacturer claimed a deduction on his income tax of $429, as 6% depreciation on the original cost of a machine. What was the original cost of the machine?

95. The population of a town moved up from 3427 residents to 12,567. What was the percent increase in residents?

96. How many dollars in sales must you generate to earn $96 at 15% commission?

97. The cost of living today is 57% more than it was a few years ago. What $20 would buy then would cost how much today?

98. A man saves $35 a month from his annual salary of $7200. His savings is what percent of his salary?

99. A picture is priced at $89.60, which is 40% above cost. What was the cost of the picture?

100. A merchant gives 3% discount for cash in 10 days. How much can be saved on the purchase of 8 shirts at $3.95 each, if cash is paid within ten days?

101. A bus has 40 seats. On one trip all seats were taken and 8 people were standing. What was the percent of capacity for that trip?

102. A baseball team has a winning streak of 12 games, which was 40% of their total wins to date. How many games have they won up to now?

103. A man wants to pay off a note in 90 days. It is for $350, plus simple interest at an annual rate of 6%. How many dollars interest must he pay?

104. In one community, the assessed value of real estate, for tax purposes, is 25% of market value. What is the assessed value on a $38,450 home built on a lot valued at $14,770?

105. One inch is approximately .0254 meter. A foot is what percent of a meter?

REVIEW

Express these numbers in words.

1. 46,231 **2.** 700,000 **3.** 4,326,511

4. 2,000,000,000 **5.** 36,000,000

6. 41,000,207

7. In 3,261,475, what is the
(a) tens digit?
(b) thousands digit?
(c) hundreds digit?
(d) ten-thousands digit?
(e) ones digit?

8. Which of these numbers:

432	4391	2765	27,651
672	5680	4391	38,000

(a) is divisible by 2?
(b) is divisible by 3?
(c) is divisible by 4?
(d) is divisible by 5?
(e) is divisible by 6?
(f) is divisible by 9?

Arrange in order, from least to greatest.

9. $\frac{3}{4}, \frac{2}{3}, \frac{5}{8}, .5$ **10.** $2\frac{2}{3}, 1.6, 2.5, 1\frac{7}{8}$

11. $\frac{4}{11}, \frac{3}{10}, \frac{2}{7}, \frac{3}{.8}$ **12.** $\frac{8}{3}, 2\frac{1}{4}, 2.3, \frac{15}{7}$

Add.

13. $\frac{3}{4} + \frac{5}{6} + \frac{7}{8}$ **14.** $\frac{1}{8} + \frac{7}{12} + \frac{2}{3}$

15. $\frac{3}{5} + \frac{2}{3} + \frac{5}{7} + \frac{5}{6}$ **16.** $\frac{2}{9} + \frac{7}{3} + \frac{3}{4} + \frac{5}{12}$

Subtract.

17. $\frac{2}{3} - \frac{7}{12}$ **18.** $\frac{7}{8} - \frac{3}{11}$

19. $1\frac{6}{7} - \frac{9}{14}$ **20.** $8\frac{1}{2} - 2\frac{5}{6}$

Multiply.

21. $\frac{5}{9} \times \frac{3}{4}$ **22.** $\frac{2}{5} \times \frac{5}{12}$

23. $7\frac{1}{2} \times 1\frac{1}{6}$ **24.** $4\frac{3}{8} \times 2\frac{4}{5}$

Divide.

25. $5 \div \frac{5}{8}$ **26.** $\frac{3}{4} \div \frac{5}{8}$

27. $\frac{2}{3} \div \frac{8}{9}$ **28.** $3\frac{3}{8} \div 4\frac{1}{2}$

29. A satellite orbits earth in $1\frac{2}{3}$ hours. How many orbits is that a day?

30. Three heirs inherit $\frac{1}{7}$, $\frac{2}{5}$, and $\frac{1}{3}$ of an estate, and the rest goes to charity. What part of the estate goes to charity?

31. What is the cost of 35 shares of stock quoted at $\$16\frac{5}{8}$ a share?

32. If a basket holds $1\frac{3}{4}$ bu. of apples, how many bushels will there be in $2\frac{1}{2}$ baskets?

Name.. Date.............................. Score......................

ACHIEVEMENT TEST NO. 1 (Chapters 1–3)

Solve and place answers in the blank provided. (The number in parentheses after each blank indicates the points for that question; total 100.)

1. $63{,}327 + 420 + 83{,}002 + 76{,}555 + 27 =$ ....(3)

2. Does the addition $4{,}286 + 3{,}795 + 8{,}632 + 427 = 17{,}239$ check by the excess of nines? Answer "yes" or "no." ....(2)

3. $3{,}007 \times 84{,}002 =$ ....(3)

4. Does the multiplication $687 \times 407 = 279{,}509$ check by the excess of nines? Answer "yes" or "no." ....(2)

5. How much must be added to 1,863 to make 2,427? ....(3)

6. What number multiplied by 637 equals 542,724? ....(3)

7. What number, divided by 803, has a quotient of 422 and a remainder of 436? ....(3)

8. What digit must replace the 0 in 82,470 to make the number exactly divisible by 4, 6, and 9? ....(3)

9. Replace the □ with either < or >.

$\frac{7}{24}$ □ $\frac{11}{36}$ ....(2)

$\frac{15}{16}$ □ .938 ....(2)

$\frac{1}{3}\%$ □ .003 ....(2)

$1\frac{7}{9}$ □ 178% ....(2)

10. Arrange the following sets of numbers in order, left to right, from least to greatest.

$\frac{7}{8}$, $\frac{4}{5}$, $\frac{8}{7}$, $\frac{17}{14}$ ...., ..., ..., ...(2)

.4, 39%, $\frac{3}{8}$, .042 ...., ..., ..., ...(2)

11. $28\frac{1}{4} + 17\frac{3}{5} + 21\frac{3}{10} + 6\frac{1}{2} =$ ....(3)

12. $3{,}002\frac{5}{8} - 1{,}712\frac{3}{4} =$ ....(3)

13. $3\frac{2}{3} - 1\frac{1}{4} + 7\frac{1}{12} - 2\frac{1}{8} =$ ....(3)

14. $3\frac{3}{8} \times 5\frac{2}{9} \times 7\frac{1}{5} =$ ....(3)

15. $528\frac{3}{4} \times 68\frac{2}{3} =$ ....(3)

16. $6\frac{3}{8} \div 3\frac{2}{5} =$ ....(3)

17. How many $1\frac{3}{8}$'s are there in $5\frac{1}{24}$? ....(3)

18. Simplify $\dfrac{8\frac{2}{5}}{\frac{3}{10}}$ ....(3)

19. Simplify $\dfrac{2 - \frac{1}{4}}{1 + \dfrac{2}{1 + \frac{1}{2}}}$ ....(3)

20. $.826 - .0346 =$ ....(3)

21. $.965 \times .0007 =$ ....(3)

22. How many .062's are there in .32767? ....(3)

23. Supply the missing equivalents. (1 point each entry.)

Fraction	*Decimal*	*Per cent*
$\frac{3}{40}$	....	....
....	1.45	....
....	....	7.25%

24. 1.9 is what per cent of 7.6? ....(3)

25. 40% of $1\frac{1}{2}$ is what? ....(3)

26. 12.6 is 35% of what? ....(3)

27. 175% of what is 3.01? ....(3)

28. If a chicken and a half costs a dollar and a half, how much would one and a half dozen chickens cost? ....(3)

29. At the beginning of the month there was $462.18 in a checking account. Checks drawn during the month: $29.32, $162.88, $405.72, $118.00. Deposits during the month: $75.00, $325.52, $197.55. What was the balance at the end of the month? ....(3)

30. How many dollars' worth of merchandise must a salesman sell to earn $162 if he works on a 15% commission? ....(3)

31. A man sold two lots for $900 each. For one he received 25% more than it cost him and for the other he received 25% less than it cost him. How much did he gain or lose on the deal? ....(3)

PART II

ESSENTIALS OF ALGEBRA

4

FROM ARITHMETIC TO ALGEBRA

1. SETS OF NUMBERS

An important concept in mathematics is that of a *set* or collection of numbers. A set may have only a few numbers in it as members, e.g.,

$\{1, 3\}$

or many members, e.g.,

$\{1, 2, 3, 4, 5, \ldots\}$

or no members:

$\{\ \}$

The set having no members is called *the empty set*.

The standard symbol for denoting a set is a pair of braces, $\{\ldots\}$.

The set of numbers normally used in arithmetic are the whole numbers and numbers that are the quotients of pairs of whole numbers. This set may be illustrated by a number line on which each number of the set corresponds to a point. It would be impossible in a single drawing to label *every* point, but *any* point may be properly labeled, as some are in Fig. 4.1.

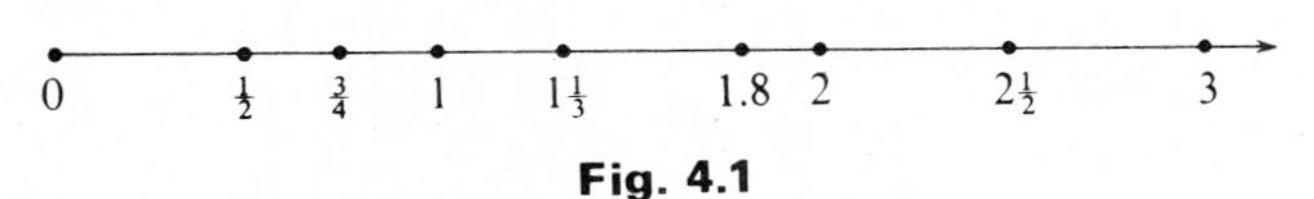

Fig. 4.1

In algebra more extensive sets of numbers are used. One important set consists of the numbers of arithmetic and their opposites. In this set the numbers of arithmetic (except 0) are referred to as *positive numbers*. For each positive number in the set there is also an opposite number, a *negative number*, whose value is such that when it is added to its corresponding positive number the sum is 0. Numerals for positive numbers are sometimes prefixed with a plus sign (+); numerals for negative numbers are usually prefixed with a minus sign (−). Thus:

$(+6) + (-6) = 0$ (read: positive 6 plus negative 6 equals 0)

$(+\frac{3}{4}) + (-\frac{3}{4}) = 0$

The set consisting of the numbers of arithmetic and their respective negatives (0 may be considered to be its own negative) make up the full set of *rational numbers*. The number line at the bottom of the page illustrates.

The rational numbers are said to be *ordered*. That is, for any partial collection of them (usually called a *subset*), or for the whole set itself, the numbers can be arranged by size. To put this another way, for any pair of rational numbers, one number will be greater (and one will be less) than the other. On the number line, as illustrated in Fig. 4.2, numbers located to the right of a given number are greater than (>) the given number; numbers located to the left of a given number are less than (<) the given number. Thus, $+2$ is located to the right of $+1$, and $+2 > +1$; -1 is located to the left of 0, and $-1 < 0$; also:

$+.25 < +3 \qquad +1 > -2$

$-\frac{2}{3} > -1 \qquad -3 < 0$

The whole numbers and their opposites or negatives make up another important set of numbers, called

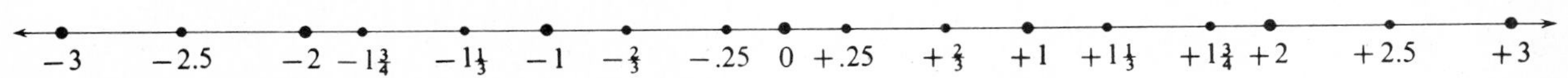

Fig. 4.2

the *integers*:

$$\{\ldots, -3, -2, -1, 0, +1, +2, +3, +4, \ldots\}$$

Thus the integers, and the numbers of arithmetic, are different subsets of the set of rational numbers. (Another way to refer to the numbers of arithmetic is as the *non-negative rationals*.) In fact, the *ratio*nals are usually defined as numbers that can be expressed as the *ratio* (fraction) of two integers. Thus $\frac{1}{2}$ and $\frac{5}{7}$ are clearly rational numbers, as is $6 = \frac{6}{1} = \frac{12}{2}$ and $0 = \frac{0}{5}$, etc.

2. SUMS AND DIFFERENCES OF RATIONAL NUMBERS

The operations of addition, subtraction, multiplication, and division hold also for the set of rational numbers. As was the case for numbers of arithmetic, computing the results of operations on the rational numbers involves a processing of their numerals. The procedures for doing so are more readily understood by introducing the concept of absolute value.

The *absolute value* of a number may be thought of as the "distance" a number lies from 0 on the number line, irrespective of direction. Note in Fig. 4.2 that $+3$ and -3 both lie at a distance of 3 units from 0. Using the symbol $|\ldots|$ to indicate absolute value, we have as an example:

$$|+3| = 3 \quad \text{and} \quad |-3| = 3$$

Similarly:

$$|+1\tfrac{1}{4}| = |-1\tfrac{1}{4}| = 1\tfrac{1}{4}$$

$$|+26\tfrac{1}{2}| = |-26\tfrac{1}{2}| = 26\tfrac{1}{2}$$

In practice, we can read the absolute value for a number by ignoring the plus or minus sign of its numeral. Computations with absolute values of numbers are essentially like those for the numbers of arithmetic.

(4.1) *To compute the sum of rational numbers when their signs are the same:*

Step 1. Find the sum of the absolute values of the rational numbers.

Step 2. Prefix the sum found in Step 1 with the common sign of the addends.

EXAMPLES

1. Compute the sum of -2 and -5.
   Step 1. Absolute value of -2: 2
   Absolute value of -5: 5
   Sum of absolute values: $2 + 5 = 7$
   Step 2. Common sign of the addends: $-$ (minus)
   Sum of -2 and -5: -7
2. Add $+4$ and $+6$.
   Step 1. $|+4| + |+6| = 4 + 6 = 10$
   Step 2. $(+4) + (+6) = +10$
3. Add: $(+2) + (+3.1) + (+5)$.
   Step 1. $|+2| + |+3.1| + |+5| = 2 + 3.1 + 5 = 10.1$
   Step 2. $(+2) + (+3.1) + (+5) = +10.1$
4. Add: $(-\frac{2}{3}) + (-\frac{1}{2}) + (-\frac{1}{4})$.
   $|-\frac{2}{3}| + |-\frac{1}{2}| + |-\frac{1}{4}| = \frac{2}{3} + \frac{1}{2} + \frac{1}{4} = \frac{17}{12} = 1\frac{5}{12}$
   $(-\frac{2}{3}) + (-\frac{1}{2}) + (-\frac{1}{4}) = -1\frac{5}{12}$
5. Add: $(-4.1) + (-3.2) + (-.7)$.
   $|-4.1| + |-3.2| + |-.7| = 4.1 + 3.2 + .7 = 8.0$
   $(-4.1) + (-3.2) + (-.7) = -8.0$

When a positive and negative number are added (e.g., $+8$ and -5), the computation could be carried out in this way (read down):

$$(+8) \quad + (-5) = ?$$
$$\overbrace{(+3) + (+5)} + (-5) = ?$$
$$(+3) + \underbrace{(+5) + (-5)}$$
$$(+3) + \quad 0 \quad = +3$$

Similarly for the sum of $+9$ and -14:

$$(+9) + \quad (-14) \quad = ?$$
$$(+9) + \overbrace{(-9) + (-5)} = ?$$
$$\underbrace{(+9) + (-9)} + (-5)$$
$$0 \quad + (-5) = -5$$

However, a more direct procedure is given in the following program.

(4.2) *To compute the sum of two rational numbers when their signs differ:*

Step 1. Find the difference between the absolute values of the two rational numbers.

Step 2. Prefix the difference found in Step 1 with the sign of the addend having the greater absolute value.

EXAMPLES

1. Add $+8$ and -5.
   Step 1. Absolute value of $+8$: 8
   Absolute value of -5: 5
   Difference of absolute values: $8 - 5 = 3$
   Step 2. Sign of the addend having the greater absolute value $(8 > 5)$ is $+$; so: $(+8) + (-5) = +3$.
2. Add $+9$ and -14.
   Step 1. $|+9| = 9$
   $|-14| = 14$ (greater absolute value)
   $14 - 9 = 5$
   Step 2. $(+9) + (-14) = -5$

3. Add: $(-\frac{3}{5}) + (+\frac{2}{3})$.
Step 1. $|-\frac{3}{5}| = \frac{3}{5} = \frac{9}{15}$
$|+\frac{2}{3}| = \frac{2}{3} = \frac{10}{15}$ (greater absolute value)
$\frac{10}{15} - \frac{9}{15} = \frac{1}{15}$
Step 2. $(-\frac{3}{5}) + (+\frac{2}{3}) = +\frac{1}{15}$

4. Add: $(+.362) + (-1.427)$.
Step 1. $|+.362| = .362$
$|-1.427| = 1.427$ (greater absolute value)
$1.427 - .362 = 1.065$
Step 2. $(+.362) + (-1.427) = -1.065$

When there are several rational numbers to be added, some positive and some negative, the following applies.

(4.3) *To compute the sum of three or more rational numbers when their signs differ:*
Step 1. Find the sum of all of the positive rational numbers.
Step 2. Find the sum of all of the negative rational numbers.
Step 3. Add the sums of Steps 1 and 2 for the desired sum.

EXAMPLES

1. Add: $(+3) + (-2) + (+5) + (+6) + (-13)$.
Step 1. $(+3) + (+5) + (+6) = +14$ (Program 4.1)
Step 2. $(-2) + (-13) = -15$ (Program 4.1)
Step 3. $(+14) + (-15) = -1$ (Program (4.2)
2. Add: $(+\frac{1}{2}) + (-\frac{1}{3}) + (+5) + (-\frac{1}{6}) + (-1)$.
Step 1. $(+\frac{1}{2}) + (+5) = +5\frac{1}{2}$
Step 2. $(-\frac{1}{3}) + (-\frac{1}{6}) + (-1) = -1\frac{3}{6} = -1\frac{1}{2}$
Step 3. $(+5\frac{1}{2}) + (-1\frac{1}{2}) = +4$
3. Add: $(+.3) + (-.4) + (-1.5) + (+1.8)$.

Step 1	*Step 2*	*Step 3*	
+ .3	− .4	+2.1	
+1.8	−1.5	−1.9	
+2.1	−1.9	+ .2	(sum)

There is an important difference between a negative number and the *negative of* a number. A negative number is one whose standard numeral carries a minus sign. The negative of a given number is the "opposite" of the given number. For example:
the negative of $+3$ is -3
the negative of -3 is $+3$
the negative of $-\frac{1}{2}$ is $+\frac{1}{2}$
This distinction is important to the following procedure, and elsewhere in mathematics.

(4.4) *To compute the difference of two rational numbers, add the negative of the subtrahend (number to be subtracted) to the minuend.**

EXAMPLES

1. Subtract: $(+8) - (+5)$.
The negative of $+5$ is -5:
$(+8) - (+5) = (+8) + (-5) = +3$ (Program 4.2)
2. Subtract: $(-4) - (-5)$.
The negative of -5 is $+5$:
$(-4) - (-5) = (-4) + (+5) = +1$ (Program 4.2)
3. Subtract: $(-8) - (+3)$.
The negative of $+3$ is -3:
$(-8) - (+3) = (-8) + (-3) = -11$ (Program 4.1)
4. Subtract: $(-\frac{3}{4}) - (+\frac{1}{4})$.
$(-\frac{3}{4}) - (+\frac{1}{4}) = (-\frac{3}{4}) + (-\frac{1}{4}) = -\frac{4}{4} = -1$
5. Subtract: $(+1.63) - (-4.25)$.
$(+1.63) - (-4.25) = (+1.63) + (+4.25) = +5.88$

(Complete Exercise 4.1)

3. PRODUCTS AND QUOTIENTS OF RATIONAL NUMBERS

Programs for computing products and quotients of rational numbers, like those for computing sums and differences, are influenced by the basic definitions of the numbers.

(4.5) *To compute the product of two or more rational numbers:*

Step 1. Find the product of the absolute values of the factors.
Step 2. Prefix the product found in Step 1 with a plus sign if the number of negative factors entering the product is even (i.e., none, two, four, etc.), and with a minus sign if the number of negative factors entering the product is odd (i.e., one, three, five, etc.).

EXAMPLES

1. Multiply: $(-6) \times (-4) \times (+10)$.
Step 1. $|-6| \times |-4| \times |+10| = 6 \times 4 \times 10 = 240$
Step 2. The number of negative factors (two) is even, so the product is positive:
$(-6) \times (-4) \times (+10) = +240$
2. Multiply: $(-3) \times (-2) \times (-1) \times (-5) \times (-6)$.
Step 1. $3 \times 2 \times 1 \times 5 \times 6 = 180$
Step 2. The number of negative factors (five) is odd, so the product is negative:
$(-3) \times (-2) \times (-1) \times (-5) \times (-6) = -180$

* At the numeral level, this might be stated: *Reverse the sign of the subtrahend, and add.*

3. Multiply: $(-\frac{1}{2}) \times (+\frac{2}{5}) \times (-\frac{2}{3}) \times (+\frac{5}{7})$.
Step 1. $\frac{1}{2} \times \frac{2}{5} \times \frac{2}{3} \times \frac{5}{7} = \frac{2}{21}$
Step 2. The number of negative factors is even, so the product is positive:
$(-\frac{1}{2}) \times (+\frac{2}{5}) \times (-\frac{2}{3}) \times (+\frac{5}{7}) = +\frac{2}{21}$
4. Multiply: $(-1.2) \times (-3.4) \times (-.5)$.
Step 1. $1.2 \times 3.4 \times .5 = 2.04$
Step 2. $(-1.2) \times (-3.4) \times (-.5) = -2.04$

(4.6) *To compute the quotient of two rational numbers:*
Step 1. Find the quotient for the absolute values of the two rational numbers.
Step 2. Prefix the quotient found in Step 1 with a plus sign if the two rational numbers are both positive or both negative, and with a minus sign if one is positive and the other is negative.

EXAMPLES

1. Divide: $(-32) \div (-8)$.
Step 1. $|-32| \div |-8| = 32 \div 8 = 4$
Step 2. -32 and -8 are both negative, so the quotient is positive:
$(-32) \div (-8) = +4$
2. Divide: $(-42) \div (+6)$.
Step 1. $42 \div 6 = 7$
Step 2. -42 and $+6$ are of different signs, so the quotient is negative: $(-42) \div (+6) = -7$
3. Divide: $(+\frac{3}{5}) \div (-\frac{3}{2})$.
Step 1. $\frac{3}{5} \div \frac{3}{2} = \frac{3}{5} \times \frac{2}{3} = \frac{2}{5}$
Step 2. $(+\frac{3}{5})$ and $(-\frac{3}{2})$ differ in sign; quotient is negative: $(+\frac{3}{5}) \div (-\frac{3}{2}) = -\frac{2}{5}$
4. Divide: $(-.27) \div (-.9)$.
Step 1. $.27 \div .9 = .3$
Step 2. Both terms negative, quotient is positive: $(-.27) \div (-.9) = +.3$

4. RULE OF SIGNS FOR FRACTIONS

Since a fraction may be interpreted as expressing the quotient of numerator divided by denominator, i.e., $n \div d = \frac{n}{d}$, it follows from Program 4.6 that the sign of the fraction would have to be positive when both numerator and denominator are of like sign, and negative when numerator and denominator differ in sign. That is:

$$\frac{+n}{+d} = (+n) \div (+d) = +\frac{n}{d}$$

$$\frac{-n}{-d} = (-n) \div (-d) = +\frac{n}{d}$$

$$\frac{-n}{+d} = (-n) \div (+d) = -\frac{n}{d}$$

$$\frac{+n}{-d} = (+n) \div (-d) = -\frac{n}{d}$$

Consequently, $\frac{+n}{+d}$, $\frac{-n}{-d}$, and $+\frac{n}{d}$ are equivalent; and $\frac{-n}{+d}$, $\frac{+n}{-d}$, and $-\frac{n}{d}$ are equivalent. From this we can generalize to a procedure that is very useful, particularly when we become more deeply involved with the manipulations of algebraic symbolism:

> If we think of any fraction as possessing three signs—the sign of the numerator, the sign of the denominator, and the sign of the complete fraction —then *any pair of signs may be reversed without changing the value of the fraction.*

EXAMPLES

$$+\frac{(+2)}{(+3)} = +\frac{(-2)}{(-3)} = -\frac{(+2)}{(-3)} = -\frac{(-2)}{(+3)}$$

$$-\frac{(+a)}{(+b)} = -\frac{(-a)}{(-b)} = +\frac{(-a)}{(+b)} = +\frac{(+a)}{(-b)}$$

Reversing one or all three signs reverses the value of the number symbolized, i.e., it becomes the negative of its former value.

(Complete Exercise 4.2)

5. SOME ALGEBRAIC TERMINOLOGY

In algebra *variables*—generally symbolized by letters of the alphabet—are used to represent numbers, but not necessarily a specific number, as does a numeral. Rather, a variable represents the members of a set of numbers called the *domain of the variable.* If, for example, x (variable) represents any number in the set of whole numbers (domain of the variable), then it is true that

$$2 \times x = x + x$$

and

$$x - x = 0$$

These two statements, and many others like them, are true for any whole number that might be substituted for the variable x. To illustrate: if we replace x with 5, it is true that $2 \times 5 = 5 + 5$, and that $5 - 5 = 0$; or if we replace x with 13, it is true that $2 \times 13 = 13 + 13$, and that $13 - 13 = 0$; and so on.

A numeral, a variable, or combinations of them (e.g., $a + 5$, $x - 2$, $3x$, $m - n$) are called *algebraic expressions*. Included also are exponential expressions, called *powers*, which are a condensed way of stating products of the equal factors (e.g., $a \times a \times a = a^3$, $3 \times 3 = 3^2$, $t \times t \times t \times t = t^4$).

An algebraic expression that is either a numeral, a variable, or a product expressed by a numeral and one or more variables (e.g., $3a$, p^2, $5xy$, $3m^2n$) is called a *monomial*, or more often, a *term*. Every monomial has a *numerical coefficient*. In the case of $3a$ (which means $3 \times a$), the numerical coefficient is 3; in the case of $6xy$, the numerical coefficient is 6; in the case of x, the numerical coefficient is understood to be 1. That is, $x = 1 \times x$; also $n^3 = 1 \times n^3$, etc. In the case of a negative expression such as $-y$, the numerical coefficient is understood to be -1; i.e., $-y = -1 \times y$.

Terms that differ only in their numerical coefficients are called *similar* or *like* terms. For example, $4x$, $5x$, and x are similar terms because they are alike in variable; $2x$ and $2y$ are not similar terms even though they have the same numerical coefficient. Further, $2x$ and $2xy$ are not similar, but $2xy$ and $5xy$ are.

Sums and differences of two monomials are called *binomials* (e.g., $m + n$, $2a + 3$, $a^2 + ab^2$). Sums and differences of three monomials are called *trinomials*. Collectively, monomials, binomials, trinomials, and sums and differences of four or more terms are called *polynomials*.

6. ADDITION AND SUBTRACTION WITH POLYNOMIALS

The basic operations in algebra are the same as those in arithmetic: addition, subtraction, multiplication, and division. As in arithmetic, procedures for computing sums, differences, products, and quotients are carried out at the symbol level.

(4.7) *To compute the sum (difference) of like terms:*

Step 1. Add (subtract) the numerical coefficients of the terms.

Step 2. Affix to the sum (difference) of Step 1 the common variable part of the terms added (subtracted).

EXAMPLES

1. Add: $4x + x + 7x$.
   Step 1. Add: $4 + 1 + 7 = 12$
   Step 2. Affix the common variable part (x) to the sum of Step 1: $12x$ is the desired sum.
2. Add: $3y + (-2y) + 4y + \frac{1}{2}y$.
   Step 1. Add: $3 - 2 + 4 + \frac{1}{2} = 5\frac{1}{2}$
   Step 2. The common variable part is y: $5\frac{1}{2}y$ is the desired sum.
3. Subtract: $6y - 2y$.
   Step 1. $6 - 2 = 4$
   Step 2. The common variable part is y: $4y$ is the desired difference.
4. Subtract: $(-3p) - (2p)$.
   Step 1. $(-3) - (2) = (-3) + (-2) = -5$
   Step 2. The common variable part is p: $-5p$ is the desired difference.
5. Subtract: $(-6ab) - (-8ab)$.
   Step 1. $(-6) - (-8) = (-6) + (+8) = 2$
   Step 2. The common variable part is ab: $2ab$ is the difference.
6. Add: $4p + 3q$.
   The terms are unlike; their sum is expressed: $4p + 3q$.
7. Subtract: $(6a) - (5b)$.
   The terms are unlike; their difference is expressed: $6a - 5b$.

(4.8) *To compute the sum of several polynomials:*

Step 1. Arrange each polynomial addend so that like terms are in the same column.

Step 2. Add each column separately.

EXAMPLES

1. Add: $(3a + b - c) + (4a - 3b + 6c) + (2a + b + c)$.

Step 1. Arrange addends:
$$\begin{array}{rrr} 3a & + b & - c \\ 4a & - 3b & + 6c \\ 2a & + b & + c \\ \hline 9a & - b & + 6c \end{array} = 9a - b + 6c \text{ (sum)}$$
Step 2. Add by columns.

2. Add: $(4p + 6q + 11r) + (3p - 2r) + (3q - 4r)$.

Step 1. Arrange addends:
$$\begin{array}{rrr} 4p & + 6q & + 11r \\ 3p & & - 2r \\ & + 3q & - 4r \\ \hline 7p & + 9q & + 5r \end{array} = 7p + 9q + 5r \text{ (sum)}$$
Step 2. Add by columns.

Subtraction in algebra is accomplished by adding the negative of the subtrahend to the minuend (recall Program 4.4). Reversing the signs of the terms of a polynomial produces the negative of that polynomial. For example, the negative of $3a - 2b + 5c - d$ is $-3a + 2b - 5c + d$. This can be verified by adding the two according to Program 4.8, and observing that the sum is 0.

(4.9) *To compute the difference of two polynomials:*

Step 1. Arrange the minuend and subtrahend so that like terms are in the same column.

Step 2. Reverse the signs of the terms of the subtrahend.
Step 3. Add the polynomial of Step 2 to the minuend.

EXAMPLES

1. Subtract: $(6x - 4y) - (3x + 8y)$.
   Step 1. Arrange minuend and subtrahend: $\begin{array}{r} +6x - 4y \\ \underline{+3x + 8y} \end{array}$
   Step 2. Reverse the signs of subtrahend: $\begin{array}{r} +6x - 4y \\ \underline{-3x - 8y} \end{array}$
   Step 3. Add by columns: $\begin{array}{r} +6x - 4y \\ \underline{-3x - 8y} \\ \text{(difference) } +3x - 12y \end{array}$
2. Subtract $(4a - 2b + c)$ from $(3a - 4b + 8c)$.
   $$\begin{array}{r} +3a - 4b + 8c \\ \underline{(-) + 4a - 2b + c} \end{array} = \begin{array}{r} +3a - 4b + 8c \\ \underline{(+) - 4a + 2b - c} \\ \text{(difference) } -a - 2b + 7c \end{array}$$
3. Subtract from $(4x - 3y)$ the binomial $(3x - 2z)$.
   $$\begin{array}{r} +4x - 3y \\ \underline{(-) + 3x \quad - 2z} \end{array} = \begin{array}{r} +4x - 3y \\ \underline{(+) - 3x \quad + 2z} \\ \text{(difference) } +x - 3y + 2z \end{array}$$

(Complete Exercise 4.3)

7. MULTIPLICATION WITH POLYNOMIALS

A power, we have noted, is a product of equal factors. It may be represented by writing the equal factor with a numeral above and to its right to show the number of times the factor occurs in the product. For example:

$$a \cdot a \cdot a = a^3, \quad b \cdot b \cdot b \cdot b = b^4$$

In such expressions the repeated factor is called the *base*; the numeral written above and to the right is called the *exponent*. As with numerical coefficients, when no exponent appears in an expression, the exponent is assumed to be one (1). Thus:

$$a = a^1; \qquad b = b^1; \qquad ab = a^1b^1$$

Since a^3 means $a \cdot a \cdot a$, and a^4 means $a \cdot a \cdot a \cdot a$, when we multiply these two powers, a^3 and a^4, we get:

$$a^3 \cdot a^4 = (a \cdot a \cdot a) \cdot (a \cdot a \cdot a \cdot a)$$
$$= a \cdot a \cdot a \cdot a \cdot a \cdot a \cdot a = a^7$$

We could have obtained the same result by simply adding the exponents of the two factors:

$$a^3 \cdot a^4 = a^{3+4} = a^7$$

This illustrates one of the basic laws of exponents, and leads to the following computational procedure.

(4.10) *To compute the product of powers having the same base:*
Step 1. Add the exponents of the factors.
Step 2. Write as the desired product the common base with an exponent equal to the sum of the exponents found in Step 1.

EXAMPLES

1. Find the product of $a^4 \times a^3 \times a^7$.
   Step 1. $4 + 3 + 7 = 14$
   Step 2. The product is a^{14}.
2. Find the product of $a^3b^2 \cdot a^2b^2 \cdot a^4b$.
   [Note: Here all factors are not of the same base. When this is so, powers having the same base are multiplied together independently.]

$$a^3b^2 \cdot a^2b^2 \cdot a^4b^1 = a^3 \cdot a^2 \cdot a^4 \cdot b^2 \cdot b^2 \cdot b^1$$
$$= a^{3+2+4}b^{2+2+1}$$
$$= a^9b^5$$

Actually, a multisymboled monomial expresses a product in itself. For instance, the monomial $6ab$ may be thought of as the product of $6 \cdot a \cdot b$. Therefore the product of two monomials, say $(6ab) \times (2ab^2c)$, may be thought of as $6 \cdot a \cdot b \cdot 2 \cdot a \cdot b^2 \cdot c$ or $12a^2b^3c$.

(4.11) *To compute the product of several monomials:*
Step 1. Multiply the numerical coefficients of the factors for the numerical coefficient of the product.
Step 2. Multiply the variable parts of the factors exponentially for the variable part of the product.

EXAMPLES

1. Multiply: $(4a^2b) \times (-2ac^2) \times (bc^3)$.
   Step 1. $(4) \times (-2) \times (1) = -8$.

   $$(a^2b) \times (ac^2) \times (bc^3) = a^2 \times b^1 \times a^1 \times c^2 \times b^1 \times c^3$$
   $$= a^{2+1}b^{1+1}c^{2+3} = a^3b^2c^5$$
   The product is $-8a^3b^2c^5$.
2. Multiply: $(-3a^2b) \times (-2a^3b) \times (4)$.
   Step 1. $(-3) \times (-2) \times (+4) = +24$.
   Step 2. $(a^2b) \times (a^3b) = a^2 \cdot b^1 \cdot a^3 \cdot b^1 = a^5b^2$.
   The product is $24a^5b^2$.
3. Multiply: $(6xy^3) \times (-3xy) \times (-x^3y)$.
   Step 1. $(+6) \times (-3) \times (-1) = +18$
   [Note: $-x^3y$ may be thought of as $-1x^3y$.]
   Step 2.
   $$(xy^3) \times (xy) \times (x^3y) = x^1 \cdot y^3 \cdot x^1 \cdot y^1 \cdot x^3 \cdot y^1$$
   $$= x^5y^5$$
   The product is $18x^5y^5$.

Among the most important properties of numbers in mathematics is the *distributive property*. In symbols it can be expressed as

$$a \times (b + c) = (a \times b) + (a \times c)$$

In effect, the distributive property holds that if we add b and c together and then multiply that sum by a, or if we multiply b and c each by a and then add the resulting products, the final result will be the same in both cases. For example:

$$3 \times (2 + 5) \overset{?}{=} (3 \times 2) + (3 \times 5)$$

$$3 \times \quad 7 \overset{?}{=} \quad 6 \quad + \quad 15$$

$$21 = 21$$

The following program is essentially an expression of the distributive property.

(4.12) *To compute the product of a monomial and a polynomial:*

Step 1. Multiply each term of the polynomial by the monomial.

Step 2. Add the products of Step 1 for the desired product.

EXAMPLES

1. Multiply: $6a \times (3b - 2a)$, or $6a(3b - 2a)$.
   Step 1. $(6a) \times (3b) = 18ab$
   $(6a) \times (-2a) = -12a^2$
   Step 2. Add the products of Step 1: $18ab - 12a^2$.
2. Multiply: $5a^3b(3a^2 - 2b + 4)$.
   Step 1. $(5a^3b)(3a^2) = 15a^5b$
   $(5a^3b)(-2b) = -10a^3b^2$
   $(5a^3b)(+4) = 20a^3b$
   Step 2. Add products: $15a^5b - 10a^3b^2 + 20a^3b$.
3. Multiply: $-3a^3(4c - 2a^4 - 6ab^2)$.
   Step 1. $(-3a^3)(4c) = -12a^3c$
   $(-3a^3)(-2a^4) = 6a^7$
   $(-3a^3)(-6ab^2) = 18a^4b^2$
   Step 2. Add products: $-12a^3c + 6a^7 + 18a^4b^2$.
4. Multiply: $-c(a^3 - 2b^2 + d)$.
   Step 1. $(-c)(a^3) = -ca^3$
   $(-c)(-2b^2) = 2b^2c$
   $(-c)(+d) = -cd$
   Step 2. Add products: $-ca^3 + 2b^2c - cd$.

The general program for computing the product of any two polynomials is very similar to that for computing the product of two multidigit numbers in arithmetic: we build up partial products between one factor and each term of the other factor, then add the partial products.

(4.13) *To compute the product of two polynomials:*

Step 1. Write one of the polynomial factors directly above the other.

Step 2. Multiply each term of the bottom factor with all terms of the top factor and arrange the terms of the resulting partial products in columns according to likeness.

Step 3. Add the terms of the partial products of Step 2 for the desired product.

EXAMPLES

1. Find the product of $(2x + 1) \times (3x + 2)$.

$$\text{Step 1.} \begin{cases} 3x + 2 \\ 2x + 1 \end{cases}$$

$$\text{Step 2.} \begin{cases} 6x^2 + 4x \\ \qquad\quad 3x + 2 \end{cases}$$

$$\text{Step 3.} \quad 6x^2 + 7x + 2$$

[Note: Since there is no "carrying" in algebraic multiplication, the computation may be performed from left to right. This usually helps in organizing the partial product terms by likeness.]

2. Multiply $(3a - 4c)$ by $(2a + 3y)$.

$$\text{Step 1.} \begin{cases} 3a - 4c \\ 2a + 3y \end{cases}$$

$$\text{Step 2.} \begin{cases} 6a^2 - 8ac \\ \qquad\qquad + 9ay - 12cy \end{cases}$$

$$\text{Step 3.} \quad 6a^2 - 8ac + 9ay - 12cy$$

3. Multiply: $(4x^2 - 3x)(2x + 7)$.

$$\text{Step 1.} \begin{cases} 4x^2 - 3x \\ \quad 2x + 7 \end{cases}$$

$$\text{Step 2.} \begin{cases} 8x^3 - 6x^2 \\ \qquad + 28x^2 - 21x \end{cases}$$

$$\text{Step 3.} \quad 8x^3 + 22x^2 - 21x$$

4. Multiply: $(4a - 2b + 3c)(2a - b - 4c)$.

$$\text{Step 1.} \begin{cases} 4a - 2b + 3c \\ 2a - b - 4c \end{cases}$$

$$\text{Step 2.} \begin{cases} 8a^2 - 4ab + 6ac \\ \qquad - 4ab \qquad\qquad + 2b^2 - 3bc \\ \qquad\qquad - 16ac \qquad\qquad + 8bc - 12c^2 \end{cases}$$

$$\text{Step 3.} \quad 8a^2 - 8ab - 10ac + 2b^2 + 5bc - 12c^2$$

8. SYMBOLS OF GROUPING

Parentheses ( ), brackets [ ], and braces { }, are the usual symbols of grouping used in algebra. They indicate that certain algebraic expressions are to be treated as a single expression. For instance, if we wish to multiply the sum of $a + b$ by c, we express the complete operation as $c \cdot (a + b)$ or $c(a + b)$. In effect, these symbols of grouping can be looked upon as the punctuation marks of our algebraic language.

In the English language we know it is possible for the same sequence of words to have two totally different meanings depending upon how the words are grouped. Consider the two statements, which differ only by a comma:

"Let us eat, John."
"Let us eat John."

For an arithmetic parallel:

(a) Three times five, plus six.
(b) Three times five plus six.

Without punctuating parentheses $3 \times 5 + 6$ must serve both (a) and (b). With punctuating parentheses it is possible to make clear the distinction between the two:

(a) $(3 \times 5) + 6 = 15 + 6 = 21.$
) $3 \times (5 + 6) = 3 \times 11 = 33.$

Often in the course of our algebraic work we have need for either removing or inserting symbols of grouping. The case in which a minus sign precedes a grouped expression, as $-(a - b)$, deserves special comment. In this we have another instance of the "unwritten numeral 1" (others, e.g., $6 = 1 \times 6 = \frac{6}{1}$, $1 \times a = a^1 = a$, etc.). Here, too, the 1 goes unwritten, but its minus sign *is* written; thus:

$$\begin{aligned} -(a - b) &= -1 \times (a - b) \\ &= (-1)(a) + (-1)(-b) \\ &= -a + b \end{aligned}$$

When there are several symbols of grouping involved, the convention is to begin with the innermost, and work outward.

EXAMPLES

1. $\{-3a[4b - a(c - d)] + 14ab\}$
   $\{-3a[4b - ac + ad] + 14ab\}$
   $\{-12ab + 3a^2c - 3a^2d + 14ab\}$
   $3a^2c - 3a^2d + 2ab$
2. $-3\{a + [6(a - 2b)] + b - [7(a + 3b)]\}$
   $-3\{a + [6a - 12b] + b - [7a + 21b]\}$
   $-3\{a + 6a - 12b + b - 7a - 21b\}$
   $-3\{-32b\} = 96b$

The insertion of parentheses or other symbols of grouping is but the inverse of the foregoing. It must be remembered, however, that *prefixing a grouped expression with a minus sign negates the enclosed expression; therefore the sign of each term of the enclosed expression must be reversed if it is to remain equivalent to the original expression.*

EXAMPLES

1. $-6 + 2x - 4y = -(6 - 2x + 4y).$
   To check, apply the distributive property:
   $$\begin{aligned} -(6 - 2x + 4y) &= (-1)(+6) + (-1)(-2x) \\ &\quad + (-1)(+4y) \\ &= -6 + 2x - 4y \end{aligned}$$
2. $$\begin{aligned} -3x^2 + 4x - 7y &= -(3x^2 - 4x + 7y) \\ &= -[3x^2 - (4x - 7y)]. \end{aligned}$$

(Complete Exercise 4.4)

9. DIVISION WITH POLYNOMIALS

The relationship between the fraction symbol and division allows us to express the quotient for $a^7 \div a^3$ $(a \neq 0)$, for example, as a^7/a^3, which in turn may be expressed as:

$$\frac{a \cdot a \cdot a \cdot a \cdot \cancel{a} \cdot \cancel{a} \cdot \cancel{a}}{\cancel{a} \cdot \cancel{a} \cdot \cancel{a}} = a^4$$

On the other hand, had the two terms been reversed, that is $a^3 \div a^7$, we would have:

$$a^3 \div a^7 = \frac{a^3}{a^7} = \frac{\cancel{a} \cdot \cancel{a} \cdot \cancel{a}}{a \cdot a \cdot a \cdot a \cdot \cancel{a} \cdot \cancel{a} \cdot \cancel{a}} = \frac{1}{a^4}$$

In both cases the exponent in the final form of the quotient could have been obtained by subtracting the smaller exponent of the two terms from the larger exponent. Later in mathematics we shall find it advantageous to define negative exponents, but until that time we use a two-part program for division of powers of the same base.

(4.14) *To compute the quotient of two powers having the same base:*

(a) *When the exponent of the dividend (numerator) is greater than the exponent of the divisor (denominator):*

Step 1. Subtract the exponent of the divisor from the exponent of the dividend.

Step 2. Write as the desired quotient the common base (zero excepted) with an exponent equal to the difference of the exponents found in Step 1.

EXAMPLES

1. Divide b^4 by b^3 $(b \neq 0)$.
   Step 1. $4 - 3 = 1$
   Step 2. The desired quotient is b^1 or b.
2. Divide: $a^9 \div a^3$ $(a \neq 0)$.
   Step 1. $9 - 3 = 6$
   Step 2. The desired quotient is a^6.

(b) *When the exponent of the divisor (denominator) is greater than that of the dividend (numerator):*

Step 1. Subtract the exponent of the dividend from the exponent of the divisor.

Step 2. Write as the desired quotient a fraction whose numerator is one and whose denominator is the common base (zero excepted) with an exponent equal to the difference of the exponents found in Step 1.

EXAMPLES

1. Divide k^5 by k^8 $(k \neq 0)$.
   Step 1. $8 - 5 = 3$
   Step 2. The desired quotient is $\frac{1}{k^3}$.
2. Divide: $x^3 \div x^7$ $(x \neq 0)$.

$$x^3 \div x^7 = \frac{1}{x^{7-3}} = \frac{1}{x^4}$$

These relationships among exponents under division suggest a plausible and useful definition for the exponent 0. For example, note what happens when we apply Program 4.14(a):

$$x^4 \div x^4 = x^{4-4} = x^0$$

But $x^4 \div x^4$ may be interpreted as a number divided by itself; and a number divided by itself (0 excepted) invariably has 1 for a quotient. Thus the general definition: *Any power having a nonzero base and an exponent of* 0 *is equal to* 1.

EXAMPLES

1. $b^0 = 1$ $(b \neq 0)$
2. $8^0 = 1$
3. $(-367\tfrac{1}{2})^0 = 1$

The quotient of two monomials is expressed most simply by a fraction in lowest terms. Recall that a monomial may be thought of as the product of its several parts as factors, e.g., $6a^3b = 2 \cdot 3 \cdot a \cdot a \cdot a \cdot b$.

(4.15) *To compute the quotient of two monomials:*

Step 1. Express dividend and divisor as numerator and denominator, respectively.

Step 2. Divide out common factors.

EXAMPLES

1. Divide: $36a^3b^2c$ by $9ab^2$.

Step 1. $\frac{36a^3b^2c}{9ab^2}$

Step 2. $\frac{2 \cdot 2 \cdot \cancel{3} \cdot \cancel{3} \cdot \cancel{a} \cdot a \cdot a \cdot \cancel{b} \cdot \cancel{b} \cdot c}{\cancel{3} \cdot \cancel{3} \cdot \cancel{a} \cdot \cancel{b} \cdot \cancel{b}} = 4a^2c$

2. Divide: $-6a^3bc^2$ by $2ab^4$.

$$\frac{\overset{-3\,a^2}{-\cancel{6}\,\cancel{a^3}\,\cancel{b}\,c^2}}{\underset{b^3}{\cancel{2}\,\cancel{a}\,\cancel{b^4}}} = \frac{-3a^2c^2}{b^3}$$

3. Divide: $(-25x^3) \div (-5x^2yz)$.

$$\frac{\overset{+5\,x^1}{-\cancel{25}\,\cancel{x^3}}}{-\cancel{5}\,\cancel{x^2}\,yz} = \frac{5x}{yz}$$

Division is said to be distributive from the right, that is the divisor distributes over the dividend:

$$(6 + 4) \div 2 \stackrel{?}{=} (6 \div 2) + (4 \div 2)$$
$$10 \div 2 \stackrel{?}{=} 3 + 2$$
$$5 = 5$$

[Note: The reverse is not true; e.g.:

$$2 \div (6 + 4) \stackrel{?}{=} (2 \div 6) + (2 \div 4)$$
$$2 \div 10 \stackrel{?}{=} \tfrac{1}{3} + \tfrac{1}{2}$$
$$\tfrac{1}{5} \neq \tfrac{5}{6}$$]

We take advantage of the right distributive property to develop the following computational procedure.

(4.16) *To compute the quotient of a polynomial divided by a monomial:*

Step 1. Divide each term of the polynomial by the monomial.

Step 2. Add the partial quotients of Step 1 for the desired quotient.

EXAMPLES

1. Divide $(6a^4b^3 - 3a^2b^3)$ by $3ab$.

Step 1. $\frac{\overset{2}{\cancel{6}}\,a^{\overset{3}{\cancel{4}}}\,b^{\overset{2}{\cancel{3}}}}{\cancel{3}\,\cancel{a}\,\cancel{b}} = 2a^3b^2$

$\frac{\overset{-1}{-\cancel{3}}\,a^{\overset{1}{\cancel{2}}}\,b^{\overset{2}{\cancel{3}}}}{\cancel{3}\,\cancel{a}\,\cancel{b}} = (-1)(ab^2) = -ab^2$

Step 2. The desired quotient is $2a^3b^2 - ab^2$.

2. Divide: $(18x^4y - 15x^2y^3) \div (3x^3y^2)$.

Step 1. $\frac{\overset{6}{\cancel{18}}\,x^{\overset{1}{\cancel{4}}}\,\cancel{y}}{\cancel{3}\,\cancel{x^3}\,y^{\overset{1}{\cancel{2}}}} = \frac{6x}{y}$

$\frac{\overset{-5}{-\cancel{15}}\,\cancel{x^2}\,y^{\overset{1}{\cancel{3}}}}{\cancel{3}\,x^{\overset{1}{\cancel{3}}}\,\cancel{y^2}} = \frac{-5y}{x} = -\frac{5y}{x}$

Step 2. The desired quotient is $\frac{6x}{y} - \frac{5y}{x}$

The general program for computing the quotient of any two polynomials parallels that of division in arithmetic.

(4.17) *To compute the quotient of two polynomials:*

Step 1. Arrange both polynomials in descending powers of the same variable.

Step 2. Compute the first term of the quotient by dividing the first term of the dividend by the first term of the divisor.

Step 3. Multiply the divisor by the quotient term of Step 2.

Step 4. Subtract the product of Step 3 from the dividend; bring down the next term of the original dividend to form the new dividend.

Step 5. Repeat the loop or sequence of Steps 2, 3, and 4 on the new dividend; keep repeating this sequence of steps until the exponent of the first term of any remainder is less than the exponent of the first term of the divisor.

EXAMPLES

1. Divide $(8x + 6x^2 + 2)$ by $(2x + 2)$.

Step 1. $(6x^2 + 8x + 2) \div (2x + 2)$

Step 2. $\frac{6x^2}{2x} = 3x$

Step 3.

$$\begin{array}{r} 3x + 1 \\ 2x + 2 \overline{)6x^2 + 8x + 2} \\ (3x)(2x+2) \longrightarrow 6x^2 + 6x \\ \hline 2x + 2 \\ 2x + 2 \\ \hline \end{array}$$

Step 4. Subtract and bring down $+2$

Step 5.
$$\begin{cases} \frac{2x}{2x} = 1 \\ (1) \times (2x + 2) = 2x + 2 \\ (2x + 2) - (2x + 2) = 0 \end{cases}$$

2. Divide: $(2x + 6x^2 - 20) \div (7 + 3x)$.

Step 1. $(6x^2 + 2x - 20) \div (3x + 7)$

Step 2. $\frac{6x^2}{3x} = 2x$

Step 3.

$$\begin{array}{r} 2x - 4 \\ 3x + 7 \overline{)6x^2 + 2x - 20} \\ 6x^2 + 14x \\ \hline -12x - 20 \\ -12x - 28 \\ \hline +8 \end{array}$$

Step 4. Subtract and bring down -20

Step 5. $\frac{-12x}{3x} = -4$

$(-4)(3x + 7) = -12x - 28$

$(-12x - 20) - (-12x - 28) = +8$ (remainder)

As in arithmetic, when a remainder occurs it is treated simply as a remainder or it is added to the quotient as a fraction whose denominator is the divisor. Thus:

$$(6x^2 + 2x - 20) \div (3x + 7) = \begin{cases} 2x - 4, \text{R}(+8) \\ \text{or} \\ 2x - 4 + \frac{8}{3x + 7} \end{cases}$$

3. Divide $(x^4 - y^4)$ by $(x - y)$.

$$\begin{array}{r} x^3 + x^2y + xy^2 + y^3 \\ x - y \overline{)x^4 - y^4} \\ x^4 - x^3y \\ \hline x^3y - y^4 \\ x^3y - x^2y^2 \\ \hline x^2y^2 - y^4 \\ x^2y^2 - xy^3 \\ \hline xy^3 - y^4 \\ xy^3 - y^4 \\ \hline \end{array}$$

(Complete Exercise 4.5)

Name.. Date.......................... **EXERCISE 4..**

1. Locate $-1\frac{1}{2}$, $+.5$, $+\frac{5}{4}$, -2, $-\frac{5}{8}$, $+2.3$, -3.1, -2.9, $+3.0$, $-\frac{1}{3}$, at an appropriate place on the following number line.

0 +1

2. Insert the appropriate symbol: $<$, $>$, $=$.

$+4$........$+6$	-8........$+10$	$+32$........$+23$	-12........-3	$+6$........-7
-2........-3	$+3$........$-\frac{27}{9}$	-3.2........-4.1	$+\frac{1}{4}$........$-\frac{5}{8}$	0........$-.1$

3. Use set notation to specify the set whose members are:

(a) Integers greater than $+2$ but less than $+7$. { ...., ..., ..., ... }

(b) Integers between (does not include) $+4$ and $+8$.

(c) Integers between -5 and $+2$.

(d) Positive integers between -3 and $+6$.

(e) Integers between -3 and -4.

(f) The four nonzero integers nearest 0 in value.

Write the equivalent (arithmetic) number for each.

4. $|+12| =$ $|+3| =$ $|-8| =$ $|-91| =$ $|-\frac{1}{4}| =$ $|+\frac{2}{3}| =$

5. $|+\frac{5}{8}| =$ $|-\frac{7}{3}| =$ $|0| =$ $|+3.06| =$ $|-3.06| =$ $|-.407| =$

Add.

6. $+13$ / $+\ 6$

7. -86 / -13

8. $+362$ / $+168$

9. -86.3 / -17.5

10. $+163.28$ / $+\ 7.43$

11. $+\frac{7}{8}$ / $+\frac{2}{3}$

12. $-\frac{17}{3}$ / $-\frac{14}{5}$

13. $+63$ / -15

14. $+128$ / $-\ 67$

15. -384 / $+126$

16. $+17.63$ / $-\ 8.25$

17. $+376$ / -954

18. -1836 / $+2748$

19. $+\ 63.24$ / -187.57

20. $+\frac{3}{4}$ / $-\frac{1}{2}$

21. $\begin{array}{r} +\frac{87}{8} \\ +\frac{42}{5} \\ \hline \end{array}$

22. $\begin{array}{r} -16\frac{3}{4} \\ +12\frac{1}{5} \\ \hline \end{array}$

23. $\begin{array}{r} +76\frac{1}{8} \\ +82\frac{3}{5} \\ \hline \end{array}$

24. $\begin{array}{r} -42.63 \\ +\ 1.78 \\ \hline \end{array}$

25. $\begin{array}{r} +\ .036 \\ -4.071 \\ \hline \end{array}$

26. $\begin{array}{r} -14 \\ -\ 6 \\ -\ 8 \\ \hline \end{array}$

27. $\begin{array}{r} -42 \\ +36 \\ -51 \\ \hline \end{array}$

28. $\begin{array}{r} -86 \\ +12 \\ -48 \\ +63 \\ \hline \end{array}$

29. $\begin{array}{r} +1.8 \\ -1.4 \\ +2.7 \\ -1.8 \\ \hline \end{array}$

30. $\begin{array}{r} +32.6 \\ -\ 1.4 \\ +26.2 \\ \hline \end{array}$

31. $\begin{array}{r} +6.37 \\ -4.89 \\ -7.56 \\ \hline \end{array}$

32. $\begin{array}{r} -\frac{2}{3} \\ +\frac{3}{4} \\ -\frac{1}{6} \\ \hline \end{array}$

33. $\begin{array}{r} -6\frac{2}{3} \\ +5\frac{7}{8} \\ -1\frac{2}{3} \\ +6\frac{1}{6} \\ \hline \end{array}$

Subtract the lower rational number from the upper.

34. $\begin{array}{r} +14 \\ +\ 8 \\ \hline \end{array}$

35. $\begin{array}{r} -18 \\ -\ 6 \\ \hline \end{array}$

36. $\begin{array}{r} -27 \\ -13 \\ \hline \end{array}$

37. $\begin{array}{r} +1.8 \\ +1.4 \\ \hline \end{array}$

38. $\begin{array}{r} +81.7 \\ +42.2 \\ \hline \end{array}$

39. $\begin{array}{r} -32.64 \\ -83.46 \\ \hline \end{array}$

40. $\begin{array}{r} -54.827 \\ -29.378 \\ \hline \end{array}$

41. $\begin{array}{r} +632 \\ -817 \\ \hline \end{array}$

42. $\begin{array}{r} +7739 \\ -4832 \\ \hline \end{array}$

43. $\begin{array}{r} -43{,}206 \\ +17{,}302 \\ \hline \end{array}$

44. $\begin{array}{r} +\frac{3}{5} \\ -\frac{1}{5} \\ \hline \end{array}$

45. $\begin{array}{r} +\frac{2}{7} \\ -\frac{5}{14} \\ \hline \end{array}$

46. $\begin{array}{r} -\frac{13}{21} \\ -\frac{11}{18} \\ \hline \end{array}$

47. $\begin{array}{r} -\frac{82}{9} \\ -\frac{17}{18} \\ \hline \end{array}$

48. $\begin{array}{r} -\frac{31}{40} \\ +\frac{99}{8} \\ \hline \end{array}$

49. $\begin{array}{r} +36\frac{2}{5} \\ -16\frac{1}{3} \\ \hline \end{array}$

50. $\begin{array}{r} +15\frac{2}{3} \\ -71\frac{5}{11} \\ \hline \end{array}$

51. $\begin{array}{r} -12.03 \\ -\ 1.67 \\ \hline \end{array}$

52. $\begin{array}{r} -12.38 \\ -52.47 \\ \hline \end{array}$

53. $\begin{array}{r} +20\frac{2}{9} \\ -19\frac{8}{9} \\ \hline \end{array}$

54. How many degrees difference is there between these pairs of temperatures?

(a) +47° and +96° (b) −6° and +8° (c) +62° and −12° (d) −47° and −8°

55.

	High	Low	Last	Net change
Comdat	$50\frac{1}{8}$	$47\frac{1}{2}$	$48\frac{1}{4}$	$+\frac{5}{8}$
New Prod.	$23\frac{1}{2}$	$22\frac{1}{4}$	$22\frac{3}{4}$	$-1\frac{7}{8}$

Interpretation: During the day reported, a share of Comdat traded for as high as $50\frac{1}{8}$ (\$$50\frac{1}{8}$ or \$50.12$\frac{1}{2}$) and as low as $47\frac{1}{2}$ (\$$47\frac{1}{2}$ or \$47.50). The last trade, at the close of the market, was at $48\frac{1}{4}$. This was \$$\frac{5}{8}$ (or 62$\frac{1}{2}$¢) better than the price of the last sale on the day before.

(a) What must have been the price of the last sale of Comdat the day before? ....

(b) How wide a price range was there in Comdat on the day reported? ....

(c) Comdat closed at how much higher than its low for the day? ....

(d) How wide a price range was there in New Products that day? ....

(e) What must have been the closing price for New Products on the day before? ....

Name.. Date.................. **EXERCISE 4.2**

Multiply.

1. $(-2) \times (+3) =$

2. $(+6) \times (-8) =$ -48

3. $(-3) \times (-7) =$

4. $(+5) \times (-32) =$

5. $(-\frac{2}{3}) \times (+6) =$

6. $(-27) \times (-\frac{1}{3}) =$ +9

7. $(-\frac{3}{5}) \times (-\frac{5}{8}) =$

8. $(-1.62) \times (+.03) =$

9. $(+.0042) \times (+.007) =$

10. $(+6.027) \times (-.03) =$

11. $(-8) \times (-3) \times (+6) =$

12. $(-2) \times (-5) \times (-7) =$

13. $(+\frac{2}{3}) \times (-\frac{3}{5}) \times (+\frac{9}{11}) =$

14. $(+.02) \times (-.06) \times (-.7) =$

15. $(-2) \times (-1) \times (+3) \times (-4) =$

16. $(-8.1) \times (-\frac{2}{3}) \times (+.3) =$

17. $(-.003) \times (-.02) \times (-.1) \times (-\frac{1}{5}) =$

18. $(-2) \times (+1) \times (+3) \times (-6) \times (-4) \times (-3) =$

19. $(+627.83) \times (-6.2) =$

20. $(-42.34) \times (-60.02) =$

Divide.

21. $(-4) \div (-2) =$ +2

22. $(-16) \div (+8) =$ -2

23. $(-42) \div (-6) =$

24. $(-32{,}256) \div (-48) =$ +672

25. $(+20{,}794) \div (-37) =$

26. $(+6783) \div (+42) =$ + 161.5

27. $(-\frac{3}{4}) \div (+\frac{2}{5}) =$

28. $(+\frac{3}{5}) \div (+\frac{4}{5}) =$ $+\frac{3}{4}$

29. $(-\frac{7}{22}) \div (+\frac{2}{11}) =$

EVEN

Divide.

30. $(+8) \div (+\frac{3}{4}) =$

31. $(-1\frac{2}{3}) \div (-6\frac{2}{3}) =$

32. $(-51) \div (+3\frac{2}{5}) =$

33. $(-8.40) \div (+.35) =$

34. $(+29{,}664) \div (-.36) =$

35. $(-.005832) \div (-3.6) =$

Carry out the operations, and simplify.

36. $\frac{(-16)}{(-4)} =$

37. $\frac{(32)}{(-2)} =$

38. $\frac{(-25)}{(-10)} =$

39. $\frac{(-36)}{(12)} =$

40. $\frac{(6)(-3)}{(-9)} =$

41. $\frac{(-4)+(-3)}{14} =$

42. $\frac{4(-2)+6}{2} =$

43. $\frac{(-24) \div (-3)}{(-2)} =$

44. $\frac{(7)(-28)}{(4)(-7)} =$

45. $\frac{(4) - 3(0)}{-5} =$

46. $\frac{(5) - (25)}{(-1)(-15)} =$

47. $\frac{16 \times (-1)}{(-2)(4)(-1)} =$

Insert the proper sign to make the pairs of fractions equivalent.

	(a)	(b)	(c)	(d)
48.	$(-)\frac{-5}{+6} = (\ \ )\frac{5}{6}$	$(+)\frac{15}{10} = (\ \ )\frac{-3}{-2}$	$(+)\frac{-3}{-7} = (\ \ )\frac{-3}{7}$	$(-)\frac{+a}{-b} = (\ \ )\frac{-a}{-b}$
49.	$(+)\frac{-10}{-7} = (\ \ )\frac{10}{7}$	$(-)\frac{-k}{+m} = (\ \ )\frac{-k}{-m}$	$(-)\frac{1}{-4} = (\ \ )\frac{4}{-16}$	$(+)\frac{-s}{+c} = (\ \ )\frac{+s}{-c}$

50. Rainfall for a 6-month period varied from the monthly average of 5 inches as follows: -3, $+2$, 0, $+4$, -4, $+6$. How many inches of rain fell during the 6-month period?

....

51. How many -16's are there in -432?

....

Name.. Date.......................... **EXERCISE 4.3**

Add.

	(a)	(b)	(c)	(d)	(e)	(f)	(g)	(h)	(i)
1.	$4x$	$3x$	$-2r$	$-6d$	$5x^2$	$3xy$	$3p^2$	$2m$	$3.2s$
	$3x$	$-2x$	$-5r$	$2d$	$3x^2$	$-5xy$	$-5p^2$	$-m$	$-5.6s$
					$-8x^2$	$-7xy$	$4k$	$-\frac{1}{2}m$	$-.7s$

2. (a) $3x + 2x - 7x =$ (b) $5y + .5y + .05y =$ (c) $3x + .3x - .03x =$

3. (a) $\frac{1}{2}x - 7x + \frac{1}{4}x =$ (b) $\frac{1}{5}ap - \frac{1}{10}ap - \frac{1}{100}ap =$ (c) $-3q - 6q^2 + 9q =$

4.
$$\begin{array}{r} 6xy - 2xy^2 \\ -3xy - 17xy^2 \\ 4xy + 12xy^2 \\ - xy - xy^2 \\ \hline \end{array}$$

5.
$$\begin{array}{r} 3x - 4p \\ -2x \\ 3x - 3p \\ - 4p \\ \hline \end{array}$$

6.
$$\begin{array}{r} 7xy - 3y + 2z \\ - 4y + 7z \\ 3xy + 5y + 2z \\ -2xy - 8z \\ \hline \end{array}$$

7.
$$\begin{array}{r} x + 3y - 12p^2 \\ -x + 5y \\ - 5y + 17p^2 \\ 3x - 4p^2 \\ \hline \end{array}$$

8. $(4x - 3y); (3d + 4x);$
$(3y + 2z); (5z - 8x)$

9. $(3x - 2y); (4x - 3y + 2z);$
$(4z - 6x); (2y + 5z)$

10. $(.3x - 2y); (.6x + .7y);$
$(-.8x - 1.4y)$

11. $(\frac{2}{3}x - \frac{1}{2}y); (\frac{3}{4}x - \frac{1}{3}y);$
$(-\frac{5}{6}x - \frac{3}{2}y)$

Subtract the lower term from the upper term.

	(a)	(b)	(c)	(d)	(e)	(f)	(g)	(h)
12.	$+14r$	$-18m$	$+1.2k$	$+16a$	$+17c$	$-\frac{3}{5}t$	$-27d$	$+2.6d$
	$+7r$	$-6m$	$+1.4k$	$-37b$	$-12c$	$+\frac{1}{3}t$	$-34e$	$-1.8d$

Subtract the right polynomial from the left.

13. $(3x - 2y); (6x - 7y)$

14. $(3x - 2y); (2y - 3x)$

15. $(3x - 4y); (6y + 12)$

16. $(x + 3p); (3p + 2)$

17. $(x - \frac{1}{2}y); (\frac{1}{2}x - \frac{3}{4}y)$

18. $(.6a - .3b); (-.4b + .7a)$

19. $(3x - 7y); (2p + 3x - 15y)$

20. $(.1a - 3.4b - 2.7c); (6d - 3.2a - .5c)$

Simplify.

21. $3x - 2y + 3x - 5x + 3y - x - 3y =$ ....

22. $5a - 3b + 2a - 6c - 7a - 15b - 4c =$ ....

23. $6ab - .6b + 3.2ab - .7b + 4ab - .1b =$ ....

24. $\frac{1}{2}x - \frac{3}{4}y + \frac{2}{3}x + \frac{1}{4}y - \frac{1}{5}x =$ ....

25. $2g - 5h - .3g + \frac{1}{2}h + \frac{1}{5}g - .4h =$ ....

26. Add $(3x - 2y)$ to itself, and subtract the sum from $(6y - 7x)$.

....

27. Increase $(4x - 7)$ by 3 and subtract that from $(5x - 12)$.

....

28. (a) Subtract $6x - 3y$ from 1.
(b) Subtract $4x - 3y$ from 0.

(a) ....

(b) ....

29. What must be added to $(3x - 2y)$ to make $(4y + z)$?

....

Name.. Date.......................... **EXERCISE 4.4**

Express the products exponentially.

1. $a^3 \cdot a^2 =$ $\qquad$ $b^4 \cdot b^3 \cdot b =$ $\qquad$ $d^2 \cdot d \cdot d^{10} =$ $\qquad$ $x^3 \cdot x^5 \cdot x^7 =$

2. $c^2 \cdot c^2 \cdot d =$ $\qquad$ $a^2b^2 \cdot a^4b^2 \cdot ab^3 =$ $\qquad$ $a^3b^2 \cdot a^5 \cdot b^7 \cdot a =$ $\qquad$ $x^3y \cdot x^2y^3 \cdot x^4 =$

Multiply.

3. $\begin{array}{r} 5x \\ -2 \\ \hline \end{array}$ $\qquad$ $\begin{array}{r} 5n \\ 2n \\ \hline \end{array}$ $\qquad$ $\begin{array}{r} -7x \\ -3x \\ \hline \end{array}$ $\qquad$ $\begin{array}{r} 4y \\ -2x \\ \hline \end{array}$ $\qquad$ $\begin{array}{r} 5xy \\ 6x \\ \hline \end{array}$ $\qquad$ $\begin{array}{r} -3xy^2 \\ 5x^2y \\ \hline \end{array}$ $\qquad$ $\begin{array}{r} -8a^2b \\ -3bx \\ \hline \end{array}$

4. $(4x)(-3x) =$ $\qquad$ $(7x)(-3y) =$ $\qquad$ $(-12x)(3a) =$ $\qquad$ $(-4x)(-3y)(2x) =$

5. $(-3x)(-2b)(+3x) =$ $\qquad$ $(4m)(-3m)(-7m^2) =$ $\qquad$ $(3st)(3st)(3st) =$ $\qquad$ $(-x^3)(-y^5)(x^2) =$

Multiply the polynomial by the monomial.

6. $(3x)(4x + 2y - z) =$

7. $(-x)(4a - 2x + 3y) =$

8. $(-3ax)(-2a + 5x - 6) =$

9. $(2a)(a - 2 + a + 3) =$

10. $(-2a^2c)(3a - 4c - 2ac) =$

11. $(-8a^2)(3b^2 - 4a - 2b) =$

12. $(-pqr)(3p + 2qr + 7) =$

13. $(-6ab^2c^3)(3a^2bc - 3ab + 3ac) =$

Multiply.

14. $\begin{array}{r} 2x - 7 \\ 5x + 2 \\ \hline \end{array}$ $\qquad$ **15.** $\begin{array}{r} 5a - 2x \\ a + 3x \\ \hline \end{array}$ $\qquad$ **16.** $\begin{array}{r} m - n \\ 2p - m \\ \hline \end{array}$ $\qquad$ **17.** $\begin{array}{r} 3x - 2y + 6 \\ x - y \\ \hline \end{array}$

18. $\begin{array}{r} 3x^2 - 2x + 7 \\ 3x - 5 \\ \hline \end{array}$ $\qquad$ **19.** $\begin{array}{r} 6x^2 - 3xy + 5y^2 \\ x - 2y \\ \hline \end{array}$ $\qquad$ **20.** $\begin{array}{r} 5a - 2a^2 + 3a^3 \\ a^2 - 2a \\ \hline \end{array}$

Multiply.

21. $\begin{array}{r} -8d^2 + 4cd - 6c^2 \\ -d + 2c \\ \hline \end{array}$

22. $\begin{array}{r} 3x^2 - 2x + 1 \\ 5x^2 - 3x + 7 \\ \hline \end{array}$

Remove all signs of grouping and simplify.

23. $6 + (3 + 4) - 5 =$

24. $8 - (3 - 2) + (4 - 6) =$

25. $5(4 - 6) - 2(3 + 1) =$

26. $5 - [6(3 - 5) + 2] =$

27. $-[3(2 - 5) - (-6 + 2)] =$

28. $4\{[6 - 7(3 - 2 - 4) + 8] + 3\} =$

29. $4p - (k - 2x) + 8k =$

30. $7s - [4s + (4s + 7)] + 7 =$

31. $-(3a - 5b + 3c) - (a + 7b - 4c) =$

32. $(2x - y) + (4x + 2y) - (6x + 3y) =$

33. $(5x + 3y) - [(2y - 6x) - (2x + 8y)] =$

34. $-[(2a - b) + (5b + 6a)] - (8a + 3b) =$

35. $6x - \{2x - [5x - (3x - 8) - 8] + 4\} =$

36. $4(2y + 6) + 3(4y - 5) =$

Write the equivalent of the following expressions, enclosing the last three terms in parentheses, preceded by a minus sign.

37. $2x + 6y - 3t - 8 =$

38. $4x - 6y + 3z + 5 =$

39. $ab - c + gk - h =$

40. $-st + q - a - c - d =$

Name.. Date.......................... **EXERCISE 4.5**

Divide exponentially.

1. $2^5 \div 2^3 =$ $2^2 \div 2^3 =$ $2^4 \div 2^7 =$ $2^6 \div 2^6 =$

2. $x^8 \div x^4 =$ $x^8 \div x^2 =$ $x^7 \div x^9 =$ $x^5 \div x^5 =$

3. $(xy)^3 \div (xy)^5 =$ $(ab)^2 \div (ab) =$ $(-a)^7 \div (-a)^5 =$ $(-2)^3 \div (-2)^6 =$

4. $(-5)^3 \div (-5)^3 =$ $(4)^0 \div (4)^2 =$ $x^0 \div x^3 =$ $(-3)^2 \div (-3)^0 =$

Divide, and express in simplest terms.

5. $14a \div (-2) =$ $-15x \div 5 =$ $-24a^2 \div 6a =$ $18x^4 \div 2x =$

6. $\frac{25p^2q}{-5pq} =$ $\frac{-84a^3b^2}{-7a^2b^2} =$ $\frac{26a^3b}{-13a^3b} =$ $\frac{40x^2y^3z}{16x^2y} =$

7. $(12a^2 - 18a^4) \div 6a^2 =$

8. $(t^3 - t^2) \div (-t^2) =$

9. $(12k^2n^3 - 28k^3n^2) \div 4k^2n^2 =$

10. $(-12a^2b^3 + 48a^4b^4) \div (-6a^2b^3) =$

11. $\frac{-18x^3y + 33x^2y - 27xy^3}{3x^2} =$

12. $\frac{28x^2yz^3 - 21x^3y + 14x^3y^2z}{-7x^2y} =$

13. $\frac{25a^3b^2c^2 - 20a^2b^3c^2 - 15a^2b^2c^4}{-5a^2b^2c^2} =$

14. $\frac{30x^4y - 5x^2y^2 + 12xy^2}{-5x^2y^2} =$

15. $\frac{m^2 - \frac{1}{2}mn - \frac{3}{4}m^2n^2}{\frac{1}{4}m} =$

16. $\frac{3.6x^2y - 5.2xy^3 + 3.0xy}{-.4x^2y^2}$

Divide.

17. $(6a^2 + 11a + 3) \div (3a + 1)$

18. $(12x^2 + 7xy - 12y^2) \div (4x - 3y)$

19. $(53a^2 + 15a^3 - 8 - 30a) \div (3a - 2)$

20. $(4x^3 + 1 - 3x) \div (2x - 1)$

21. $(10x^2 + 1 + 11x) \div (2x + 3)$

22. $(a^3 - a^2 + a - 1) + (a + 1)$

23. How many $(x - 2)$'s are there in $(x^3 - 8)$?

24. If $(2x - 4)$ similar machines cost $(4x^2 - 9x + 2)$ dollars, what will one machine cost?

1. Draw a number line and locate on it:
$-\frac{3}{4}$; $+.6$; 0; $-2\frac{1}{2}$; $+3.1$; $+\frac{3}{5}$; $-.25$.

Add.

2. $\begin{array}{r} -468 \\ +397 \\ \hline \end{array}$ **3.** $\begin{array}{r} +\frac{49}{16} \\ -\frac{39}{12} \\ \hline \end{array}$ **4.** $\begin{array}{r} -18\frac{9}{10} \\ -13\frac{7}{8} \\ \hline \end{array}$

5. $\begin{array}{r} -38 \\ +47 \\ +19 \\ \hline \end{array}$ **6.** $\begin{array}{r} +492 \\ -381 \\ -117 \\ +162 \\ \hline \end{array}$ **7.** $\begin{array}{r} -17.7 \\ -\ \ 8.3 \\ +20.3 \\ +\ \ 6.2 \\ \hline \end{array}$

8. $\begin{array}{r} -12.8 \\ +15.7 \\ -\ \ 1.9 \\ \hline \end{array}$ **9.** $\begin{array}{r} +\ \frac{4}{5} \\ -1\frac{1}{16} \\ +\ \frac{7}{15} \\ \hline \end{array}$ **10.** $\begin{array}{r} -1\frac{5}{16} \\ -\ \frac{8}{11} \\ -\ \frac{5}{22} \\ \hline \end{array}$

Subtract the lower term from the upper.

11. $\begin{array}{r} -\ \ 6 \\ -18 \\ \hline \end{array}$ **12.** $\begin{array}{r} +1.7 \\ -1.5 \\ \hline \end{array}$ **13.** $\begin{array}{r} +83.7 \\ -64.2 \\ \hline \end{array}$

14. $\begin{array}{r} -37.672 \\ -54.791 \\ \hline \end{array}$ **15.** $\begin{array}{r} -53{,}627 \\ +42{,}874 \\ \hline \end{array}$ **16.** $\begin{array}{r} +\frac{5}{8} \\ -\frac{2}{5} \\ \hline \end{array}$

17. $\begin{array}{r} -\frac{8}{9} \\ -\frac{3}{4} \\ \hline \end{array}$ **18.** $\begin{array}{r} -\frac{34}{45} \\ +\frac{11}{12} \\ \hline \end{array}$ **19.** $\begin{array}{r} -47\frac{3}{8} \\ +24\frac{4}{5} \\ \hline \end{array}$

Compute the products.

20. $(+10) \times (-13)$ **21.** $(-47) \times (-23)$
22. $(-\frac{2}{9}) \times (+81)$ **23.** $(-14) \times (-\frac{10}{7})$
24. $(-0.4) \times (+0.421)$ **25.** $(-4283) \times (-0.004)$
26. $(+98.76) \times (+0.02)$
27. $(-0.003) \times (+0.06) \times (-4.6)$
28. $(-\frac{1}{2}) \times (-7) \times (-8)$
29. $(+\frac{11}{9}) \times (-\frac{3}{2}) \times (-\frac{7}{4}) \times (+\frac{12}{7})$

Compute the quotients.

30. $(-75) \div (-5)$ **31.** $(-4.5) \div (+12)$
32. $(+24.568) \div (-296)$ **33.** $(-389{,}906) \div (-439)$
34. $(-\frac{7}{6}) \div (+\frac{2}{5})$ **35.** $(-\frac{13}{9}) \div (+\frac{7}{3})$
36. $(-6\frac{1}{3}) \div (+10\frac{4}{9})$ **37.** $(-18.62) \div (+4.9)$

Simplify.

38. $\dfrac{(-56) - (+20)}{(-9) \times (-4)}$ **39.** $\dfrac{(+39) - (+44)}{(-5) \times (-5)}$

40. $\dfrac{(+40) \div (+\frac{2}{3})}{(-5)}$ **41.** $\dfrac{(-\frac{5}{8}) \div (+\frac{1}{6})}{(+\frac{14}{15}) \div (-\frac{9}{10})}$

Insert the proper sign in the empty parentheses to make the following pairs of fractions equivalent.

42. $(-)\dfrac{-28}{+40} = (\ \ )\dfrac{+28}{+40}$ **43.** $(+)\dfrac{+39}{+40} = (\ \ )\dfrac{-39}{-40}$

44. $(+)\dfrac{-11}{+77} = (\ \ )\dfrac{+1}{-7}$ **45.** $(\ \ )\dfrac{+49}{-21} = (\ \ )\dfrac{-7}{-3}$

Add.

46. $\begin{array}{r} -3.9x \\ +4.6x \\ \hline \end{array}$ **47.** $\begin{array}{r} -\frac{5}{8}m \\ +\frac{3}{5}m \\ \hline \end{array}$ **48.** $\begin{array}{r} +42x \\ -18x \\ +\ \ 7x \\ \hline \end{array}$

49. $\begin{array}{r} -6.3a \\ -7.2a \\ +4.1a \\ \hline \end{array}$ **50.** $\begin{array}{r} -103y \\ +\ \ 27y \\ -\ \ 63y \\ \hline \end{array}$ **51.** $\begin{array}{r} -\frac{3}{8}a \\ +\frac{2}{3}a \\ -\frac{3}{5}a \\ \hline \end{array}$

Subtract the lower term from the upper term.

52. $\begin{array}{r} +623a \\ -418a \\ \hline \end{array}$ **53.** $\begin{array}{r} -6.37x \\ +2.42x \\ \hline \end{array}$ **54.** $\begin{array}{r} +\ \ 9.62m \\ +10.37m \\ \hline \end{array}$

55. $\begin{array}{r} +62p \\ +37q \\ \hline \end{array}$ **56.** $\begin{array}{r} -\ \ \frac{5}{8}x \\ -3\frac{1}{4}x \\ \hline \end{array}$ **57.** $\begin{array}{r} +2\frac{3}{4}d \\ -1\frac{5}{8}d \\ \hline \end{array}$

Simplify.

58. $7.9x - 3.4x + 2 - 8.7x + 1$
59. $96x - 362x - 421x + 37x$
60. $\frac{3}{4}x - \frac{2}{3}x + \frac{5}{8}x + \frac{1}{2}x - \frac{5}{12}x$
61. $7x - 0.7x + 0.07x$
62. $4\frac{1}{4}x - 3\frac{1}{5}x - 2\frac{1}{2}x + 16$

Compute the products.

63. $a^7 \cdot a^3 \cdot a^5$
64. $(-6) \times (14xy)$
65. $(\frac{1}{3}ab)(63a^2b^2)$

66. $(.03a^2b)(-.02ab)$
67. $(0.3m)(-6m)(0.01m)(-0.02)$
68. $(\frac{3}{4}xy)(\frac{2}{3}xyz)(-\frac{1}{5}x)(\frac{5}{7})$
69. $(x^4)(-y^3)(-x^2)(y)$
70. $(0.02x)(-0.01y)(-0.0003x^2)$

Compute the quotients.

71. $x^9 \div x^3$
72. $(xy)^3 \div (xy)$
73. $-18x \div 6$
74. $(-76a^2b^3) \div (19ab^2)$
75. $\frac{5}{8}a^2b \div \frac{1}{4}ab$
76. $0.08ab^4 \div 0.02ab^2$
77. $\frac{3}{5}x^2y^3 \div \frac{7}{10}xy^3$
78. $(-3\frac{3}{8}m^2n) \div 3\frac{1}{3}mn^3$

Add.

79.
$$\begin{array}{r} 32xy - 6xy^2 \\ 14xy + 9xy^2 \\ -3xy - 7xy^2 \\ xy + xy^2 \\ \hline \end{array}$$

80.
$$\begin{array}{rrr} x + 2y & - & z \\ -x & + & z \\ 3y & - & 4z \\ 2x - 2y & & \\ \hline \end{array}$$

81. $(4x - 2y); (3x - 7y + 2z); (5z - 3y); (2x + 2y)$
82. $(\frac{3}{4}x - \frac{1}{3}y); (\frac{2}{5}x + \frac{3}{7}y); (-\frac{1}{2}x + \frac{3}{4}y); (\frac{2}{3}y - \frac{1}{2}x)$

Subtract the right polynomial from the left.

83. $(4x - 7y); (9y - 4x)$
84. $(0.3a + 0.7b); (0.5a - 0.2b)$
85. $(0.3a - 4.1b + 2.6c); (7.1c - 3.2d)$

Simplify.

86. $14ab - 0.3b + 0.62ab - 0.4b + 2.1ab$
87. $\frac{3}{4}x - \frac{2}{3}y + \frac{3}{8}xy + \frac{1}{2}x - \frac{1}{5}y$
88. Subtract $5x - 8y$ from -1.
89. Subtract $9a - 2b$ from 0.
90. What must be added to $(7m - 3k)$ to make the sum $(5a + 2k)$?

Multiply.

91. $(-3xy)(4a - 2x + 7)$
92. $(3m)(a - 3m + 2 + 4a)$
93. $(-1.6a^2b)(-0.1a + 4b - 0.3c)$
94. $(3x - 2y)(3x - 3y + 7)$
95. $(-a + 4d)(-6a^3 + 3a^2 + 2a)$
96. $(3x^2 + 4x - 2)(12x^2 - 3x + 1)$
97. What would be the sum if $(3a - \frac{1}{2}c)$ were added repeatedly $(2a - d)$ times?
98. What is the product if one factor is x increased by y, and the other is y increased by x?

Remove all signs of grouping and simplify.

99. $9 - [3(5 + 2) + 6]$
100. $3\{[8 - 2(6 - 3 + 7) + 4] - 7\}$
101. $(3m - 6t) - (m + 2t)$
102. $8a - [5a + (3a - 2)] + 7$
103. $(5a - b) + (3a - 7b) - (2a + 3b)$
104. $-[(5x - 3y) + (4y - 2x)] - (2y - 3x)$
105. $4m(3a + 2b) - a(4m - 6b)$
106. $0.5(4a - 6b) - 0.8(5a + 1.5b)$
107. $12m - \{3m - [7 - (4m + 2) + 3m] - 8\}$
108. Write the equivalent of $2z + 6m - 4n + 2$ with the last three terms in parentheses, preceded by a minus sign.

Divide; express quotients in simplest terms.

109. $(15a^2b - 18a^4) \div (3a^2)$
110. $(m - 12m^2n^3) \div (mn)$
111. $(24a^2b^3 - 16ab^4) \div (4ab^2)$
112. $(-14x^2y + 21xy^2 - 63x^2y^3) \div (-7xy)$
113. $(36x^2y^3 + 12x^2y - 15x^2y^2) \div (-3xy)$
114. $(102x^4y^4z^2 - 51x^3y^2z^3 + 85x^2y^2z^2) \div (-17x^2y^2z^2)$
115. $(x^5 - 2x^2 - 3) \div \left(-\frac{1}{3a}\right)$
116. $(a^4 - a^3 - 3a^2) \div (-\frac{1}{2}a)$
117. $(21x^2 - xy - 10y^2) \div (3x + 2y)$
118. $(8x^2 + 10x - 65) \div (2x + 7)$
119. How many $(x - 3)$'s are there in $(x^3 - 27)$

REVIEW

1. $\frac{3}{4} + \frac{7}{8} + \frac{2}{9} + \frac{1}{3}$
2. $3\frac{1}{4} + 4\frac{1}{2} + \frac{5}{8} + 2\frac{3}{7}$
3. $\frac{7}{8} - \frac{2}{3}$
4. $123\frac{1}{5} - 67\frac{4}{7}$
5. $.042 \times .036 \times 1.5$
6. $\frac{3}{4} \times \frac{5}{7} \times \frac{14}{19}$
7. $3\frac{1}{4} \times 6\frac{2}{3} \times 4\frac{7}{13}$
8. $\frac{8}{11} \div \frac{2}{7}$
9. $5\frac{6}{7} \div 4\frac{1}{2}$
10. $2.5 - 3\frac{1}{4} + 6\frac{7}{8}$
11. $4.16 - 2\frac{3}{4} + 1.62 - \frac{1}{2}$
12. (a) Round to the nearest thousandth.
    7.3241 .61582 3.4445
    (b) Round to the nearest ten.
    62.417 248.32 56,715.21
13. Express as a percent:
    (a) 34 out of 100. (b) 30 out of 200.
    (c) 7 compared to 5.
14. Find 18% of 650.
15. What number is $3\frac{1}{2}\%$ of 260?
16. What number is 120% more than 260?
17. Find $\frac{3}{4}\%$ of 96.
18. 57 is what percent of 76?
19. 81 is what percent of 27?
20. 27 is what percent of 81?
21. 45 is $37\frac{1}{2}\%$ of what number?
22. 70% of what number is 43.4?
23. $\frac{1}{2}$ is 75% of what number?
24. A team won 18 and lost 7 games. What percent of its games has it lost?

5

FIRST-DEGREE EQUATIONS AND INEQUALITIES

1. EQUATIONS

An equation is a mathematical sentence which states that two number expressions, called *members*, are equal. When one or both members of an equation contain a variable, and when the equation is true for all numbers in the domain of the variable,* the equation is called an *identity*. For example,

$$x^2 + 2x + 1 = (x + 1)^2$$

is an identity. No matter which number in the domain replaces the variable, x, a true sentence results. For 3 and -2, say:

$$x^2 + 2x + 1 = (x + 1)^2$$

$$x = 3: \quad 3^2 + 2(3) + 1 \stackrel{?}{=} (3 + 1)^2$$

$$9 + 6 + 1 \stackrel{?}{=} (4)^2$$

$$16 = 16$$

$$x = -2: \quad (-2)^2 + 2(-2) + 1 \stackrel{?}{=} (-2 + 1)^2$$

$$4 + (-4) + 1 \stackrel{?}{=} (-1)^2$$

$$1 = 1$$

When an equation is true for some but not all numbers in the domain of the variable, the equation is called a *conditional equation*. For example,

$$x + 2 = 7$$

is true for only one replacement of the variable: 5. Any other replacement produces a false sentence, for example:

$$x + 2 = 7$$

$x = -4$: $(-4) + 2 = 7$ (false)

$x = 0$: $(0) + 2 = 7$ (false)

$x = 3$: $(3) + 2 = 7$ (false)

$x = 5$: $(5) + 2 = 7$ (true)

$x = 6$: $(6) + 2 = 7$ (false)

Those numbers in the domain of the variable that make an equation a true sentence are called *solutions* of the equation. Solutions are said to *satisfy* an equation. For an identity, all numbers in the domain satisfy the equation. For a conditional equation, only some numbers in the domain satisfy the equation.

When an equation involves but a single variable and the exponent is one, the equation is known as a *first-degree* equation in a single variable. Because the graph of a first-degree equation is a straight line, as will be explained in Section 8, first-degree equations are also called *linear* equations. Provided the domain is sufficiently extensive, it can be shown that there are as many solutions for a conditional equation in a single variable as the number (degree) of its highest exponent.

2. SOLVING FIRST-DEGREE EQUATIONS IN A SINGLE VARIABLE

Two equations are said to be *equivalent* if they have the same solutions. The following operations always produce an equation that is an equivalent of the original equation:

(a) *Addition or subtraction of the same number (which may be in the form of the variable) to both members.*

(b) *Multiplication or division of both members by the same number (except zero and terms involving the variable).*

* Unless stated otherwise, in this chapter the domain of the variable is assumed to be the set of rational numbers.

EXAMPLES

1. Given the equation: $x - 4 = 6$
   Add 4 to each member: $x - 4 \mathbf{+ 4} = 6 \mathbf{+ 4}$
   Simplify each member: $x = 10$
   The equation $x = 10$ is an equivalent equation of $x - 4 = 6$. Note that the solution of the equation is immediately evident in $x = 10$ while in $x - 4 = 6$ it is not.
2. Given the equation: $x + 2 = 7$
   Subtract 2 from each member:
   $x + 2 \mathbf{- 2} = 7 \mathbf{- 2}$
   Simplify each member: $x = 5$
   The equation $x = 5$ is an equivalent equation of $x + 2 = 7$. The solution of the equation is immediately evident in $x = 5$ while in $x + 2 = 7$ it is not.
3. Given the equation: $\frac{1}{2}x = 6$
   Multiply each member by 2: $\mathbf{2}(\frac{1}{2}x) = \mathbf{2}(6)$
   Simplify each member: $x = 12$
   The equation $x = 12$ is an equivalent equation of $\frac{1}{2}x = 6$. The solution of the equation is immediately evident in $x = 12$ while in $\frac{1}{2}x = 6$ it is not.
4. Given the equation: $4x = 12$
   Divide each member by 4: $\frac{4x}{\mathbf{4}} = \frac{12}{\mathbf{4}}$
   Simplify each member: $x = 3$
   The equation $x = 3$ is an equivalent equation of $4x = 12$. The solution of the equation is immediately evident in $x = 3$ while in $4x = 12$ it is not.

From these examples the technique for solving first-degree equations can be summarized: *Transform the given equation into an equivalent equation in which one member contains only the variable with a numerical coefficient of* $+1$ *and the other member contains all the other terms which do not involve the variable.* The "other" member is the solution of the equation.

Sometimes this technique requires that we go through a series of steps—in effect a series of equivalent equations—before arriving at the final stage where the solution becomes clearly evident. There is no fixed pattern for such a series of steps except that each involves one or more of the operations stated in (a) and (b) above.

EXAMPLES

1. Solve for x: $7x - 2 = 4x + 7$.
   Subtract $4x$ from both members:
   $7x - 2 \mathbf{- 4x} = 4x + 7 \mathbf{- 4x}$
   $3x - 2 = 7$
   Add 2 to both members:
   $3x - 2 \mathbf{+ 2} = 7 \mathbf{+ 2}$
   $3x = 9$
   Divide both members by 3:
   $$\frac{3x}{\mathbf{3}} = \frac{9}{\mathbf{3}}$$
   $x = 3$ (the solution)
2. Solve for x: $2x - 12 = 4x - 8$.
   Subtract $4x$ from both members:
   $2x - 12 \mathbf{- 4x} = 4x - 8 \mathbf{- 4x}$
   $-2x - 12 = -8$
   Add 12 to both members:
   $-2x - 12 \mathbf{+ 12} = -8 \mathbf{+ 12}$
   $-2x = 4$
   Divide both members by -2:
   $$\frac{-2x}{\mathbf{-2}} = \frac{4}{\mathbf{-2}}$$
   $x = -2$ (the solution)

   [Note: The variable need not be kept on the left side of the equality sign. In this example it might have been easier to keep the x on the right side of the equality sign:
   $2x - 12 = 4x - 8$
   Subtract $2x$: $-12 = 2x - 8$
   Add 8: $-4 = 2x$
   Divide by 2: $-2 = x$]

(5.1) *To solve a first-degree equation in a single variable:*

Step 1. *If the equation involves fractional coefficients, multiply both members of the equation by the least common denominator of the fractions; remove all parentheses and simplify.*
Step 2. *As necessary, add to, subtract from, divide both members equally to produce an equivalent equation in which one member contains only the variable with a numerical coefficient of* $+1$. *The other member is the solution of the equation.*
Step 3. *Substitute the solution of Step 2 for the variable in the given equation to verify that the left member equals the right member.*

EXAMPLES

1. Solve for x: $5x - \frac{2}{15} = \frac{2}{3} + 3x$.
   Step 1. LCD = 15:
   $15(5x - \frac{2}{15}) = 15(\frac{2}{3} + 3x)$
   $75x - 2 = 10 + 45x$
   Step 2. Add 2 to each member; subtract $45x$ from each member; simplify:
   $75x - 2 + 2 - 45x = 10 + 45x + 2 - 45x$
   $30x = 12$

Divide each member by 30:

$$\frac{30x}{30} = \frac{12}{30}$$

$$x = \frac{12}{30} = \frac{2}{5} \text{ (the solution)}$$

Step 3. Verify by substituting $\frac{2}{5}$ for x wherever it appears in the given equation:

$$5x - \frac{2}{15} = \frac{2}{3} + 3x$$

$$x = \frac{2}{5}: \quad 5(\tfrac{2}{5}) - \frac{2}{15} \stackrel{?}{=} \frac{2}{3} + 3(\tfrac{2}{5})$$

$$2 - \frac{2}{15} \stackrel{?}{=} \frac{2}{3} + \frac{6}{5}$$

$$\frac{30}{15} - \frac{2}{15} \stackrel{?}{=} \frac{10}{15} + \frac{18}{15}$$

$$\frac{28}{15} = \frac{28}{15}$$

2. Solve for y: $\frac{5}{6}(y + 1) = \frac{1}{2}y - \frac{1}{4}(2y + 5)$.

Step 1. LCD = 12:

$$12[\tfrac{5}{6}(y + 1)] = 12[\tfrac{1}{2}y - \tfrac{1}{4}(2y + 5)]$$
$$10(y + 1) = 6y - 3(2y + 5)$$
$$10y + 10 = 6y - 6y - 15$$
$$10y + 10 = -15$$

Step 2. Subtract 10 from both members:

$$10y + 10 - 10 = -15 - 10$$
$$10y = -25$$

Divide each member by 10:

$$\frac{10y}{10} = \frac{-25}{10}$$

$$y = \frac{-25}{10} = \frac{-5}{2}$$

Step 3. Verify by substituting $\frac{-5}{2}$ for y in the given equation:

$$\frac{5}{6}(y + 1) = \frac{1}{2}y - \frac{1}{4}(2y + 5)$$

$$y = \frac{-5}{2}: \quad \frac{5}{6}\left(\frac{-5}{2} + 1\right) \stackrel{?}{=} \frac{1}{2}\left(\frac{-5}{2}\right) - \frac{1}{4}\left[2\left(\frac{-5}{2}\right) + 5\right]$$

$$\frac{5}{6}\left(\frac{-3}{2}\right) \stackrel{?}{=} \frac{1}{2}\left(\frac{-5}{2}\right) - \frac{1}{4}(-5 + 5)$$

$$\frac{-5}{4} \stackrel{?}{=} \frac{-5}{4} - \frac{1}{4}(0)$$

$$\frac{-5}{4} = \frac{-5}{4}$$

The term *transpose* is often used in connection with the solution of equations. It describes an *apparent* effect of adding to or subtracting from both members of an equation. For instance, subtracting 4 from both members of

$$x + 4 = 5$$

has the apparent effect of transposing the $+4$ from the left side of the equation to the right side, with a reversal of sign en route:

$$x + 4 = 5$$
$$\text{(subtract 4)} \quad x + 4 - 4 = 5 - 4$$
$$x = 5 - 4$$

Also in

$$5x = -2x - 9$$
$$2x + 5x = -2x - 9 + 2x$$
$$2x + 5x = -9$$

adding 2x to both sides (second line) makes it appear that the $-2x$ has moved from the right side in the top equation to the left side in the bottom equation, reversing its sign en route. The 4 and the $-2x$ terms in these two illustrations are said to have been transposed. We may generalize on this with a

Rule of transposition: *An equivalent equation results when a term is "moved" from one side of an equation to the other, provided its sign is reversed.*

[Note: The word "term" here means a *complete* term. For instance, in $3x + 7 = 5 - 2x$ the $-2x$ may be transposed to the left side as $+2x$, but its parts, -2 and x, may not be transposed separately.]

EXAMPLE

Solve: $x - 4 = 3x + 6$.

Transpose the -4 to the right side and the $3x$ to the left side:

$$x - 4 = 3x + 6$$
$$x - 3x = 6 + 4$$
$$-2x = 10$$
$$x = -5$$

(Complete Exercise 5.1)

3. TRANSLATING PROBLEMS INTO ALGEBRAIC LANGUAGE

Algebra is often compared to a language, a language stripped down to absolute essentials. From this point of view its use in the solution of problems involves a two-step procedure:

(1) Translate the problem into an algebraic sentence of equality—the equation.
(2) Solve the equation by algebraic techniques.

While the second step is easily handled, being purely a matter of mechanics, the first step is usually more difficult.

A sentence of equality, like any similarly complex sentence in English, is made up of phrases, and these phrases are made up of terms which correspond to the data of the problem. As we shall see in the next section, such problems are solved through an algebraic statement of some equivalence relationship which exists between the variable, representing the unknown, and the known, i.e., by an equation.

An important first step is to translate the information of the problem, both known and unknown, into accurate algebraic phrases. Ordinarily, most problems solvable by first-degree equations involve either comparisons of more and less or of ratio. Phrases of more and less can be handled by sums and differences. For instance, if x represents a number, then

$$x + 7 \quad \text{and} \quad 7 + x$$

represent a number "7 greater than x" or a number "x more than 7." The phrase

$$x - 12$$

represents a number which is "12 less than x" or simply the "difference between x and 12." Similarly,

$$12 - x$$

represents a number which is the "difference between 12 and x" or perhaps "that which is left of a dozen after an unspecified number have been removed."

Phrases of ratio comparisons can be stated in terms of products or quotients. For instance, if n represents a number, then

$$7n$$

represents a number "seven times the size of n." And

$$\frac{1}{7}n \quad \text{or} \quad \frac{n}{7}$$

represents a number "$\frac{1}{7}$ the size of n" or perhaps "one of seven equal parts of n."

Algebraic phrases, like those in English, can also be compounded. For instance:

$$5(x + 7)$$

represents a number "five times as great as the unknown number (x) increased by 7." Similarly,

$$\tfrac{1}{3}(x - 2)$$

represents a number which is "$\frac{1}{3}$ the size of an unknown number that has been diminished by 2." The phrase

$$\frac{5x + 10x}{100}$$

might represent the worth in dollars of a group of coins containing the same number (x) of nickels and dimes.

Translating problem situations into accurate algebraic language is absolutely essential to successful problem solving by algebra.

4. SOLVING PROBLEMS BY EQUATIONS

In problem situations, that which makes a problem a problem is the existence of an unknown whose value we seek; that which makes the problem *solvable* is availability of enough collateral information to allow us to determine the value of the unknown. Consequently the first step in solving a problem is to establish from the problem just what we wish to know. *A letter symbol (variable) is assigned to that unknown.*

The second step is to *identify from the problem some equivalence relationship which involves known data and the unknown* of the problem. Every problem capable of solution by an equation must contain—either stated or implied—such an equivalence. The following two problems illustrate.

(1) PROBLEM WITH THE EQUIVALENCE RELATIONSHIP STATED:

What number when doubled and then increased by 3 is 27?

If we substitute the word "equals" for the "is" and translate the phrase before it, using x to represent the unknown, we get

x: the unknown
$2x$: the unknown doubled
$2x + 3$: the doubled unknown increased by 3
$2x + 3 = 27$: the equation

(2) PROBLEM WITH THE EQUIVALENCE RELATIONSHIP IMPLIED:

What is the altitude of a triangle that has an area of 36 and a base of 6?

In this problem it is necessary for us to know, either as a matter of fact or by formula, that the area of a triangle is numerically equal to one-half the product of its base and altitude measurements. Thus, if we let x represent the measure of the unknown altitude, then:

$$\text{(area) } 36 = \tfrac{1}{2}(6)(x) \quad \text{(the equation)}$$

where (6) is the base and (x) is the altitude.

The third step in problem solving has already been demonstrated in the two foregoing illustrations: *The unknown and the knowns are incorporated into an algebraic statement of some basic equivalence relationship.* This is the heart of the solution of the problem, for it provides an equation by which the unknown can become known.

The fourth step is purely mechanical: *Solve the equation for the variable.*

The final step is to verify: *Substitute the solution of the equation back* **in the problem**—not in the equation. To substitute the solution in the equation simply verifies the mechanical work and will not detect any error in the setup of the equation, the major source of difficulty in solving problems.

(5.2) *To solve problems by equation:*

Step 1. Determine from the statement of the problem that which is sought—the unknown; represent it by a variable.

Step 2. Identify some equivalence relationship which involves both known data and the unknown of the problem.

Step 3. State algebraically (by equation) the equivalence relationship of Step 2.

Step 4. Solve the equation of Step 3 for the variable.

Step 5. Verify by substituting back in the problem the solution found in Step 4.

EXAMPLES

1. A board 12 feet long is cut in two pieces so that one piece is three times the length of the shorter piece. How long is the shorter piece?

   Step 1. What is sought? The length of the shorter piece. Let x represent the length of the shorter piece.

   Step 2. Identify an equivalence relationship:

   length of short piece + length of long piece = 12 ft

   Step 3. State algebraically (by equation) the equivalence relationship of Step 2: If x is the length of the shorter piece, then the length of the longer piece, which is three times the length of the shorter piece, must be $3x$. Therefore:

   length of short piece + length of long piece = 12 ft

   $$x + 3x = 12$$

   This is the basic equation.

   Step 4. Solve the basic equation for the variable:

   $$x + 3x = 12$$
   $$4x = 12$$
   $$x = 3 \quad \text{(ft—length of short piece)}$$

   Step 5. Verify by going back to the problem and reasoning: If the length of the short piece is 3 ft, then the length of the long piece, which is three times that of the short piece, must be 3×3 ft or 9 ft; the sum of the short and the long piece (3 ft + 9 ft = 12 ft) should equal the length of the board (12 ft). Since it does, the problem checks.

2. Given three consecutive integers. If the sum of the two smaller integers is decreased by three times the largest, the result is -37. What are the integers?

   Step 1. Three consecutive integers are sought. Let x represent the smallest of these.

   Step 2. Equivalence relationship:

   $$\text{Integer}_1 + \text{Integer}_2 - 3\,(\text{Integer}_3) = -37$$

   Step 3. If the smallest integer is x, the next consecutive integer will be one greater than x, or $(x + 1)$; the integer after the second will be two greater than the smallest, or one greater than the second integer:

   $$(x + 1) + 1 = (x + 2) = \text{the third integer}$$
   $$\text{Integer}_1 + \text{Integer}_2 - 3\,(\text{Integer}_3) = -37$$
   $$x + (x + 1) - 3\,(x + 2) = -37 \quad \text{(basic equation)}$$

   Step 4. Solve the basic equation:

   $$x + (x + 1) - 3(x + 2) = -37$$
   $$x + x + 1 - 3x - 6 = -37$$
   $$x + x - 3x = -37 + 6 - 1$$
   $$-x = -32$$
   $$x = 32 \quad (\text{Integer}_1)$$
   $$\text{then} \quad x + 1 = 33 \quad (\text{Integer}_2)$$
   $$\text{and} \quad x + 2 = 34 \quad (\text{Integer}_3)$$

   Step 5. The sum of the two smaller integers is $32 + 33 = 65$; if 65 is diminished by 3 times the largest $(3 \times 34 = 102)$ the result should be -37: $65 - 102 = -37$. The solution checks.

3. A merchant sells a mixture of olive oil and corn oil for salad dressing. If olive oil costs him \$.32 a unit and corn oil \$.28 a unit how much of each should be used if the cost of 10 units of the mixture is to be \$2.94?

   Let x = number of units of corn oil.
   Then $10 - x$ = number of units of olive oil.
   The equivalence relationship is:

   $$\left[\text{Units olive oil} \times \text{Cost per unit}\right] + \left[\text{Units corn oil} \times \text{Cost per unit}\right] = \$2.94$$

   $$[(10 - x) \times (\$.32)] + [\,(x) \times (\$.28)\,] = \$2.94$$
   $$(10 - x)(.32) + (x)(.28) = 2.94$$
   $$3.2 - .32x + .28x = 2.94$$
   $$-.04x = -.26$$
   $$x = \frac{-.26}{-.04} = \frac{26}{4}$$
   $$= 6\tfrac{1}{2} \text{ (units corn oil)}$$
   $$10 - x = 10 - 6\tfrac{1}{2}$$
   $$= 3\tfrac{1}{2} \text{ (units olive oil)}$$

   Verification:

$3\frac{1}{2}$ units	@ \$.32 =	\$1.12
$6\frac{1}{2}$ units	@ \$.28 =	\$1.82
10 units		\$2.94

4. A coin bank holds nickels, dimes, and quarters. If it contains three times as many dimes as nickels and two more quarters than dimes, and if the total value of its contents is \$4.90, how many of each coin are there in the bank?

Let x = the *number* of nickels;
$3x$ = the *number* of dimes;
$3x + 2$ = the *number* of quarters.
The equivalence relationship:

$$\begin{pmatrix}\text{No. of}\\ \text{nickels}\end{pmatrix} \times 5¢ + \begin{pmatrix}\text{No. of}\\ \text{dimes}\end{pmatrix} \times 10¢ + \begin{pmatrix}\text{No. of}\\ \text{quarters}\end{pmatrix} \times 25¢ = 490¢$$

$$[(x)(5)] + [(3x)(10)] + [(3x + 2)(25)] = 490$$
$$5x + 30x + 75x + 50 = 490$$
$$110x = 440$$
$$x = 4 \text{ (nickels)}$$
$$3x = 12 \text{ (dimes)}$$
$$3x + 2 = 14 \text{ (quarters)}$$

Verification: 4 nickels = \$0.20
12 dimes = 1.20
14 quarters = 3.50
\$4.90

5. A stream flows at the rate of 4 miles per hour. A launch takes 3 hours to make a trip to a certain town downstream, but it takes 7 hours to return because of adverse current. What is the speed of the launch in still water?
Let x represent the speed of the launch in still water; $x + 4$, the speed of the launch going downstream; $x - 4$, the speed of the launch going upstream.
Since (distance) = (rate) × (time), the basic equivalence for this problem may be stated in terms of distance:

distance down = distance back
(rate down) × (time down) = (rate up) × (time up)

$$(x + 4) \times (3 \text{ hr}) = (x - 4) \times (7 \text{ hr})$$
$$3(x + 4) = 7(x - 4)$$
$$3x + 12 = 7x - 28$$
$$40 = 4x$$
$$10 = x \text{ (speed in still water)}$$

Verification: If the launch moves at 10 miles an hour in still water, with the current it "makes good" $10 + 4 = 14$ miles per hour. A three-hour trip at this rate must be $3 \times 14 = 42$ miles long. Coming back, against the current, the launch makes good only $10 - 4 = 6$ miles per hour. At this rate, a seven-hour trip must be $7 \times 6 = 42$ miles long. Hence, distance down (42 miles) = distance up (42 miles) and the solution is verified.

5. NUMERICAL APPROACH TO PROBLEM SOLVING

When there is difficulty in setting up the equation for a problem, a numerical approach often proves helpful. By this technique we assume a *plausible* value for the unknown and then reason through the problem as we would in the check. Except by accident, the assumed value will of course be incorrect, though if our reasoning is correct, substituting a letter symbol for the assumed value in our so-called "check" provides us immediately with the required basic equation.

EXAMPLES

1. Bill is now twice the age of John. Seven years ago the sum of their ages was 28. What are their ages now?
Numerical Approach: Assume that John is now 12 years old: that would make Bill, who is twice John's age, 24 years old. Seven years ago John would have been $12 - 7$, or 5, and Bill would have been $24 - 7$, or 17. According to the data of the problem, the sum of their ages seven years ago would have been 28. But
$$(12 - 7) + (24 - 7) \neq 28$$
However, if we introduce x instead of the assumed 12 as John's present age (and $2x$ instead of Bill's 24) in the false equation above, we obtain the necessary basic equation:
$$(x - 7) + (2x - 7) = 28$$
$$3x - 14 = 28$$
$$3x = 42$$
$$x = 14 \text{ (John's present age)}$$
$$2x = 28 \text{ (Bill's present age)}$$

2. If half the sum of two consecutive even integers is subtracted from twice the smaller integer, the result will be 23. What are the integers?
Numerical Approach: Assume the smaller even integer is 18; the next consecutive even integer would then be 20. Half their sum would be $\frac{1}{2}(18 + 20) = 19$, which, according to the data of the problem, when subtracted from twice the smaller $(2 \times 18 = 36)$ should equal 23. But $(2 \times 18) - \frac{1}{2}(18 + 20) \neq 23$. However, if we substitute x for the smaller integer instead of the assumed 18, and $x + 2$ for the next consecutive even integer instead of the assumed 20, we are led to the equation:
$$(2x) - \tfrac{1}{2}[(x) + (x + 2)] = 23$$
$$2x - \tfrac{1}{2}[2x + 2] = 23$$
$$2x - x - 1 = 23$$
$$x = 24 \text{ (smaller integer)}$$
$$x + 2 = 26 \text{ (next consecutive even integer)}$$

(Complete Exercises 5.2 and 5.3)

6. INEQUALITIES

An *inequality* is also a mathematical sentence, one which states that two number expressions or members are unequal. For example:

$$3 + x > 5$$

Like equations, two inequalities are equivalent if they have the same solutions; unlike equations, inequalities generally have a great many solutions. For example, if the domain of the variable for the inequality above is the set of integers, then every integer greater than 2 will satisfy it. Numbers which satisfy an equation or inequality are said to belong to the *solution set*. Thus, for

$$3 + x > 5$$

the solution set is:

$$\{3, 4, 5, 6, 7, \ldots\}$$

Similarly, for the inequality

$$x + 2 < 4$$

the solution set is:

$$\{\ldots, -3, -2, -1, 0, 1\}$$

EXAMPLES

1. Give the solution set for $8 + x > 12$ if the domain of the variable is the set of integers.
   The number 4 is not in the solution set because $8 + 4$ *equals* 12, and therefore is not *greater than* 12. Numbers less than 4 are not in the solution set because when each is added to 8 the sum is less than 12. Numbers greater than 4 are in the solution set because when each is added to 8 the sum is greater than 12. So $\{5, 6, 7, 8, 9, \ldots\}$ are the integers that satisfy $8 + x > 12$.

2. Give the solution set for $6 > x + 3$ if the domain of the variable is the set of integers.
   Answer: $\{\ldots, -4, -3, -2, -1, 0, 1, 2\}$

3. Give the solution set for $5 \leq x - 1$ if the domain of the variable is the set of integers.
   The sentence "$5 \leq x - 1$" is really a compound sentence which states "$5 < x - 1$ or $5 = x - 1$." The solution set for the compound sentence contains all those numbers that make *either* sentence true. Thus, 6 makes $5 = x - 1$ true, and 7, 8, 9, 10, ... make $5 < x - 1$ true. So $\{6, 7, 8, 9, 10, \ldots\}$ is the solution set for $5 \leq x - 1$.

4. Give the solution set for $2x \geq -6$ if the domain of the variable is the set of integers.
   Answer: $\{-3, -2, -1, 0, 1, \ldots\}$

If the domain of the variable in the four previous examples had not been restricted to the set of integers, then the solution set would have contained many more members. Also, the solution set could not have been expressed as simply, using brace notation. In such cases a graph is often used, as illustrated in the following examples.

EXAMPLES

1. Graph the solution set for $8 + x > 12$.

   $8 + x > 12$:

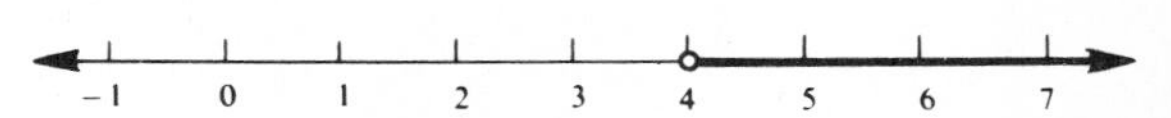

   Basis for the graph is the number line. The hollow dot at 4 means that 4 is not in the solution set. The thick stripe to the right of 4 means that all numbers greater than 4 are in the solution set.

2. Graph the solution set for $6 > x + 3$.

   $6 > x + 3$:

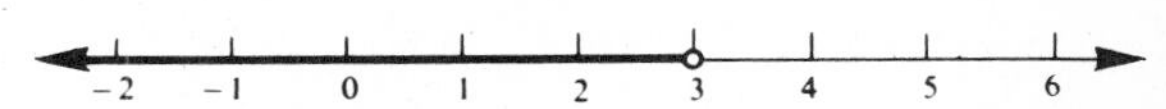

   Interpretation: 3 (hollow dot) is not in the solution set, but all numbers less than 3 are in the solution set.

3. Graph the solution set for $5 \leq x - 1$.

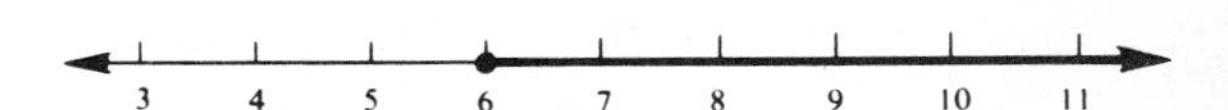

   Interpretation: 6 (solid dot) is in the solution set, and so is every number greater than 6 in the solution set.

4. Graph the solution set for $2x \geq -6$.

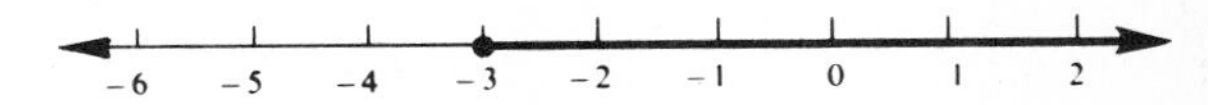

   Interpretation: -3 (solid dot) is in the solution set, and so is every number greater than -3 in the solution set.

7. SOLVING FIRST-DEGREE INEQUALITIES IN A SINGLE VARIABLE

Simple inequalities, such as those of the examples in Section 6, can often be solved by inspection. More complicated inequalities are not so easily solved. As with equations, however, it is possible to proceed through a series of equivalent but simpler inequalities until we arrive at one whose solution set is obvious. The following operations always produce an inequality that is equivalent to the original inequality.

(a) *Addition or subtraction of the same number (which may be in the form of the variable) to or from both members.*
(b) *Multiplication or division of both members by the same positive number (variable excepted).*
(c) *Multiplication or division of both members by the same negative number (variable excepted) when the sense or direction of the inequality is reversed, i.e., from $>$ to $<$, or from $<$ to $>$.*

EXAMPLES

1. Solve: $8 + x > 12$.
   Subtract 8 from both members:
   $8 + x - 8 > 12 - 8$
   Simplify: $x > 4$
   The inequality $x > 4$ is equivalent to $8 + x > 12$. The solution set for $x > 4$ is immediately evident, but not for $8 + x > 12$. (Compare with both Examples numbered 1 in Section 6.)
2. Solve: $5 \leq x - 1$.
   Add 1 to both members:
   $1 + 5 \leq x - 1 + 1$
   Simplify: $6 \leq x$
   (Compare with both Examples 3 in Section 6.)
3. Solve: $2x \geq -6$.
   Divide both members by 2:
   $$\frac{2x}{2} \geq \frac{-6}{2}$$
   $$x \geq -3$$
   (Compare with both Examples 4 in Section 6.)
4. Solve: $3 - 2x < 5$.
   Subtract 3 from both members:
   $3 - 2x - 3 < 5 - 3$
   $$-2x < 2$$
   Divide both members by -2 (reverse the sense or direction of the inequality):
   $$\frac{-2x}{-2} > \frac{2}{-2}$$
   $$x > -1$$

An alternative procedure for solving inequalities is given in the following Program.

(5.3) *To solve a first-degree inequality in a single variable:*

Step 1. Transform the given inequality into an equation by replacing the $<$ or $>$ sign with an $=$ sign.

Step 2. Solve the equation of Step 1.
Step 3. Choose two trial numbers, one greater and one less than the solution found in Step 2, and substitute each for the variable in the given inequality.
(a) If the greater trial number satisfies the given inequality, then the solution set contains those numbers in the domain of the variable that are greater than the solution found in Step 2.
(b) If the lesser trial number satisfies the given inequality, then the solution set contains those numbers in the domain of the variable that are less than the solution found in Step 2.

EXAMPLES

1. Solve: $3 - 2x < 5$.
   Step 1. Replace $<$ with $=$:
   $3 - 2x = 5$
   Step 2. Solve for x:
   $-2x = 5 - 3$
   $-2x = 2$
   $x = -1$

   Step 3. Select two trial numbers, one greater than -1 (say, 0) and one less than -1 (say, -3), and substitute each for the variable in the given inequality:

   $$3 - 2x < 5$$
   $x = 0$: $3 - 2(0) < 5$
   $3 - 0 < 5$
   $3 < 5$ (true)
   $x = -3$: $3 - 2(-3) < 5$
   $3 + 6 < 5$
   $9 < 5$ (false)

   The greater trial number satisfies the given inequality, therefore the solution set for the given inequality is $x > -1$; or:

   (Compare with Example 4 in the left column.)
2. Solve: $2x - 2 > 4x + 6$.
   Step 1. $2x - 2 = 4x + 6$
   Step 2. $2x - 4x = 6 + 2$
   $-2x = 8$
   $x = -4$
   Step 3. Trial numbers: $-5\,(< -4)$ and $1\,(> -4)$
   $$2x - 2 > 4x + 6$$
   $x = -5$: $2(-5) - 2 > 4(-5) + 6$
   $-12 > -14$ (true)
   $x = 1$: $2(1) - 2 > 4(1) + 6$
   $0 > 10$ (false)

   Solution: $x < -4$, or:

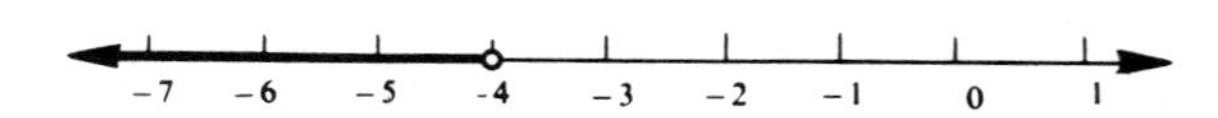

3. Solve: $3p + 2 \geq 4p + 1$.

Step 1. $3p + 2 = 4p + 1$

Step 2. $3p - 4p = 1 - 2$

$-p = -1$

$p = 1$

Step 3. Trial numbers: 0 and 2.

$3p + 2 \geq 4p + 1$

$p = 0$: $0 + 2 \geq 0 + 1$

$2 \geq 1$ (true)

$p = 2$: $3(2) + 2 \geq 4(2) + 1$

$8 \geq 8 + 1$

$8 \geq 9$ (false)

Solution: $p \leq 1$, or:

[Note: $p = 1$ is included in the solution set because 1 satisfies the equality part of the given inequality.]

(Complete Exercise 5.4)

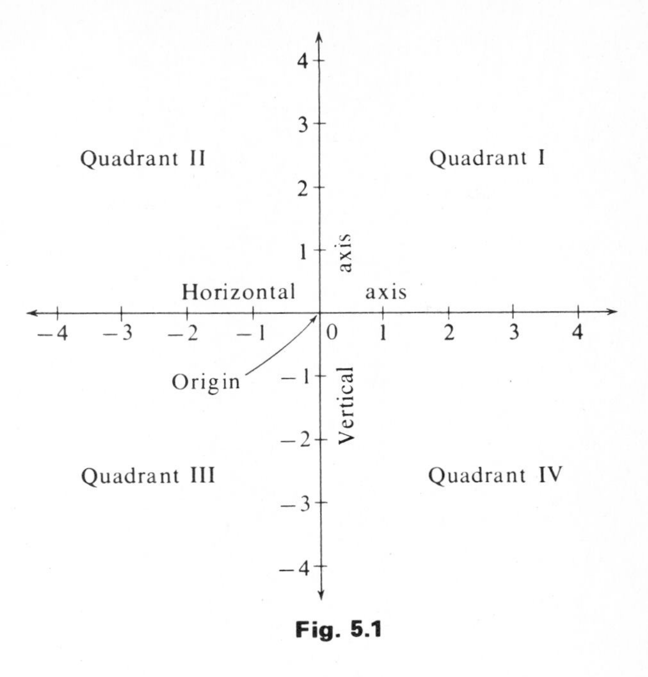

Fig. 5.1

8. GRAPHING LINEAR EQUATIONS

Up to this point we have considered first-degree equations and inequalities in a single variable. We now consider first-degree equations in two variables. For example:

$$x + 2y = 11$$

Here the solution set consists of ordered pairs of numbers, among them (1, 5), (3, 4), (5, 3), $(10, \frac{1}{2})$, (13, −1), (19, −4). By convention the first number of the pair is considered a replacement for x and the second a replacement for y. The order is important, as (3, 4), for example, satisfies $x + 2y = 11$, but (4, 3) does not:

$x + 2y = 11$

(3, 4): $(3) + 2(4) = 11$ (true)

(4, 3): $(4) + 2(3) = 11$ (false)

The solution set for an equation in two variables is often expressed more easily in the form of a graph in the *Cartesian* or *rectangular coordinate system.* This system consists of two perpendicular number lines, or *axes*, which divide the *coordinate plane* into four *quadrants.* Ordinarily, the quadrants are numbered as shown in Fig. 5.1.

Every point in the coordinate plane will lie directly opposite some numbered point on each axis. The first number is taken from the horizontal axis and the second from the vertical axis. The result is an ordered pair of numbers called the *coordinates* of the point. The first coordinate is also referred to as the *abscissa* of the point, and the second coordinate as the *ordinate* of the point.

EXAMPLES

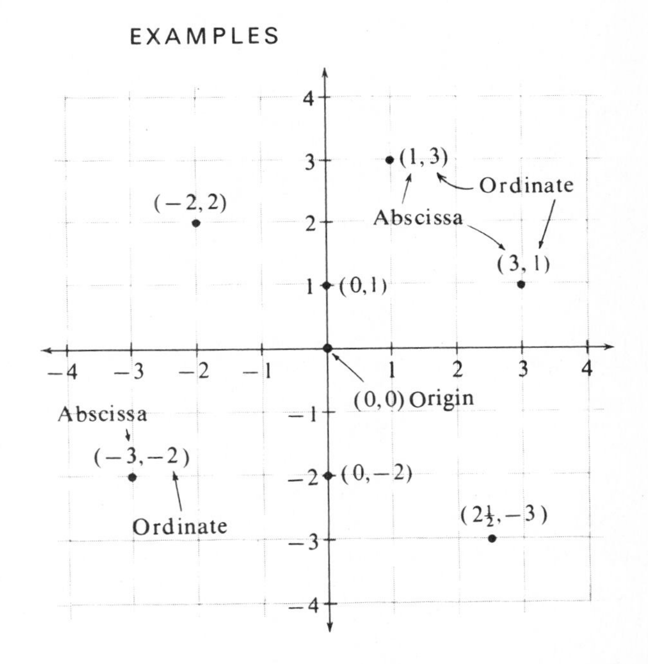

The *graph of an equation* is actually the graph of its solution set.

(5.4) *To graph a first-degree equation in two variables:*

Step 1. *Find an ordered pair of numbers that satisfies the equation by substituting some rational number for the first variable and solving the resulting equation for the second.*

Step 2. Repeat Step 1 using a different number for the first variable.
Step 3. Plot the two points corresponding to the ordered pairs of Steps 1 and 2.
Step 4. Draw a line through the two points of Step 3 for the graph of the given equation.

EXAMPLES

1. Graph the equation: $x + 2y = 7$.
Step 1. Select 1, say, as the first number; substitute it for x in the equation and solve for y:

$$x + 2y = 7$$
$$x = 1: \quad 1 + 2y = 7$$
$$y = 3$$

Ordered pair: (1, 3)

Step 2. Select -1, say, as the second number; substitute it for x and solve for y:

$x = -1$:
$$-1 + 2y = 7$$
$$y = 4$$

Ordered pair: $(-1, 4)$

Steps 3 and 4.

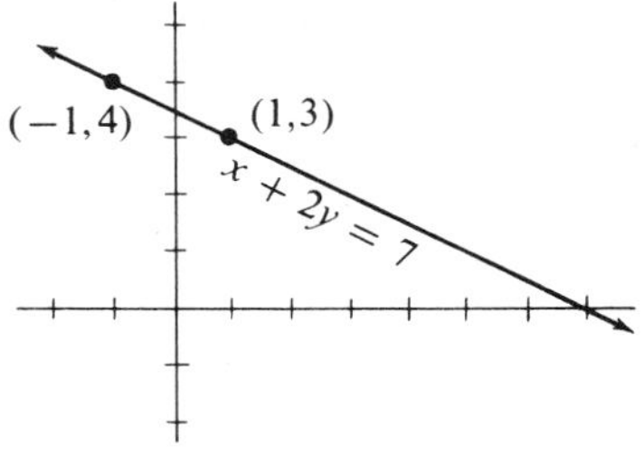

2. Graph: $3x - 2y = 12$.
Steps 1 and 2.

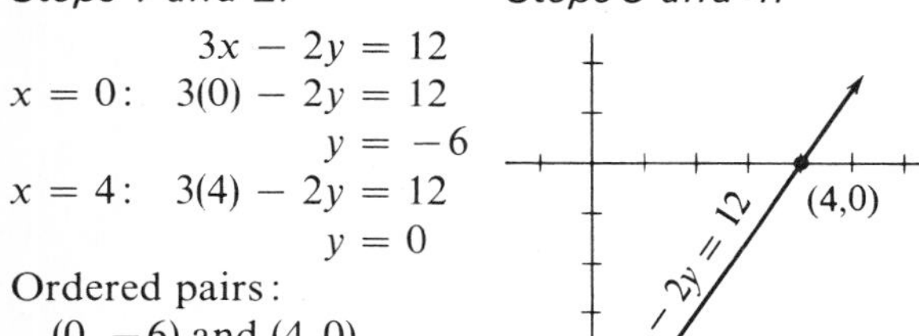

$$3x - 2y = 12$$
$$x = 0: \quad 3(0) - 2y = 12$$
$$y = -6$$
$$x = 4: \quad 3(4) - 2y = 12$$
$$y = 0$$

Ordered pairs:
$(0, -6)$ and $(4, 0)$

Steps 3 and 4.

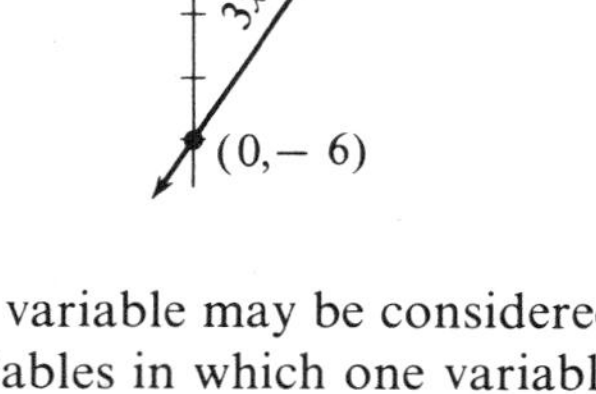

3. Graph: $y = 3$.
An equation in a single variable may be considered an equation in two variables in which one variable has 0 as numerical coefficient; e.g., $y = 3$ implies $0x + y = 3$.

From this it is evident that no matter what number replaces x, the equation will always simplify to $y = 3$. In terms of the graph, this means that for any and all abscissas, the ordinate is 3.

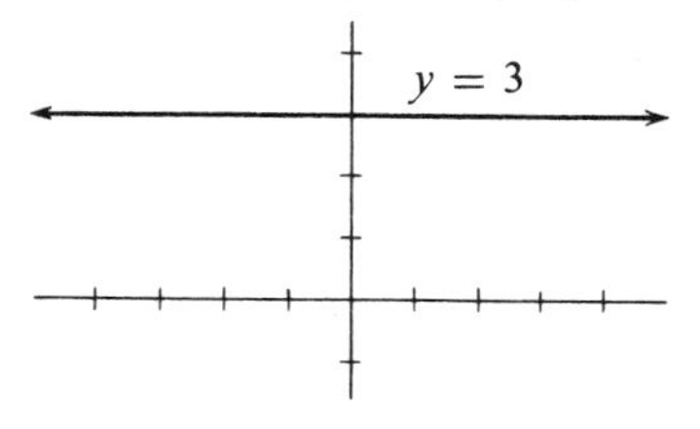

4. Graph: $x = -2$.
We might think of the equation, $x = -2$, as
$x + 0y = -2$
No matter what number replaces the variable y, x will always have a value of -2.

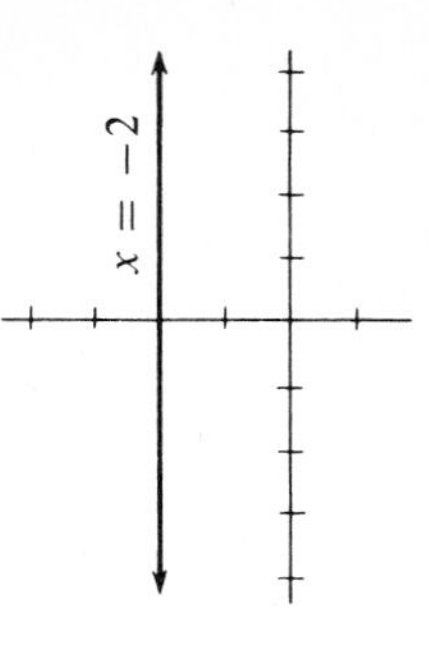

5. Graph $2x - y = 7$ and $x + y = 5$ in the same coordinate system and estimate their point of intersection.

$$2x - y = 7$$
$$x = 5: \quad 2(5) - y = 7$$
$$y = 3$$
$$x = 3: \quad 2(3) - y = 7$$
$$y = -1$$

$$x + y = 5$$
$$x = 2: \quad (2) + y = 5$$
$$y = 3$$
$$x = 6: \quad (6) + y = 5$$
$$y = -1$$

Ordered pairs:
(5, 3) and $(3, -1)$

Ordered pairs:
(2, 3) and $(6, -1)$

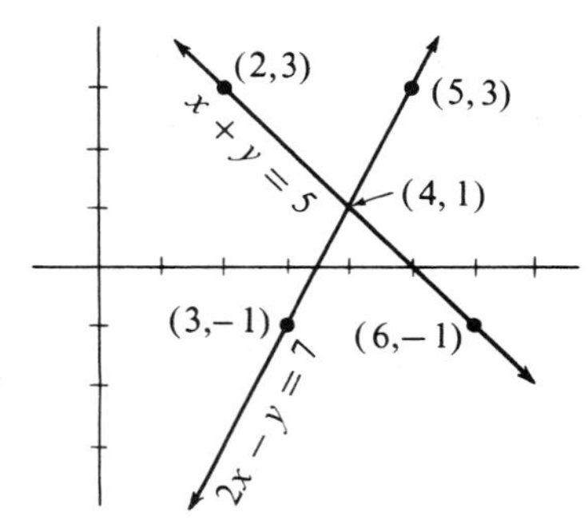

The estimated point of intersection for the graphs of $2x - y = 7$ and $x + y = 5$ is (4, 1).

(Complete Exercise 5.5)

9. SYSTEMS OF EQUATIONS

A *system of equations* is a set of equations. The simplest system is one consisting of two first-degree or linear equations in two variables. For example:

$$\begin{cases} 2x - y = 7 \\ x + y = 5 \end{cases}$$

The solution of such a system is an ordered pair of numbers that satisfies both equations. One direct way to identify the pair is to graph the equations and to note the abscissa and ordinate of their point of intersection. This was done for the system immediately above in Example 5 of Section 8; the point of intersection was estimated to be (4, 1).

$$\begin{array}{lll} & 2x - y = 7 & x + y = 5 \\ \left.\begin{array}{l} x = 4 \\ y = 1 \end{array}\right\} & 2(4) - (1) \stackrel{?}{=} 7 & (4) + (1) \stackrel{?}{=} 5 \\ & 8 - 1 = 7 & 5 = 5 \end{array}$$

Since (4, 1) satisfies both equations, (4, 1) is the solution of the system.

Solution of systems of equations by graphic means is an approximate method. More precise are the following algebraic methods.

(5.5) *To solve a system of two linear equations in two variables by substitution:*

Step 1. *Solve one of the equations of the system for one of the variables in terms of the other.*
Step 2. *Substitute the solution of Step 1 for the variable in the other equation.*
Step 3. *Solve the equation which results from Step 2.*
Step 4. *Substitute the solution of Step 3 for that variable in either of the two equations of the system and solve for the other variable.*
Step 5. *Verify by substituting the obtained solutions in the equation of the system not used in Step 4.*

EXAMPLES

1. Solve the system by substitution: $\begin{cases} 2x - y = 7 \\ x + y = 5 \end{cases}$

Step 1. Solve $2x - y = 7$ for y: $y = 2x - 7$.
Step 2. Substitute $y = 2x - 7$ for y in

$$x + y = 5$$

$$y = 2x - 7: \quad x + (2x - 7) = 5$$

Step 3. $x + 2x - 7 = 5$

$$3x = 12$$
$$x = 4$$

Step 4. Substitute 4 for x in $2x - y = 7$.

$$\begin{aligned} x = 4: \quad 2(4) - y &= 7 \\ 8 - y &= 7 \\ -y &= -1 \\ y &= 1 \end{aligned}$$

Step 5. Verify by substituting 4 for x and 1 for y in the other equation of the system:

$$\begin{array}{ll} & x + y = 5 \\ \left.\begin{array}{l} x = 4 \\ y = 1 \end{array}\right\} & 4 + 1 \stackrel{?}{=} 5 \\ & 5 = 5 \end{array}$$

The solution: (4, 1)

2. Solve by substitution: $\begin{cases} 6x - 3y = 15 \\ x - 4y = 6 \end{cases}$

Step 1. $x - 4y = 6$

$$x = 4y + 6$$

Step 2. $6x - 3y = 15$

$$x = 4y + 6: \quad 6(4y + 6) - 3y = 15$$

Step 3. $24y + 36 - 3y = 15$

$$\begin{aligned} 24y - 3y &= 15 - 36 \\ 21y &= -21 \\ y &= -1 \end{aligned}$$

Step 4. $6x - 3y = 15$

$$\begin{aligned} y = -1: \quad 6x - 3(-1) &= 15 \\ 6x + 3 &= 15 \\ 6x &= 15 - 3 \\ 6x &= 12 \\ x &= 2 \end{aligned}$$

Step 5. Verification: $x - 4y = 6$.

$$\begin{array}{ll} \left.\begin{array}{l} x = 2 \\ y = -1 \end{array}\right\} & (2) - 4(-1) \stackrel{?}{=} 6 \\ & 2 + 4 \stackrel{?}{=} 6 \\ & 6 = 6 \end{array}$$

The solution: (2, −1)

3. Solve by substitution: $\begin{cases} 4x + 3y = 7 \\ 8x - 3y = 41 \end{cases}$

Step 1. $4x + 3y = 7$

$$\begin{aligned} 4x &= 7 - 3y \\ x &= \frac{7 - 3y}{4} \\ &= (\tfrac{1}{4})(7 - 3y) \end{aligned}$$

Step 2. $8x - 3y = 41$

$$x = \tfrac{1}{4}(7 - 3y): \quad 8(\tfrac{1}{4})(7 - 3y) - 3y = 41$$

Step 3. $2(7 - 3y) - 3y = 41$

$$\begin{aligned} 14 - 6y - 3y &= 41 \\ -6y - 3y &= 41 - 14 \\ -9y &= 27 \\ y &= -3 \end{aligned}$$

Step 4. $4x + 3y = 7$

$$\begin{aligned} y = -3: \quad 4x + 3(-3) &= 7 \\ 4x - 9 &= 7 \\ 4x &= 16 \\ x &= 4 \end{aligned}$$

Step 5. $8x - 3y = 41$

$$\begin{array}{ll} \left.\begin{array}{l} x = 4 \\ y = -3 \end{array}\right\} & 8(4) - 3(-3) \stackrel{?}{=} 41 \\ & 32 + 9 \stackrel{?}{=} 41 \\ & 41 = 41 \end{array}$$

The solution: (4, −3)

In Example 3 above, solving either of the equations for one variable in terms of the other involves fractions. Practically speaking, the substitution method is worth considering only when the numerical coefficient of at least one of the variables in either equation is unity (+1 or −1). Otherwise the addition-subtraction method (Program 5.6 on page 102) is preferable.

By an appropriate addition or subtraction of the two equations, term by term, it is often possible to produce an equivalent equation in which only one of the variables is present, i.e., the other variable takes on a numerical coefficient of 0. This is referred to as *eliminating* a variable. In cases where eliminating a variable is not immediately possible, appropriate multiplication of all terms of either one or both equations will permit a subsequent addition or subtraction of the two equations to eliminate one of the variables. Thus in Example 3 on page 101, by *adding* the two equations together the y-term is eliminated and the resulting equation is solvable for x:

$$\begin{array}{r} 4x + 3y = 7 \\ (+)8x - 3y = 41 \\ \hline 12x + 0y = 48 \\ 12x = 48 \\ x = 4 \end{array}$$

(5.6) *To solve a system of two linear equations in two variables by addition or subtraction:*

Step 1. *Multiply, if necessary, one or both equations by a number that will make the numerical coefficients of one of the variables in both equations alike, except possibly for sign.*

Step 2. *Eliminate the variable whose numerical coefficients are alike by adding the two equations together (when the terms to be eliminated differ in sign) or by subtracting one equation from the other (when the terms to be eliminated agree in sign).*

Step 3. *Solve the equation which results from Step 2.*

Step 4. *Substitute the solution of Step 3 for that variable in either of the two equations of the system and solve for the other variable.*

Step 5. *Verify by substituting the obtained solutions in the equation of the system not used in Step 4.*

EXAMPLES

1. Solve the system by the addition-subtraction method:

$$\begin{cases} 3x + 2y = 12 \\ 3x + 4y = 18 \end{cases}$$

Step 1. No multiplication is necessary, since the numerical coefficients of the x-terms are alike.

Step 2. Subtract the bottom equation from the top:

$$\begin{array}{r} 3x + 2y = 12 \\ (-)3x + 4y = 18 \\ \hline 0x - 2y = -6 \end{array}$$

Step 3. Solve for y:

$$-2y = -6$$
$$y = \frac{-6}{-2}$$
$$y = 3$$

Step 4. Substitute 3 for y in one of the equations of the system and solve for x:

$$\begin{array}{rr} & 3x + 2y = 12 \\ y = 3: & 3x + 2(3) = 12 \\ & 3x + 6 = 12 \\ & 3x = 6 \\ & x = 2 \end{array}$$

Step 5. Verify by substituting 2 for x and 3 for y in the other equation of the system:

$$3x + 4y = 18$$
$$\left.\begin{array}{l} x = 2 \\ y = 3 \end{array}\right\} 3(2) + 4(3) \stackrel{?}{=} 18$$
$$6 + 12 \stackrel{?}{=} 18$$
$$18 = 18$$

Solution: $(2, 3)$

2. Solve by the addition-subtraction method:

$$\begin{cases} 6x + 2y = 11 \\ 4x + 3y = 14 \end{cases}$$

Step 1. Multiply all terms of the top equation by 2 and of the bottom equation by 3:

$$2 \cdot (6x + 2y = 11) = 12x + 4y = 22$$
$$3 \cdot (4x + 3y = 14) = 12x + 9y = 42$$

Step 2. Subtract: $-5y = -20$

Step 3. Solve for y: $y = 4$

Step 4. Substitute 4 for y in one of the equations of the system and solve for x:

$$\begin{array}{rr} & 6x + 2y = 11 \\ y = 4: & 6x + 2(4) = 11 \\ & 6x + 8 = 11 \\ & 6x = 3 \\ & x = \frac{1}{2} \end{array}$$

Step 5. Verify, using the equation of the system not used in Step 4:

$$4x + 3y = 14$$
$$\left.\begin{array}{l} x = \frac{1}{2} \\ y = 4 \end{array}\right\} 4(\tfrac{1}{2}) + 3(4) \stackrel{?}{=} 14$$
$$2 + 12 \stackrel{?}{=} 14$$
$$14 = 14$$

Solution: $(\frac{1}{2}, 4)$

3. Solve: $\begin{cases} 5x - 2y = 5 \\ 2x + 3y = 21 \end{cases}$

Step 1. Multiply the top equation through by 3 and the bottom equation by 2:

$$3 \cdot (5x - 2y = 5) = 15x - 6y = 15$$
$$2 \cdot (2x + 3y = 21) = 4x + 6y = 42$$

Step 2. Add: $19x = 57$

Step 3. Solve for x: $x = 3$

Step 4. Substitute 3 for x in one of the equations of the system and solve for y:

$$\begin{array}{rr} & 5x - 2y = 5 \\ x = 3: & 5(3) - 2y = 5 \\ & 15 - 2y = 5 \\ & -2y = -10 \\ & y = 5 \end{array}$$

Step 5. Verify, using the other equation of the system:

$$2x + 3y = 21$$

$$\left.\begin{array}{l} x = 3 \\ y = 5 \end{array}\right\} \quad 2(3) + 3(5) \stackrel{?}{=} 21$$

$$6 + 15 \stackrel{?}{=} 21$$

$$21 = 21$$

Solution: (3, 5)

Not all pairs of linear equations will have a unique solution. When one of the two equations is an equivalent equation of the other, the system is said to be *dependent*. In terms of their graphs, the two equations of a dependent system, being equivalent, will have the same straight line—which demonstrates that there are an infinite number of ordered pairs that satisfy *both* equations.

EXAMPLE

$$\text{Solve:} \begin{cases} 8x - 2y = 10 \\ 12x - 3y = 15 \end{cases}$$

Step 1. Multiply the top equation by 3 and the bottom equation by 2:

$$3 \cdot (8x - 2y = 10) \ = 24x - 6y = 30$$

$$2 \cdot (12x - 3y = 15) = 24x - 6y = 30$$

Step 2. Subtract: $0x + 0y = 0$

$$0 = 0$$

The equations of the system are dependent. Any pair of numbers that satisfies one equation will satisfy the other. The graph of both equations is shown in Fig. 5.2(a).

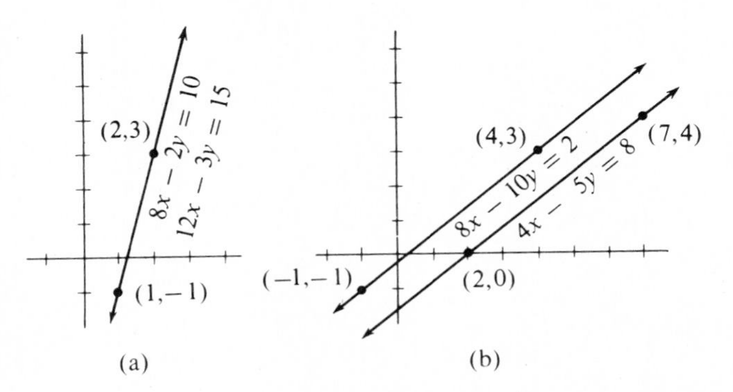

Fig. 5.2

It is also possible for a system to have no ordered-pair solution. Such systems are called *inconsistent*. The graphs of the equations of an inconsistent system are parallel lines, i.e., they have no point of intersection.

EXAMPLE

$$\text{Solve:} \begin{cases} 4x - 5y = 8 \\ 8x - 10y = 2 \end{cases}$$

Step 1. Multiply the top equation by 2:

$$2 \cdot (4x - 5y = 8) = 8x - 10y = 16$$

$$8x - 10y = 2$$

Step 2. Subtract: $0x - 0y = 14$

$$0 = 14$$

The result of Step 2: $0 = 14$, is false, and indicates that the equations of the system are inconsistent. There is no ordered pair of numbers that satisfies *both* equations. The graphs of the equations are parallel lines, as can be seen in Fig. 5.2(b).

(Complete Exercise 5.6)

10. DETERMINANTS

There is another technique for solving systems of linear equations, one that has special utility for computers, particularly when the system is relatively small. It employs a type of mathematical notation called a *determinant*—a square array of numerals between vertical lines. For example:

$$\begin{vmatrix} a_1 & b_1 \\ a_2 & b_2 \end{vmatrix}$$

A determinant is made up of:

elements—the numerals a_1, a_2, b_1, b_2;

rows—the elements in a horizontal line (same subscripts, e.g., a_2, b_2);

columns—the elements in a vertical line (e.g., b_1, b_2); and

principal diagonal—the elements occurring along a diagonal that starts at the upper left element and ends at the lower right element; e.g.,

$$\begin{vmatrix} \mathbf{a_1} & b_1 \\ a_2 & \mathbf{b_2} \end{vmatrix}$$

When there are two elements in each row and in each column, the determinant is called a *second-order determinant*. The *expansion* of a determinant is a polynomial. By definition, the expansion of

$$\begin{vmatrix} a_1 & b_1 \\ a_2 & b_2 \end{vmatrix} = (a_1)(b_2) - (a_2)(b_1)$$

(5.7) *To expand a second-order determinant:*

Step 1. Multiply the elements along the principal diagonal.

Step 2. Multiply the elements along the other diagonal.

Step 3. Subtract the product of Step 2 from that of Step 1.

EXAMPLES

1. Expand $\begin{vmatrix} a_1 & b_1 \\ a_2 & b_2 \end{vmatrix}$

Step 1. $\begin{vmatrix} a_1 & b_1 \\ a_2 & b_2 \end{vmatrix}$ $a_1 \cdot b_2 = a_1b_2$

Step 2. $\begin{vmatrix} a_1 & b_1 \\ a_2 & b_2 \end{vmatrix}$ $a_2 \cdot b_1 = a_2b_1$

Step 3. $\begin{vmatrix} a_1 & b_1 \\ a_2 & b_2 \end{vmatrix} = a_1b_2 - a_2b_1$

2. Expand $\begin{vmatrix} 2 & 7 \\ 5 & 3 \end{vmatrix}$

Step 1. $\begin{vmatrix} 2 & 7 \\ 5 & 3 \end{vmatrix}$ $(2)(3)$

Step 2. $\begin{vmatrix} 2 & 7 \\ 5 & 3 \end{vmatrix}$ $(5)(7)$

Step 3. $\begin{vmatrix} 2 & 7 \\ 5 & 3 \end{vmatrix} = (2)(3) - (5)(7) = 6 - 35$
$= -29$

3. Evaluate the determinant $\begin{vmatrix} 4 & 3 \\ -2 & 1 \end{vmatrix}$

$$\begin{vmatrix} 4 & 3 \\ -2 & 1 \end{vmatrix} = (4)(1) - (-2)(3) = 4 - (-6)$$
$$= 4 + 6 = 10$$

4. Evaluate $\begin{vmatrix} 3 & 2 \\ 0 & -5 \end{vmatrix}$

$$\begin{vmatrix} 3 & 2 \\ 0 & -5 \end{vmatrix} = (3)(-5) - (0)(2) = -15$$

There are determinants of higher orders. A third-order determinant consists of three rows and three columns of elements, a fourth-order determinant has four rows and four columns of elements, and so on. Methods for expanding them differ somewhat from that of the foregoing second-order determinants.

11. SOLUTION BY DETERMINANTS

Consider the following system of linear equations:

$$\begin{cases} a_1x + b_1y = c_1 \\ a_2x + b_2y = c_2 \end{cases}$$

If we multiply the top equation by b_2 and the bottom equation by b_1, and subtract, we obtain a value for x:

$$a_1b_2x + b_1b_2y = b_2c_1$$
$$(-)\ a_2b_1x + b_1b_2y = b_1c_2$$
$$a_1b_2x - a_2b_1x = b_2c_1 - b_1c_2$$
$$(a_1b_2 - a_2b_1)x = b_2c_1 - b_1c_2$$
$$(1) \qquad x = \frac{b_2c_1 - b_1c_2}{a_1b_2 - a_2b_1}$$

Similarly we can find a value for y by multiplying the top equation of the system by a_2 and the bottom equation by a_1, and subtracting to eliminate the x-terms:

$$(2) \quad y = \frac{a_1c_2 - a_2c_1}{a_1b_2 - a_2b_1}$$

Notice that in each instance, designated by (1) and (2) above, the denominator, $a_1b_2 - a_2b_1$, is the expansion of the determinant

$$D = \begin{vmatrix} a_1 & b_1 \\ a_2 & b_2 \end{vmatrix}$$

Moreover, the numerator of the x and y values, respectively, are expansions of the determinants D_x and D_y:

$$D_x = \begin{vmatrix} c_1 & b_1 \\ c_2 & b_2 \end{vmatrix} \qquad D_y = \begin{vmatrix} a_1 & c_1 \\ a_2 & c_2 \end{vmatrix}$$

Thus, we may express (1) and (2) as quotients of determinants:

$$(1)\quad x = \frac{b_2c_1 - b_1c_2}{a_1b_2 - a_2b_1} = \frac{D_x}{D} = \frac{\begin{vmatrix} c_1 & b_1 \\ c_2 & b_2 \end{vmatrix}}{\begin{vmatrix} a_1 & b_1 \\ a_2 & b_2 \end{vmatrix}}$$

$$(2)\quad y = \frac{a_1c_2 - a_2c_1}{a_1b_2 - a_2b_1} = \frac{D_y}{D} = \frac{\begin{vmatrix} a_1 & c_1 \\ a_2 & c_2 \end{vmatrix}}{\begin{vmatrix} a_1 & b_1 \\ a_2 & b_2 \end{vmatrix}}$$

In this we have the basis for another method for solving a system of two linear equations in two variables, known as *Cramer's Rule*.

(**5.8**) *To solve a system of two linear equations in two variables, x and y, by determinants:*

Step 1. Arrange the terms in each equation of the system so that the variables appear in the same order in the left member of each equation, and the constant terms in the right member.
Step 2. Form a determinant, D, that has for elements the numerical coefficients to the variables as they occur in the equations.
Step 3. Form a determinant, D_x, similar to the one in Step 2 except replace the elements that are the coefficients of the x-terms by the constant terms of the respective equations.
Step 4. Form a determinant, D_y, similar to the one in Step 2 except replace the elements that are the coefficients of the y-terms by the constant terms of the respective equations.

Step 5. Compute the solution set using $x = \dfrac{D_x}{D}$ *and* $y = \dfrac{D_y}{D}$.

EXAMPLES

1. Solve $\begin{cases} 2x - y - 7 = 0 \\ x + y - 5 = 0 \end{cases}$

Step 1. $\begin{cases} 2x - y = 7 \\ x + y = 5 \end{cases}$

Step 2. Form determinant D: $\begin{vmatrix} 2 & -1 \\ 1 & 1 \end{vmatrix}$

Step 3. Form determinant D_x: $\begin{vmatrix} 7 & -1 \\ 5 & 1 \end{vmatrix}$

Step 4. Form determinant D_y: $\begin{vmatrix} 2 & 7 \\ 1 & 5 \end{vmatrix}$

Step 5.

$$x = \frac{D_x}{D} = \frac{\begin{vmatrix} 7 & -1 \\ 5 & 1 \end{vmatrix}}{\begin{vmatrix} 2 & -1 \\ 1 & 1 \end{vmatrix}} = \frac{(7) - (-5)}{(2) - (-1)} = \frac{12}{3} = 4$$

$$y = \frac{D_y}{D} = \frac{\begin{vmatrix} 2 & 7 \\ 1 & 5 \end{vmatrix}}{\begin{vmatrix} 2 & -1 \\ 1 & 1 \end{vmatrix}} = \frac{(10) - (7)}{(2) - (-1)} = \frac{3}{3} = 1$$

Solution: (4, 1). Compare with Example 1, p. 101.

2. Solve $\begin{cases} 5x - 2y = 5 \\ 2x + 3y = 21 \end{cases}$

Step 1. Equations are already properly arranged.

Step 2. $D = \begin{vmatrix} 5 & -2 \\ 2 & 3 \end{vmatrix} = (15) - (-4) = 19$

Step 3. $D_x = \begin{vmatrix} 5 & -2 \\ 21 & 3 \end{vmatrix} = (15) - (-42) = 57$

Step 4. $D_y = \begin{vmatrix} 5 & 5 \\ 2 & 21 \end{vmatrix} = (105) - (10) = 95$

Step 5.

$$x = \frac{D_x}{D} = \frac{57}{19} = 3$$

$$y = \frac{D_y}{D} = \frac{95}{19} = 5$$

Solution: (3, 5). Compare with Example 3, p. 102.

3. Solve $\begin{cases} 3x + 2y = 4 \\ 3y = 6 \end{cases}$

Step 1. Equations properly arranged.

Step 2. $D = \begin{vmatrix} 3 & 2 \\ 0 & 3 \end{vmatrix} = 9 - 0 = 9$

Step 3. $D_x = \begin{vmatrix} 4 & 2 \\ 6 & 3 \end{vmatrix} = 12 - 12 = 0$

Step 4. $D_y = \begin{vmatrix} 3 & 4 \\ 0 & 6 \end{vmatrix} = 18 - 0 = 18$

Step 5.

$$x = \frac{D_x}{D} = \frac{0}{9} = 0$$

$$y = \frac{D_y}{D} = \frac{18}{9} = 2$$

Solution: (0, 2).

4. Solve $\begin{cases} 8x - 2y = 10 \\ 12x - 3y = 15 \end{cases}$

On p. 103 we found this system to be dependent. Notice what happens to D, D_x and D_y in such systems.

$$D = \begin{vmatrix} 8 & -2 \\ 12 & -3 \end{vmatrix} = (-24) - (-24) = 0$$

$$D_x = \begin{vmatrix} 10 & -2 \\ 15 & -3 \end{vmatrix} = (-30) - (-30) = 0$$

$$D_y = \begin{vmatrix} 8 & 10 \\ 12 & 15 \end{vmatrix} = (120) - (120) = 0$$

Consequently, $x = \frac{D_x}{D}$ and $y = \frac{D_y}{D}$ are undefined fractions, $\frac{0}{0}$. When $D = D_x = D_y = 0$, the system is dependent.

5. Solve $\begin{cases} 4x - 5y = 8 \\ 8x - 10y = 2 \end{cases}$

On p. 103 we found this system of equations to be inconsistent. Notice what happens to D, D_x and D_y in such systems.

$$D = \begin{vmatrix} 4 & -5 \\ 8 & -10 \end{vmatrix} = (-40) - (-40) = 0$$

$$D_x = \begin{vmatrix} 8 & -5 \\ 2 & -10 \end{vmatrix} = (-80) - (-10) = -70$$

$$D_y = \begin{vmatrix} 4 & 8 \\ 8 & 2 \end{vmatrix} = (8) - (64) = -56$$

Consequently $x = \frac{D_x}{D} = \frac{-70}{0}$ and $y = \frac{D_y}{D} = \frac{-56}{0}$, both undefined. When $D = 0$, $D_x \neq 0$, and $D_y \neq 0$, the system is inconsistent.

(Complete Exercise 5.7)

Name.. Date.............................. **EXERCISE 5.1**

Solve, and verify.

1. $x + 8 = 12$

2. $y - 4 = 7$

3. $p - 7 = -3$

4. $8 = p - 4$

5. $s + 4 = 0$

6. $3 - y = 4$

7. $-3x = -15$

8. $2x - 4 = 6$

9. $3y - 2 = 2y$

10. $3x + 7 = 6x$

11. $4x - 11 = 2x - 7$

12. $\frac{1}{2}x = 14$

13. $\frac{3}{4}x - 12 = 0$

14. $3x - 8 = 12x - 7$

15. $\frac{x}{5} = -25$

16. $4x - 3 = 3(6 - x)$

17. $7x = -2(x + 9)$

18. $(2a - 9) - (a - 3) = 0$

19. $4 - (2 - b) = 2 + 2b$

20. $(1 - t) - 6 = 3(7 - 2t)$

21. $3x + 4(3x - 5) = 12 - x$

22. $3(6a - 5) = 7(3a + 10)$

23. $5(x - 5) = 15 + 5(7 - 2x)$

24. $\frac{1}{2}x - x = 12$

25. $a - 5 = \frac{1}{4}a$

26. $\frac{p}{3} - \frac{7}{6} + \frac{2p}{5} = \frac{3p}{4}$

27. $\frac{1}{3}(x + 2) - \frac{1}{2}(x + 8) = -2$

28. $(a - 1) - (a + 4) = (2a - 5)$

29. $\frac{x}{2} + \frac{3x + 1}{5} = \frac{x + 3}{10}$

30. $\frac{2x + 3}{4} - \frac{3x + 2}{12} = 1$

31. $\frac{.06(25 - .3y)}{.3} - 3.41 = \frac{.003(5 - .1y)}{.01}$

32. $\frac{x + 1}{3} - \frac{x + 3}{5} = x - \frac{3x - 2}{3}$

33. Is $\frac{1}{2}$ a solution of the equation

$(x - 2) + \frac{x - 3}{3} = \frac{x + 7}{15}$?

....

34. (a) Is 1 the only solution for

$\frac{x - 3}{5} + 2x = \frac{11x - 3}{5}$?

....

(b) What type of equation is the equation of (a) above?

....

Name.. Date.............................. **EXERCISE 5.2**

Translate the phrases into algebraic language, using x, unless otherwise indicated, to represent the unknown.

1. A number 5 more than the unknown ....

2. A number 17 less than the unknown ....

3. A number twice the size of the unknown ....

4. A number one-fourth the size of the unknown ....

5. A number 3 more than twice the unknown ....

6. A number half the size of 3 more than the unknown ....

7. A number 40% of the size of the unknown ....

8. A number 40% larger than the unknown ....

9. That which remains after the unknown has been reduced by 6 ....

10. That which remains after the unknown has been removed from 6 ....

11. That which remains after the unknown has been reduced $\frac{1}{6}$ ....

12. 17 divided by double the unknown ....

13. The sum of two numbers a and b decreased by half their product ....

14. The other number when one number is x and their sum is p ....

15. John's age (x at present) 17 years from now ....

16. Cost of x items at p dollars per item ....

17. One-fifth of twice the unknown reduced by 20 ....

18. One-fifth of twice the unknown, reduced by 20 ....

19. The number of minutes in x hours ....

20. The number of weeks in x days ....

Translate the statements into equations. Use x to represent the unknown.

21. The unknown decreased by 7 is 15. ....

22. The unknown increased by 8 is 19. ....

23. Five less than twice the unknown is 16. ....

24. Three times the unknown is 8 greater than the unknown. ....

25. Ten divided by the unknown is 2. ....

26. Sixty is 4 greater than twice the unknown. ....

27. Ten increased by 8 and the unknown is equal to twice the unknown. ....

28. One-third of the unknown equals the unknown diminished by 6. ....

29. Two-thirds of the unknown, increased by 4, is 8. ....

30. Two-thirds of the unknown increased by 4 is 8. ....

Translate the problems into algebraic equations and solve.

31. A yardstick is broken in such a way that one piece is 5 inches longer than the other. What are the lengths of the two pieces?

32. Four milk shakes cost 30 cents more than the $1.50 I have. What is the cost of a single milk shake?

33. Three consecutive integers add to 51. What are they?

....

34. Thirty years from now John will be three times his present age. How old is he now?

....

35. The larger of two numbers exceeds the smaller by 7; their sum is 31. What are the numbers?

....

36. Three consecutive even integers are such that the sum of the two largest is 6 greater than twice the smallest. What are the integers?

....

37. A certain number exceeds 20 by as much as 46 exceeds that number. Find the number.

....

38. Two-thirds of a number is 2 more than half the number. Find the number.

....

Name.. Date.............................. **EXERCISE 5.3**

Write an equation for each, and solve.

1. One-half the sum of three consecutive odd integers is equal to $3\frac{1}{2}$ more than the middle one. Find the integers.

....

2. The sum of the digits of a standard three-digit numeral is 15. The hundreds digit is one more than the tens digit and one-half the ones digit. What is the numeral?

....

3. A clerk gave a customer 25 coins, all quarters and dimes. If the amount of these coins totaled $3.85, how many of each coin did he give?

....

....

4. How many pounds each of cashew nuts ($1.90/lb) and peanuts ($.70/lb) must be mixed together to form 20 pounds of a mixture to sell at a dollar a pound?

....

....

5. Find the dimensions of a rectangle whose perimeter is 88 ft and whose width is $\frac{3}{8}$ its length.

....

6. A triangle and rectangle have equal bases. The altitude of the triangle is 10 ft and that of the rectangle is 15 ft. If their combined area is 440 sq ft, what is the length of their bases?

....

7. A room can be painted by Bill in 12 hours or by his wife in 16 hours. How long will it take them to paint the room if they work together?

....

8. A basketball team scored 86 points in a game in which they scored 16 more field goals than foul shots (field goal = 2 points, foul shot = 1 point). How many of each were scored?

....

9. The sum of the digits of a standard two-digit numeral is 9. If the digits are reversed, the numeral represents a number that is 45 less. What is the original numeral?

....

10. John is 5 years younger than Sue. Six years ago his age was $\frac{2}{3}$ her age then. How old is John now?

....

11. Find the measures of the angles of a triangle (total 180°) where one angle is $\frac{3}{2}$ the size of the second angle and the third angle is 30° less than the second angle.

....

12. The second of three numbers is larger than the first by 16, while the third is $\frac{3}{4}$ that of the second. Find the numbers if their sum is 94.

....

13. Part of $2000 is invested at 4% and the remainder at 6%. If the total yearly income on these two investments is $104.90, how much is invested at each rate?

....

....

14. Mike and Ike are carrying loads of 60 lb and 70 lb, respectively. How many pounds must be removed from Mike's load and added to Ike's load, so that Mike will be carrying a load equal to $\frac{3}{4}$ of Ike's?

....

15. An 8-gallon tank contains a 40% solution of alcohol and water. How much of the solution must be drawn off and replaced by pure water to bring the alcohol concentration down to 25%?

....

16. A boy runs to the store at the rate of 9 miles per hour and walks back at the rate of 3 miles per hour. If his total traveling time is 10 minutes, how far is the store?

....

Name.. Date.................................. **EXERCISE 5.4**

Use set notation to indicate the solution set for each inequality. The domain of the variable is the set of integers.

1. $x > 5$

2. $x < 0$

3. $y \geq 4$

4. $p \leq -2$

5. $x + 1 < 6$

6. $6 > x - 1$

7. $p + 3 \geq 4$

8. $5 > t - 6$

9. $x + 3 \leq 0$

10. $3x < 12$

11. $144 \leq 12m$

12. $\frac{1}{2}k \leq -5$

13. $k + 3 < 2k + 1$

14. $6 + x > 2x + 5$

Graph the solution set for each inequality. There is no restriction on the domain of the variable.

15. $x < 7$

16. $x > -5$

17. $x \leq -2$

18. $3x > -15$

19. $14 \leq \frac{1}{2}x$

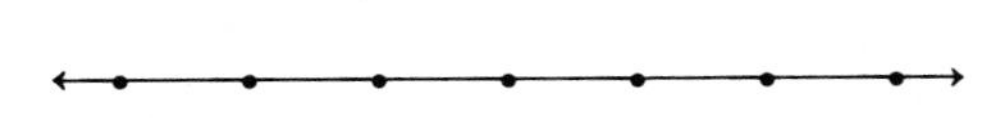

20. $2a - 3 \leq 5a$

21. $3z - 4 > 2z - 9$

22. $2p - 3p + 7 \geq 0$

23. $0 < 2x - 3 + 4x$

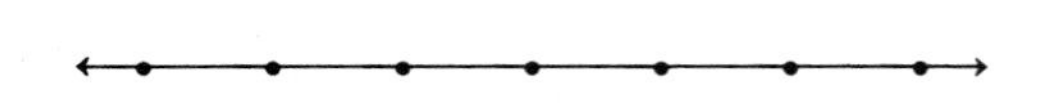

24. $3m - 2 + m \leq 5m + 6$

25. $5p - 2 + 7p > 4p - 6 + 2p$

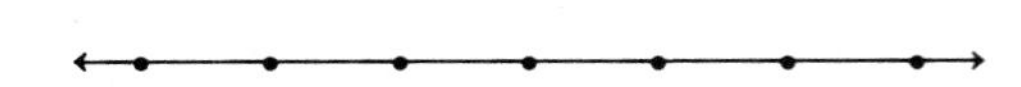

Name.. Date.............................. **EXERCISE 5.5**

Circle the ordered pairs that satisfy the equation.

1. $x + 2y = 13$

(x, y): (7, 3) (3, 7) (3, −5) (11, −1) (15, −1) (5, 4)

2. $3x - y = 2$

(x, y): (3, −7) (2, 4) (1,1) (0, −2) (5, 3) (−3, −7)

3. $2a + 3b = -2$

(a, b): (−1, 0) (0, −1) $(\frac{1}{2}, -\frac{2}{3})$ (.5, −1) (−2, 2) (−5, −4)

4. Give the coordinates of the lettered points.

a:	*b*:
c:	*d*:
e:	*f*:
g:	*h*:
i:	*j*:
k:	*l*:

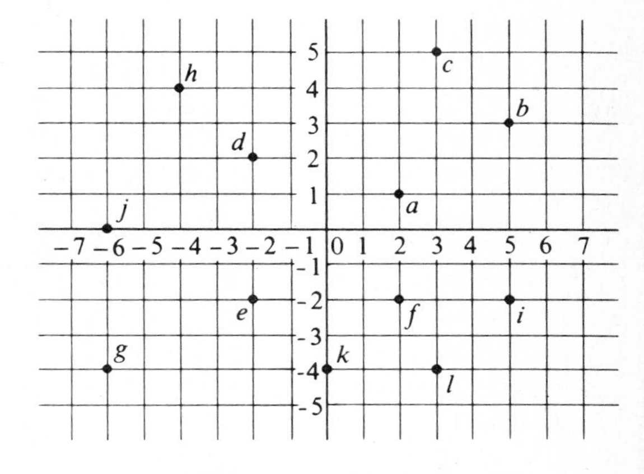

5. Locate the points in the coordinate system at the right whose abscissas and ordinates are given.

a: (4, 4)	*b*: (6, 7)
c: (−3, 5)	*d*: (3, −5)
e: (−5, 3)	*f*: (−3, −5)
g: (3, −2)	*h*: (0, 0)
i: (−5, 0)	*j*: (6, −3)
k: (−6, −2)	*l*: (2, 0)
m: $(6, 2\frac{1}{2})$	*n*: $(-4\frac{1}{2}, -2\frac{1}{2})$

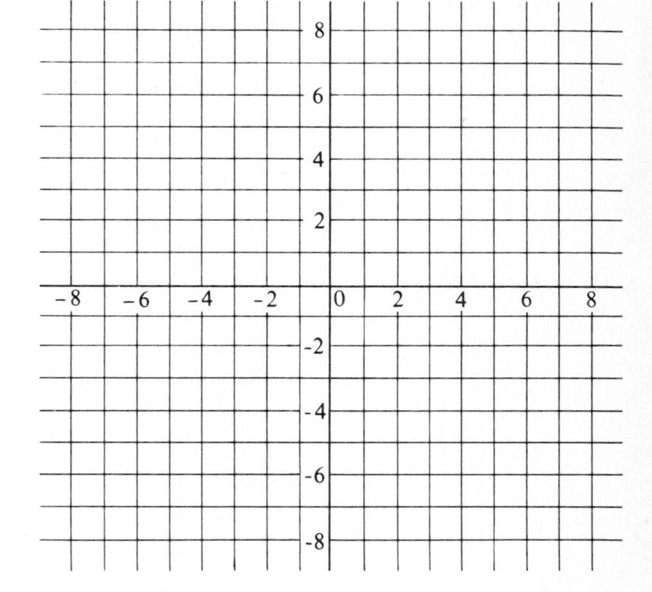

Graph each of the following pairs of equations and estimate the coordinates of their point of intersection.

6. $2x + 3y = 13$ and $x - 2y = -4$

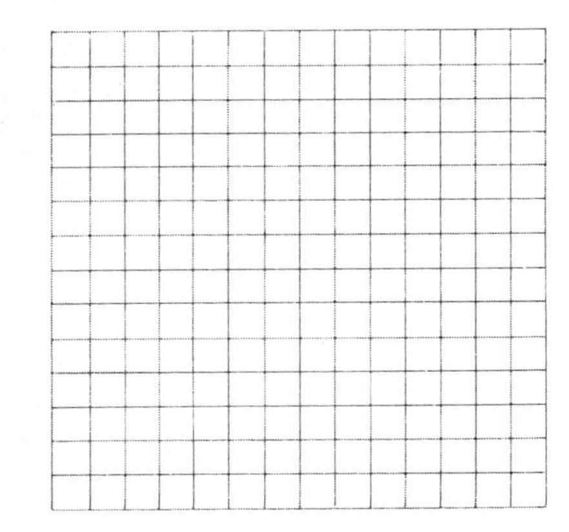

7. $2x + y = 17$ and $3x - 2y = 1$

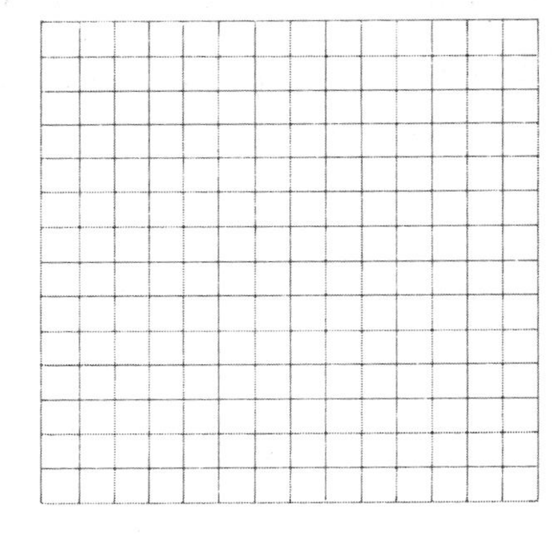

8. $3x - 2y = 4$ and $3x + 2y = 8$

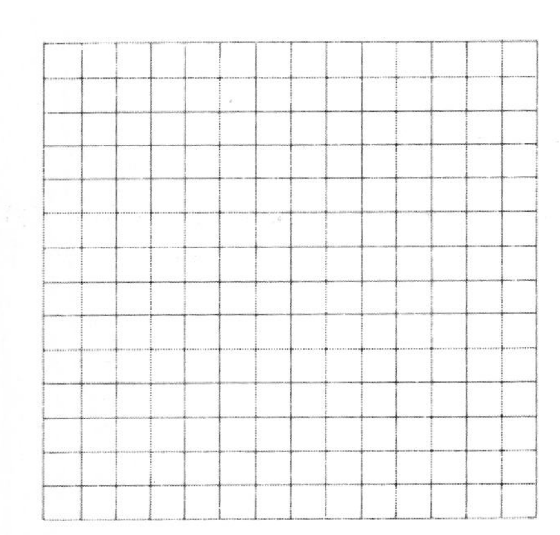

9. $3x - 2y = 9$ and $3x - 5y = 18$

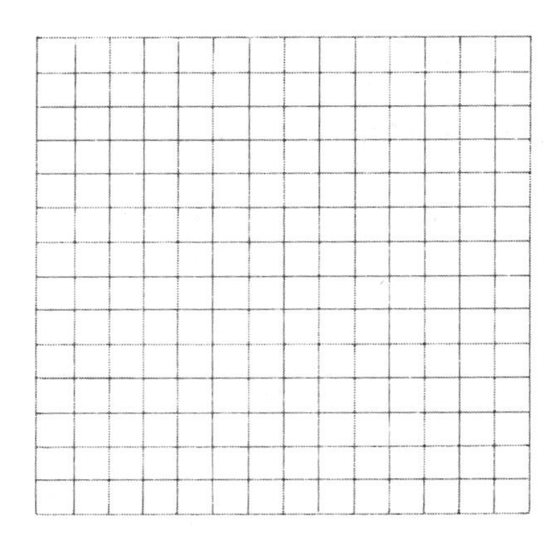

10. $3x + 6y = 9$ and $4x + 8y = 12$

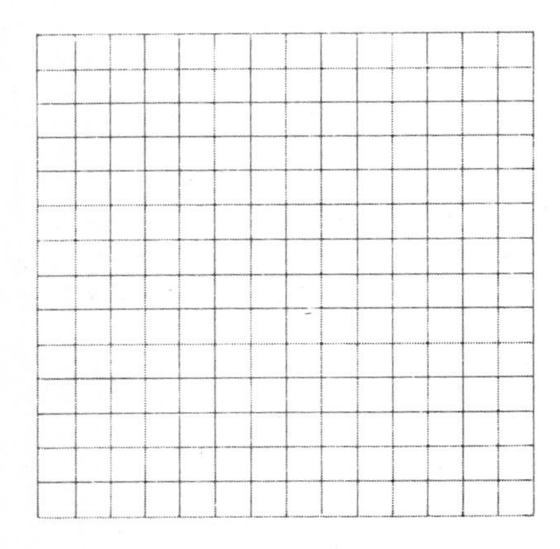

11. $2x - 3y = 5$ and $-6y + 4x = 1$

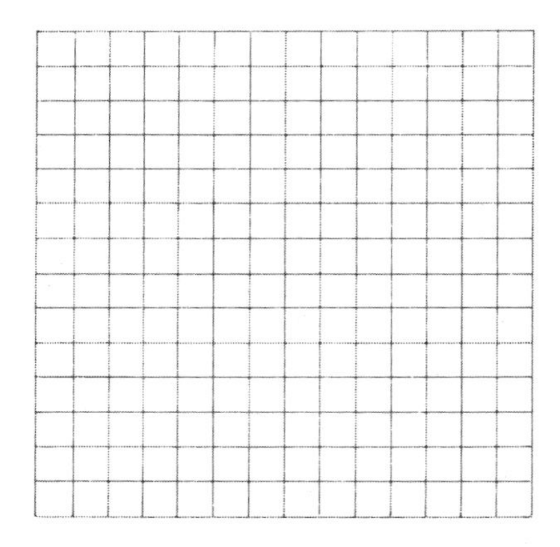

Name.. Date.............................. **EXERCISE 5.6**

Solve by substitution. Compare results of 1 and 2 with 6 and 7 of Exercise 5.5.

1. $\begin{cases} 2x + 3y = 13 \\ x - 2y = -4 \end{cases}$

2. $\begin{cases} 2x + y = 17 \\ 3x - 2y = 1 \end{cases}$

3. $\begin{cases} 4x + 2y = 5 \\ 3x + y = 2 \end{cases}$

4. $\begin{cases} x + 2y = -1 \\ 2x + 5y = 0 \end{cases}$

Solve by addition-subtraction. Compare results of 5 and 6 with 8 and 9 of Exercise 5.5.

5. $\begin{cases} 3x - 2y = 4 \\ 3x + 2y = 8 \end{cases}$

6. $\begin{cases} 3x - 2y = 9 \\ 3x - 5y = 18 \end{cases}$

7. $\begin{cases} 3x - 4y = 6 \\ 2x - y = -1 \end{cases}$

8. $\begin{cases} 2x - 3y = 20 \\ 3x + 5y = 11 \end{cases}$

Solve by either algebraic method. Compare results of 11 and 13 with 10 and 11 of Exercise 5.5.

9. $\begin{cases} 7x + 3y = 4 \\ 14x + 9y = 7 \end{cases}$

10. $\begin{cases} x - 2y = 4 \\ x + 4y = 1 \end{cases}$

11. $\begin{cases} 3x + 6y = 9 \\ 4x + 8y = 12 \end{cases}$

12. $\begin{cases} x - 4y = 1 \\ 3x - 8y = 7 \end{cases}$

13. $\begin{cases} 2x - 3y = 5 \\ -6y + 4x = 1 \end{cases}$

14. $\begin{cases} 6x + y = 4 \\ y + 8 = 0 \end{cases}$

15. $\begin{cases} 4x + 2y = 7 \\ y = -2x + 3 \end{cases}$

Solve using a system of two equations in two variables.

16. The length of a rectangle exceeds its width by 8 ft and the perimeter of the rectangle is 76 ft. Find the dimensions of the rectangle.

....

17. During one month the charge for 25 daily papers and 5 Sunday papers was $5.00. During the next month the paper bill was $4.84 for 27 daily and 4 Sunday papers. What is the unit cost for daily and Sunday papers?

....

....

18. Between two cities an 8-minute phone call costs $3.15, while a 10-minute call costs $3.85. Find the basic rate for the first three minutes and the overtime rate for each additional minute.

....

....

19. A chemistry student wishes to make 100 cubic centimeters of 27% acid solution by combining quantities of 20% and 30% solutions. How many cubic centimeters of each will be necessary?

....

....

Name.. Date.................. **EXERCISE 5.7**

Complete the expansions.

1. $\begin{vmatrix} 5 & 2 \\ 3 & 3 \end{vmatrix} = (5)(\quad) - (3)(\quad) = 9$

2. $\begin{vmatrix} 4 & 0 \\ 1 & 3 \end{vmatrix} = (4)(\quad) - (\quad)(\quad) = 12$

3. $\begin{vmatrix} 2 & 4 \\ -1 & -5 \end{vmatrix} = (\quad)(-5) - (\quad)(\quad) = -6$

4. $\begin{vmatrix} -3 & 4 \\ 2 & 3 \end{vmatrix} = (\quad)(\quad) - (\quad)(\quad) = -17$

Evaluate.

5. $\begin{vmatrix} -3 & 2 \\ 4 & 1 \end{vmatrix}$

6. $\begin{vmatrix} 2 & 4 \\ 1 & 3 \end{vmatrix}$

7. $\begin{vmatrix} 6 & 3 \\ 2 & -4 \end{vmatrix}$

8. $\begin{vmatrix} 7 & -1 \\ -3 & 2 \end{vmatrix}$

9. $\begin{vmatrix} 3 & 4 \\ 0 & 5 \end{vmatrix}$

10. $\begin{vmatrix} -2 & 4 \\ -3 & 6 \end{vmatrix}$

11. $\begin{vmatrix} 0 & -7 \\ 4 & 0 \end{vmatrix}$

12. $\begin{vmatrix} 0 & 6 \\ 0 & 7 \end{vmatrix}$

Complete the solution, using determinants.

13. $\begin{cases} x + 2y = 8 \\ 2x - y = 1 \end{cases}$

$$x = \frac{D_x}{D} = \frac{\begin{vmatrix} 8 & 2 \\ 1 & -1 \end{vmatrix}}{\begin{vmatrix} 1 & 2 \\ 2 & -1 \end{vmatrix}} = \frac{\qquad}{(-1) - (4)} =$$

$$y = \frac{D_y}{D} = \frac{\begin{vmatrix} 1 & 8 \\ 2 & 1 \end{vmatrix}}{\begin{vmatrix} 1 & 2 \\ 2 & -1 \end{vmatrix}} = \frac{\qquad}{\qquad} =$$

14. $\begin{cases} 2x - y = 4 \\ 3x + 2y = -8 \end{cases}$

$$x = \frac{D_x}{D} = \frac{\begin{vmatrix} & \end{vmatrix}}{\begin{vmatrix} & \end{vmatrix}}$$

$$y = \frac{D_y}{D} = \frac{\begin{vmatrix} & \end{vmatrix}}{\begin{vmatrix} & \end{vmatrix}}$$

Solve, using determinants (Cramer's Rule).

15. $\begin{cases} 2x + 3y = 13 \\ x + 2y = 8 \end{cases}$

16. $\begin{cases} x - 4y = 0 \\ 3x + 2y = 7 \end{cases}$

17. $\begin{cases} 4a - 3b = 1 \\ a - 2b = 4 \end{cases}$

18. $\begin{cases} 3x - 2y = 4 \\ 6x - 4y = 5 \end{cases}$

19. $\begin{cases} 7x + 4y = -12 \\ 3x - 2y = 6 \end{cases}$

20. $\begin{cases} 4a - 2b = 6 \\ 6a - 3b = 9 \end{cases}$

From the replacement set, select the member that will make the equality true.

1. $\frac{2}{x} + x = 2x - 3\frac{1}{2}$ $\{0, 1, 2, 3, 4\}$
2. $\frac{3}{x} + 2x = 3x + 2$ $\{-5, -4, -3, -2\}$
3. $\frac{2}{x} + x = \frac{3}{x} - 1.5$ $\{\frac{1}{5}, \frac{1}{4}, \frac{1}{3}, \frac{1}{2}, 1\}$
4. $\frac{2}{x^2} + 3x = x^2 + 2\frac{1}{2}$ $\{0, 1, 2, 3, 4\}$
5. $2x + 4x^2 = 204x^2$ $\{1, .1, .01, .001, .0001\}$
6. $x^2 + 2x - 3 = 0$ $\{-1, -2, -3, -4\}$

Solve for x.

7. $2x - 4 + 3x - 2x = 6 + x + 7x$
8. $4x - 2 + 3x + 7 = 4 - 3x + 2 - 5x$
9. $3x + 2 = 4(x - 7) + 6$
10. $3(x - 5) = 2(3x - 2) - 2$
11. $4(2x - 3) = 7(x - 4 + x)$
12. $2(7 - 3x) + 10 = 5(2x + 3 - 6 + 2x)$
13. $4(2x + 5) + 3(x - 7) = x - 1$
14. $2(x - 3 + x) - x + 3 = 5(x - 2 + x - 7)$
15. $3(x + 2) - (4x - 7) + 3(x - 5) = 2(x - 1)$
16. $\frac{x - 2}{5} + \frac{3x - 2}{6} = \frac{x + 8}{15}$
17. $\frac{1}{3}(4 - x) - \frac{3}{5}(5 - 2x) = \frac{2}{3}(x - 7)$
18. $\frac{2}{3}(2 - 3x) + x - \frac{3}{4}(x - 5) = 2 + \frac{1}{6}(2x + 1)$
19. $m - 3(2x - 3m) = 2(3x - m)$

Translate the following statements into algebraic language; use x for the unknown, unless otherwise indicated:

20. A number p less than the unknown.
21. A number one-ninth the size of twice the unknown.
22. A number half the size of the unknown diminished by k.
23. A number 78% larger than the unknown.
24. A number 125% greater than the unknown.
25. The remainder after the unknown has been taken from 22.
26. The other number when the larger is x and their positive difference is p.
27. The average rate of speed when the distance is $3x$ and the time is $x - 7$.
28. One-sixth of twice the unknown, reduced by m.
29. The number of dozens in x.
30. The unknown decreased by 22 is the same number as six times the unknown.
31. Three less than the unknown is a number that is three-quarters of the unknown.
32. Seven decreased by the unknown and 12 is 5 less than the negative of the unknown.
33. Two-thirds of the unknown is the equivalent of the unknown increased by p.
34. A number 3 more than 7, when divided by the unknown, is $1\frac{1}{2}$ more than the unknown.

Solve.

35. The perimeter of a rectangle is 148 in., and the length of one side is 1 in. short of being $1\frac{1}{2}$ times the length of the other side. What are the dimensions of the rectangle?
36. A line 7 ft long is to be cut into three pieces, and each finished piece is to have a knot, which takes up 2 in. of line, tied on each end. If one finished piece is to be twice the length of another and 2 in. shorter than the third, what should be the unknotted lengths of the three pieces of line?
37. Find the sizes of the angles of a triangle if one is 19° larger than the middle-sized angle, and the third is 19° less than that angle.
38. Seven consecutive integers add to 728. Which are they?
39. A numeral has two digits. The tens digit is $2\frac{1}{2}$ times the size of the ones digit, and the sum of the digits is 7. What is the numeral?
40. Reversing the digits, which differ by 2, of a two-digit numeral and subtracting the number it represents from that of the first yields a difference of 18. What are the digits?
41. Five integers, in ordered sequence, differ one from the next by 7. The sum of the four smallest is 10 less than the largest. What are the integers?
42. A confectioner has some candy that normally sells for \$1.50 a pound and some that sells for 80¢ a pound. How much of the expensive candy must he use to produce 20 lbs of a mixture to sell for \$1 a pound?
43. What volumes of 10% and 4% solutions must be mixed together to yield 81 cc of 6% solution?
44. A basketball team scored 74 points in a game in which they scored 16 more field goals than foul shots. (Field goal = 2 points; foul shot = 1 point.) How many field goals were scored?

45. At one time, an investor could borrow money on his good credit at $4\frac{1}{4}\%$ and invest these same funds at $5\frac{3}{4}\%$ elsewhere. If he earned \$600 that way one year, how much did he borrow?

46. Mr. Allen had part of \$40,000 invested at 4% and the rest at 6%. Had he switched these investments, his yield would have been \$400 more. How much must he have had invested at 6%?

47. A father, who is 36, has two sons, one twice the age of the other. In three years, the sum of all three ages will be 63. How old are the sons now?

48. An uncle is 3 times the age of his niece, and in 3 years he will be 3 times as old as his niece is now. How old is the uncle if the total of the 3 ages is now 66?

49. A woman has 3 times as many quarters as dimes in her purse. If the quarters were dimes and the dimes were dollars, she would be richer by \$1.80. How much money does she have in the purse?

Solve for the variable; the replacement set is the set of integers.

50. $x - 1 < 6$ **51.** $y + 5 \geq 6$

52. $-4 > t + 5$ **53.** $m - 4 \geq 0$

54. $3x \leq -15$ **55.** $100 \geq 10a$

56. $\frac{1}{3}p + 2 > 8$ **57.** $-6 + \frac{1}{2}x < 0$

Solve for the variable. There is no restriction on the domain of the variable.

58. $k + 1 \geq -3$ **59.** $x + 1 < 1$

60. $2x > -4$ **61.** $3m + 2 \leq -4$

62. $-x + 3 > 0$ **63.** $2m - 3 \leq 4m - 5$

64. $0 \leq 4x - 2 - 7x$ **65.** $3y - \frac{1}{2} \geq y - 4 + 3y$

Graph the following pairs of equations, and estimate the coordinates of their point of intersection.

66. $x + y = 7$ and $x - y = 5$

67. $x - 2y = 7$ and $2x + y = 9$

68. $3x + y = 3$ and $2x - 3y = 2$

69. $2x - 5y = 11$ and $4x - 3y = 1$

Solve for x and y by addition-subtraction, or by substitution.

70. $\begin{cases} 2x - y = 4 \\ 3x + 2y = 13 \end{cases}$ **71.** $\begin{cases} x - 3y = 7 \\ x + 6y = -2 \end{cases}$

72. $\begin{cases} 2x + y = -9 \\ x - 2y = 8 \end{cases}$ **73.** $\begin{cases} 3x - 2y = 6 \\ 6x - 4y = 2 \end{cases}$

74. $\begin{cases} 6x + 8y = 0 \\ 3x + 2y = -1 \end{cases}$ **75.** $\begin{cases} x + 4y = -5 \\ -x + y = 3\frac{1}{4} \end{cases}$

76. $\begin{cases} 2x - 4y = 12 \\ \frac{1}{2}x - y = 3 \end{cases}$ **77.** $\begin{cases} 2x - 3y = 9 \\ y = -3 \end{cases}$

Evaluate the determinants.

78. $\begin{vmatrix} 2 & 3 \\ 3 & 5 \end{vmatrix}$ **79.** $\begin{vmatrix} -4 & 2 \\ 3 & 5 \end{vmatrix}$

80. $\begin{vmatrix} 6 & 0 \\ 4 & -9 \end{vmatrix}$ **81.** $\begin{vmatrix} 0 & 3 \\ 0 & 7 \end{vmatrix}$

82. $\begin{vmatrix} 8 & 5 \\ 9 & -3 \end{vmatrix}$ **83.** $\begin{vmatrix} 5 & 0 \\ 0 & -3 \end{vmatrix}$

Solve, using determinants.

84. $\begin{cases} 3x + 2y = 7 \\ 2x + 3y = 3 \end{cases}$ **85.** $\begin{cases} 2x + 3y = 7 \\ 7x + 8y = 12 \end{cases}$

86. $\begin{cases} x - 6y = 0 \\ x - 4y = 1 \end{cases}$ **87.** $\begin{cases} 3x - 2y = 3 \\ -7x + 4y = -1 \end{cases}$

88. Problem 73 above.

89. Problem 76 above.

90. Problem 77 above.

91. The sum of two numbers is 6, and their difference is 10. What are the numbers?

92. What volumes of 15% and 5% solutions must be mixed together to yield 20 cc of 12% solution?

REVIEW

Simplify

1. $(-8\frac{3}{4}) + (+4\frac{7}{8}) + (-3\frac{1}{2}) + (+2\frac{3}{8})$
2. $(+26) \times (-5) \times (-31)$
3. $(+.42) \times (-.6) \times (+2.1)$
4. $(+2\frac{3}{4}) \div (-1\frac{5}{6})$
5. $3.2x - 4.1x + 3 - 2.7x - 1$
6. $a^3b^2 \cdot ab^3 \cdot ab \cdot a^2b \cdot a$
7. $(-6x)(-2y)(3x)(-y)$
8. $(a - 3b + c)(a - 1)$
9. $(2x^2 - x - 1)(x^2 + x + 1)$
10. $(8a^6 - 8a^5 + 4a^4) \div (4a^3)$
11. $(15x^7 - 10x^5 + 10x^3) \div (-5x^2)$
12. $(m^2n^2 - 3m^2n^3 - 2m^4n^2) \div (mn)$
13. $(y^2 + 8y + 12) \div (y + 2)$
14. $(m^2 + 12m + 27) \div (m + 3)$
15. $(3x^2 + 7x - 6) \div (3x - 2)$

Name.. Date........................ Score....................

ACHIEVEMENT TEST NO. 2 (Chapters 4 and 5)

Place answers, in simplest form, in the blanks provided. (4 points each correct answer, total 100)

Perform the operations and simplify:

1. $(3x + 2xy + 4y) + (7x - 3xy - 3y)$
$+ (3y - 9x) + (5xy + 12x) =$ ....

2. $\left(\frac{x}{3} - \frac{3y}{4}\right) - \left(\frac{3x}{2} - \frac{y}{8}\right) =$ ....

3. $4x - 3y + .2x - \frac{1}{2}y - \frac{1}{4}x + .3y =$ ....

4. $(-3ab^2)(-6a^3 + 4ab - 2b^2) =$ ....

5. $(2x^2 - x + 1)(3x - 5) =$ ....

6. $\{2 + |-5|\}\{3 - [4 + (2 - 1) + 3]\}$
$\{-5[3 - (6 - 6)]\} =$ ....

7. $[(3m - s) - (2m - 5s)] - 2(m - 4s) =$ ....

8. $-2[3(5a - 3b) - 6a] - 3b =$ ....

9. $\frac{(16)(-2)}{(2)(-4)(-1)} =$ ....

10. $\frac{20a^3b^2c^2 - 25a^2b^3c^2 - 15a^2b^2c^4}{-5a^2b^2c^0} =$ ....

11. $(53x^2 + 15x^3 - 8 - 30x) \div (3x - 2) =$ ....

12. $(x^4 - 16) \div (x - 2) =$ ....

Solve for x.

13. $6(x - 5) = 25 + 5(5 - 2x) =$ ....

14. $\frac{x}{3} - \frac{7}{6} + \frac{2x}{5} = \frac{3x}{4}$ ....

15. $3x - 4 < 2x - 8$ ....

16. $2 - 3x \geq 6 - x$ ....

17. Is 2 a solution for

$$\frac{1}{2}x = \frac{x + 3}{10} - \frac{1}{5}(3x + 1)?$$

(Answer "yes" or "no.") ....

18. Evaluate.

$$\begin{vmatrix} 7 & 6 \\ -3 & 10 \end{vmatrix}$$

Solve.

19. $\begin{cases} 4x - 5y = 13 \\ 3x + 4y = 2 \end{cases}$ ....

20. $\begin{cases} 4y = 2x \\ 3x - 2y = 5 \end{cases}$ ....

21. At what point will the graphs of $y = -3$ and $3x - 2y = 12$ intersect?

22. If the sum of two numbers is 28 and the larger is 5 less than twice the smaller, find the two numbers.

....

23. What four consecutive odd integers have a sum of 88?

....

24. A group of four neighbors agree to contribute equally to the cost of a ladder. If two more neighbors join in the venture, the cost for each person will be $2 less. What is the cost of the ladder?

....

25. What volumes of 10% and 4% solutions must be mixed together to yield 81 cc of 6% solution?

....

....

6

SPECIAL PRODUCTS, FACTORING, AND FRACTIONS

1. INTRODUCTION

At the computational level of algebra, it is often useful to "rename" an algebraic expression, to express it equivalently but in different form. Some of the procedures developed in Chapter 4 accomplish this. For example, the product of two algebraic factors may be *expanded* by Program 4.13 into an equivalent sum of monomials; and a sum of monomials may be *factored* by Program 4.17 into the product of two factors (i.e., the divisor and quotient).

Many algebraic expressions may be expanded or factored on sight. Learning how to do so is best done through a study of "special products." These provide direct means for expanding frequently used types of products and also afford helpful insight into the more difficult process of factoring.

It is important to note that whether a polynomial is *factorable* or not depends upon the domain of the numerical coefficients. In this chapter a polynomial with integral coefficients is considered factorable if it can be expressed as the product of two or more factors in which the numerical coefficients are integers and the exponents are whole numbers. A polynomial is said to be *prime* if, for the conditions established, it is not factorable except for the trivial factors of itself and 1, or the negative of itself and -1.

2. PRODUCT OF A MONOMIAL AND A POLYNOMIAL

Program 4.12 offers a procedure for computing the product of a polynomial and a monomial (multiply each term of the polynomial by the monomial and add the products). Invariably the product will be a sum of terms, each of which contains a common factor, namely the monomial. Inversely, this suggests a way to rename such sums as a product of factors.

(6.1) *To factor a polynomial whose terms contain a common factor:*

Step 1. *Inspect the prime factors of each term and identify all factors common to all terms.*

Step 2. *Divide the polynomial by the product of the common factors identified in Step 1.**

Step 3. *The quotient and divisor of Step 2 are a pair of factors of the given polynomial.*

EXAMPLES

1. Factor: $8x^2 + 12xy + 4x$.

   Step 1. $8x^2 = 2 \cdot 2 \cdot 2 \cdot x \cdot x$
   $12xy = 2 \cdot 2 \cdot 3 \cdot x \cdot y$
   $4x = 2 \cdot 2 \cdot x$

   The common prime factors are 2, 2, and x; hence the HCF $= 2 \cdot 2 \cdot x = 4x$.

   Step 2. $4x\overline{)8x^2 + 12xy + 4x}$ with quotient $2x + 3y + 1$

   Step 3. $8x^2 + 12xy + 4x = (4x)(2x + 3y + 1)$

2. Factor: $4x^2 - 2x + 6xy^2$.

   Step 1. $4x^2 = 2 \cdot 2 \cdot x \cdot x$
   $2x = 2 \cdot x$
   $6xy^2 = 2 \cdot 3 \cdot x \cdot y \cdot y$
   HCF $= 2 \cdot x = 2x$

   Step 2. $2x\overline{)4x^2 - 2x + 6xy^2}$ with quotient $2x - 1 + 3y^2$

   Step 3. $4x^2 - 2x + 6xy^2 = (2x)(2x - 1 + 3y^2)$

3. Factor: $-15x^3 + 45x^2y - 30x^2$.

   Step 1. $15x^3 = 3 \cdot 5 \cdot x \cdot x \cdot x$
   $45x^2y = 3 \cdot 3 \cdot 5 \cdot x \cdot x \cdot y$
   $30x^2 = 2 \cdot 3 \cdot 5 \cdot x \cdot x$
   HCF $= 3 \cdot 5 \cdot x \cdot x = 15x^2$

---

* The product of these common factors is the highest common factor (HCF)—in effect, the greatest common divisor (GCD)—of the terms under consideration.

Step 2. (When the numerical coefficient of the first term of the given polynomial is negative, the numerical coefficient of the common factor is usually made negative, which makes the coefficient of the first term in the other factor positive.)

$$-15x^2\overline{)-15x^3 + 45x^2y - 30x^2}^{\;x \;-\; 3y \;+\; 2}$$

Step 3. $-15x^3 + 45x^2y - 30x^2$
$= (-15x^2)(x - 3y + 2)$

3. SQUARE OF A BINOMIAL

A binomial is a two-term expression, such as $(3a + b)$. To square a binomial means to multiply it by itself. Thus:

$$(3a + b)^2 = (3a + b)(3a + b)$$

which expands to $9a^2 + 6ab + b^2$:

$$\begin{array}{r} 3a + b \qquad \\ \times\; 3a + b \qquad \\ \hline 9a^2 + 3ab \qquad \\ 3ab + b^2 \\ \hline 9a^2 + 6ab + b^2 \end{array}$$

If we analyze the final product above, we can see that it is the sum of the square of the first term of the binomial, plus twice the product of the two terms, plus the square of the second term of the binomial. With a little practice, the computations can be done mentally and the product written directly.

(6.2) *To square a binomial:*

Step 1. Square the first term.
Step 2. Double the product of the first and second terms (consider the sign between the two terms as belonging to the second term).
Step 3. Square the second term.
Step 4. Express the results of Steps 1, 2, and 3 as a sum for the desired product.

EXAMPLES

1. Square: $4a + 3c$.
   Step 1. Square the first term:
   $(4a)^2 = 16a^2$
   Step 2. Double the product of the two terms:
   $2[(4a)\cdot(3c)] = 2[12ac] = 24ac$
   Step 3. Square the second term:
   $(3c)^2 = 9c^2$
   Step 4. Add the results of Steps 1, 2, and 3 for the desired product:
   $(4a + 3c)^2 = 16a^2 + 24ac + 9c^2$
2. Expand: $(2a - b)^2$.
   Step 1. $(2a)^2 = 4a^2$
   Step 2. $2[(2a)(-b)] = 2[-2ab] = -4ab$
   Step 3. $(-b)^2 = b^2$
   Step 4. Add: $4a^2 - 4ab + b^2$
3. Expand: $(ab - c)^2$.
   Step 1. $(ab)^2 = a^2b^2$
   Step 2. $2[(ab)(-c)] = 2[-abc] = -2abc$
   Step 3. $(-c)^2 = c^2$
   Step 4. $a^2b^2 - 2abc + c^2$
4. Square 53.
   Express 53 as $(50 + 3)$:
   $53^2 = (50 + 3)^2$
   $= (50)^2 + 2(50)(3) + (3)^2$
   $= 2{,}500 + 300 + 9 = 2{,}809$

The square of a binomial is always a *perfect square trinomial*: a three-term polynomial in which two of the terms—usually the first and third—are perfect squares, and the other—middle—term is (disregarding the sign) twice the product of the square roots of the other two terms. It is easily verified that when the binomial is the *sum* of two terms, the sign of the middle term is invariably positive; when the binomial is the *difference* of two terms, the sign of the middle term is invariably negative.

(6.3) *To factor a perfect square trinomial:*

Step 1. Find the square roots of the two terms which are perfect squares.
Step 2. Write two identical binomial factors whose terms are the square roots of Step 1 separated by the sign of the remaining trinomial term.

EXAMPLES

1. Factor: $4x^2 + 12x + 9$.
   $4x^2 + 12x + 9$ is a perfect square trinomial.
   Step 1. $\sqrt{4x^2} = 2x$
   $\sqrt{9} = 3$
   Step 2. The remaining trinomial term is $+12x$ (positive). Therefore, $(2x + 3)(2x + 3)$ is the factored form of $4x^2 + 12x + 9$.
2. Factor: $25x^2 - 40x + 16$.
   $25x^2 - 40x + 16$ is a perfect square trinomial.
   Step 1. $\sqrt{25x^2} = 5x$
   $\sqrt{16} = 4$
   Step 2. The remaining trinomial term is $-40x$ (negative). Therefore, $(5x - 4)(5x - 4)$ is the factored form of $25x^2 - 40x + 16$.
3. Factor: $16a^2 - 24ab + 9b^2$.
   $16a^2 - 24ab + 9b^2$ is a perfect square trinomial.

Step 1. $\sqrt{16a^2} = 4a$

$\sqrt{9b^2} = 3b$

Step 2. $(4a - 3b)(4a - 3b)$ is the factored form of $16a^2 - 24ab + 9b^2$.

4. Factor: $4x^2 + 24xy + 36y^2$.
$4x^2 + 24xy + 36y^2$ is a perfect square trinomial and is also a type covered by Program 6.1 (contains a common factor 4). Remove the factor 4 first:
$4(x^2 + 6xy + 9y^2)$
But $x^2 + 6xy + 9y^2$ is also a perfect square trinomial, so:

Step 1. $\sqrt{x^2} = x$

$\sqrt{9y^2} = 3y$

Step 2. $x^2 + 6xy + 9y^2 = (x + 3y)(x + 3y)$.
Therefore the factored form of $4x^2 + 24xy + 36y^2$ is:
$(4)(x + 3y)(x + 3y)$
[Note: The same result could have been obtained by factoring $4x^2 + 24xy + 36y^2$ (a perfect square trinomial) first as
$(2x + 6y)(2x + 6y)$
then noting that each of the factors has a common factor of 2, and can be further factored:

$$\begin{aligned}[2x + 6y][2x + 6y] &= [(2)(x + 3y)][(2)(x + 3y)] \\ &= (2)(x + 3y)(2)(x + 3y) \\ &= (2)(2)(x + 3y)(x + 3y) \\ &= 4(x + 3y)^2\end{aligned}$$

Usually, however, it simplifies computation to factor out first any common factor that is present among the terms.]

(Complete Exercise 6.1)

4. PRODUCT OF THE SUM AND DIFFERENCE OF TWO TERMS

Let the binomials $(x + y)$ and $(x - y)$ represent, respectively, the sum and difference of the two terms x and y. The product of these two binomials can be found by Program 4.13:

$$\begin{array}{l} x \quad - y \\ \underline{x \quad + y} \\ x^2 - xy \\ \underline{\quad + xy \quad - y^2} \\ x^2 + 0xy - y^2 = x^2 - y^2 \end{array}$$

Note that this product amounts to the square of the first term of the binomials (x) minus the square of the second term of the binomials (y); the so-called middle term of the product of the sum and difference of two terms adds to zero, and does not appear in the final expression.

(6.4) *To compute the product of the sum and difference of two terms:*

Step 1. *Square the two terms.*

Step 2. *Write the product as the square of the first term minus the square of the second term.*

EXAMPLES

1. Expand: $(3x + 4)(3x - 4)$.
Step 1. $(3x)^2 = 9x^2$
$(4)^2 = 16$
Step 2. $9x^2 - 16$
2. Expand: $(4 + 3x)(4 - 3x)$.
Step 1. $(4)^2 = 16$
$(3x)^2 = 9x^2$
Step 2. $16 - 9x^2$
3. Expand: $(a - b)(a + b)$.
Step 1. $(a)^2 = a^2$
$(b)^2 = b^2$
Step 2. $a^2 - b^2$
4. Multiply: 98×102.
Express 98 as $(100 - 2)$, and 102 as $(100 + 2)$.

$$\begin{aligned}98 \times 102 &= (100 - 2)(100 + 2) \\ &= (100)^2 - (2)^2 \\ &= 10{,}000 - 4 = 9{,}996\end{aligned}$$

The foregoing examples illustrate that the product of the sum and difference of two terms is the difference of the squares of the two terms. Inversely, then, an algebraic expression involving the difference of two squares will have for factors the sum and difference of the square roots of these two terms.

(6.5) *To factor the difference of two squared terms:*

Step 1. *Find the square root of each of the terms.*

Step 2. *Write one factor as the sum of the two square roots of Step 1 and the other factor as the difference of the two square roots.*

EXAMPLES

1. Factor: $a^2 - 4b^2$.
Step 1. $\sqrt{a^2} = a$
$\sqrt{4b^2} = 2b$
Step 2. $(a + 2b)(a - 2b)$
2. Factor: $1 - 16x^2y^2$.
Step 1. $\sqrt{1} = 1$
$\sqrt{16x^2y^2} = 4xy$
Step 2. $(1 + 4xy)(1 - 4xy)$
3. Factor: $12x^2y - 27y^3$.
$12x^2y - 27y^3$ is an example of the type covered by

Program 6.1. Remove the common factor $3y$:
$12x^2y - 27y^3 = 3y(4x^2 - 9y^2)$
Since the factor
$(4x^2 - 9y^2) = (2x + 3y)(2x - 3y)$
the complete factorization of
$12x^2y - 27y^3 = (3y)(2x + 3y)(2x - 3y)$

5. PRODUCT OF ANY TWO BINOMIALS

The two preceding types of special products (Programs 6.2 and 6.4) have been singled out because their forms are readily recognized. In this section we shall develop a general approach for finding the product of *any* two binomials directly (which includes instances covered by Programs 6.2 and 6.4).

Consider the following product of two binomials computed by Program 4.13:

$$\begin{array}{r} 2x + y \\ x - 3y \\ \hline 2x^2 + xy \\ -6xy - 3y^2 \\ \hline 2x^2 - 5xy - 3y^2 \end{array}$$

Note in the computation above that: (a) the first term of the product ($2x^2$) is the product of the first terms of the binomials; (b) the last term of the product ($-3y^2$) is the product of the second terms of the binomials; (c) the middle term of the product ($-5xy$) is the result of adding the two products formed by multiplying the first term of each binomial with the second term of the other binomial. Part (c) is often referred to as cross multiplying or computing the cross-products. Shown below is a horizontal pattern for writing the product directly, with the computation above repeated to illustrate corresponding steps. Note that the middle term in the horizontal method is computed mentally:

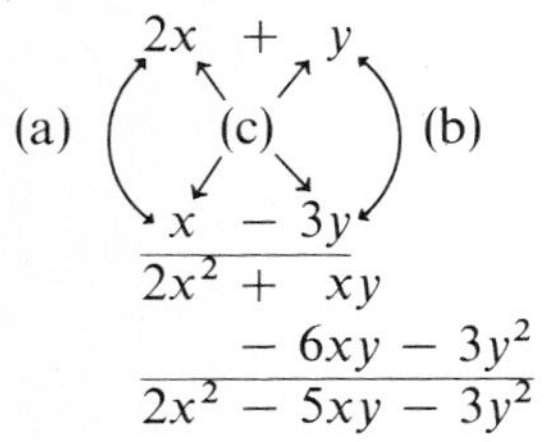

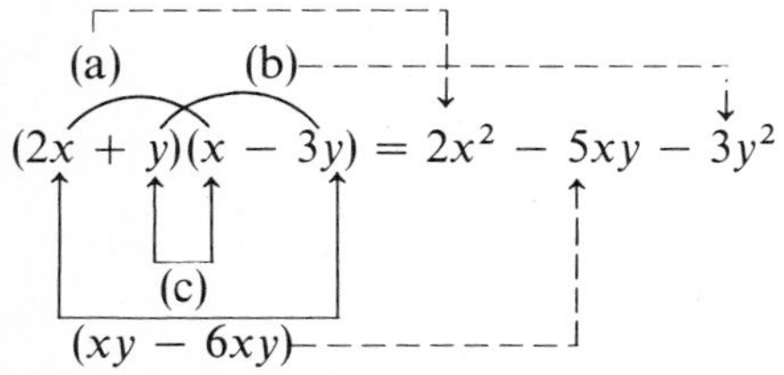

(a) (c) (b)

$$\begin{array}{r} 2x + y \\ x - 3y \\ \hline 2x^2 + xy \\ -6xy - 3y^2 \\ \hline 2x^2 - 5xy - 3y^2 \end{array}$$

(a) (b)

$(2x + y)(x - 3y) = 2x^2 - 5xy - 3y^2$

(c)

$(xy - 6xy)$

(6.6) *To compute the product of any two binomials directly:*

Step 1. Find the product of the first terms of the two binomials.

Step 2. Find the sum of the two cross-products.

Step 3. Find the product of the second terms of the two binomials.

Step 4. Express as a sum the results of Steps 1, 2, and 3 for the product of the two binomials.

EXAMPLES

1. Write the product directly for $(3x - 2y)(x - 4y)$.
   Step 1. Product of the first terms:
   $(3x)(x) = 3x^2$
   Step 2. Sum of cross-products:
   $\left.\begin{array}{r}(3x)(-4y) = -12xy \\ (-2y)(x) = -2xy\end{array}\right\} -14xy$
   Step 3. Product of the second terms:
   $(-2y)(-4y) = 8y^2$
   Step 4. $(3x - 2y)(x - 4y) = 3x^2 - 14xy + 8y^2$
2. Write the product directly for $(4x - 5)(x + 6)$. 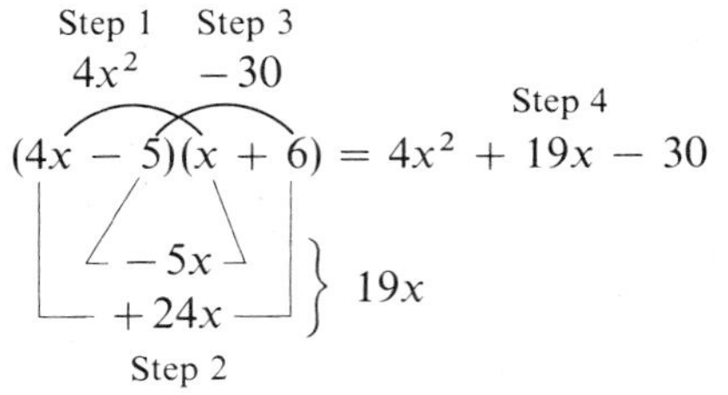
   Step 1 Step 3
   $4x^2$ -30
   Step 4
   $(4x - 5)(x + 6) = 4x^2 + 19x - 30$
   $\left.\begin{array}{r}-5x \\ +24x\end{array}\right\} 19x$
   Step 2
3. Write the product directly for $(3x - 2y)(3x - 2y)$. 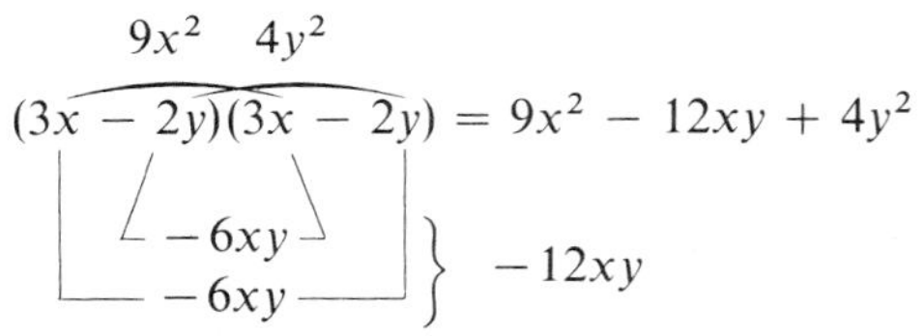
   $9x^2$ $4y^2$
   $(3x - 2y)(3x - 2y) = 9x^2 - 12xy + 4y^2$
   $\left.\begin{array}{r}-6xy \\ -6xy\end{array}\right\} -12xy$
   [Note: $(3x - 2y)(3x - 2y) = (3x - 2y)^2$ and, as can be expected, the product is a perfect square trimonial.]

Finding two binomial factors whose product is a trimonial of the form $ax^2 + bxy + cy^2$, in which a, b, c are numerical coefficients, is by no means always possible, and more often than not requires trial and error tactics. However, a few generalizations coupled with conscious experience in computing binomial products directly will tend to minimize the number of necessary trials.

From our procedure for finding the products of two binomials directly, we know that the first term of the trinomial to be factored must contain as its factors the first terms of the desired binomials; that the last term of the trinomial must contain as its factors the second terms of the desired binomials, and that the middle term of the trinomial must be the sum of the cross-products of the first and second terms of the binomial factors.

Another aid to factoring factorable trinomials is an awareness of the distribution of signs in the trinomial. In the products of the pairs of binomials given below, note how the signs are distributed among the terms:

(a) $(3x + 2y)(x + y) = 3x^2 + 5xy + 2y^2$
(b) $(3x - 2y)(x - y) = 3x^2 - 5xy + 2y^2$
(c) $(3x + 2y)(x - y) = 3x^2 - xy - 2y^2$
(d) $(3x - 2y)(x + y) = 3x^2 + xy - 2y^2$

We may generalize with the following sign analysis:

1. When the sign of the last term of a factorable trinomial is positive, the signs of the second terms in both binomial factors are the same as that of the middle term of the trinomial.
2. When the sign of the last term of a factorable trinomial is negative, the signs of the second terms of the binomial factors differ, and the prevailing sign of the sum of the resulting cross-products is the same as that of the middle term of the trinomial.

(6.7) *To factor a trinomial that is the product of two binomials:*

Step 1. Determine two factors of the first term of the trinomial.
Step 2. Determine two factors of the last term of the trinomial.
Step 3. Combine these to form two binomials such that each binomial contains one factor from each of Steps 1 and 2. Insert signs according to the sign analysis of the trinomial.
Step 4. Check the sum of the resulting cross-products in Step 3. If it agrees with the middle term of the trinomial, the binomials of Step 3 are the desired factors. If it does not agree, repeat Steps 1–3 with new factor combinations, and with new sign arrangements if the binomial factors differ in sign.

EXAMPLES

1. Factor: $x^2 - 5x + 6$.
   Step 1. The factors of the first term of the trinomial (x^2) are x and x.
   Step 2. The factors of the last term of the trinomial (6) are 6 and 1, also 3 and 2.
   Step 3. Coupling the x and x with the 6 and 1 we get $(x \quad 6)(x \quad 1)$. From analysis of the signs of the trinomial, the signs of the second terms of the binomials will be negative. Thus: $(x - 6)(x - 1)$.
   Step 4. The sum of the cross-products for $(x - 6)(x - 1)$ is $-7x$. This does not agree in sign with the middle term of the trinomial $(-5x)$, therefore $(x - 6)$ and $(x - 1)$ are *not* the factors of $x^2 - 5x + 6$.
   Step 3. Couple the x and x factors of the first term with the factors 3 and 2 of the last term for $(x - 3)(x - 2)$.
   Step 4. The sum of the cross-products for $(x - 3)(x - 2)$ is $-5x$; consequently $(x - 3)(x - 2)$ is the factored form of $x^2 - 5x + 6$.

   [Note: When the coefficient of the first term of the trinomial is one (1), as in this example, the correct set of coefficient factors of the last term can be quickly determined by the following.
   (a) If the sign of the third term of the trinomial is positive, the correct set of coefficient factors will be those whose sum is the coefficient of the middle term of the trinomial.
   (b) If the sign of the third term of the trinomial is negative, the correct set of coefficient factors will be those whose difference is the coefficient of the middle term of the trinomial.]

2. Factor: $x^2 + 2x - 15$.
   Step 1. The factors of x^2 are x and x.
   Step 2. The factors of 15 are 15 and 1, also 5 and 3. The sign of the third term is negative. The difference between the factors 15 and 1 is 14; the difference between the factors 5 and 3 is 2, which is the coefficient of the middle term. We choose 5 and 3.
   Step 3. Couple the x and x with the 5 and 3 in two binomials: $(x \quad 5)(x \quad 3)$. Since the sign of the third term in the trinomial is negative, the signs of the second terms of the two binomials will differ; and since the middle term is positive we place the $(+)$ sign before the 5 and the $(-)$ sign before the 3: $(x + 5)(x - 3)$.
   Step 4. Check the sum of the cross-products:
   $5x + (-3x) = 2x$
   Hence the factored form of $x^2 + 2x - 15$ is $(x + 5)(x - 3)$.

3. Factor: $2x^2 + xy - 3y^2$.
   Step 1. Factor $2x^2$: $2x \cdot x$
   Step 2. Factor $3y^2$: $3y \cdot y$
   Step 3. Couple these factors into binomial form. Because the last term is negative, the signs of the second terms of the binomials will differ; because the sign of the middle term (xy) is positive, the $(+)$ sign should go with the greater cross-product:
   $(2x - y)(x + 3y)$
   Step 4. Check the sum of the cross-products:
   $6xy + (-xy) \neq xy$
   Step 3. Try a new combination:
   $(2x + 3y)(x - y)$
   Step 4. Check the sum of the cross-products:
   $3xy + (-2xy) = xy$

Hence $(2x + 3y)(x - y)$ is the factored form of $2x^2 + xy - 3y^2$.

4. Factor: $6x^2 - x - 12$.
   The factors of $6x^2$: $6x$ and x, $3x$ and $2x$.
   The factors of 12: 12 and 1; 6 and 2; 4 and 3.
   From an analysis of the signs in the trinomial, establish the distribution of signs in the binomials: $(\ldots + \ldots)(\ldots - \ldots)$
   Test the various factor combinations of 12 with the $6x$ and x factors of $6x^2$, keeping in mind that the sum of the cross-products must be negative (because the middle term in the trinomial is negative).

Factors(?)	*Middle term check*
$(6x + 1)(x - 12)$	$(x) + (-72x) \neq -x$
$(6x + 2)(x - 6)$	$(2x) + (-36x) \neq -x$
$(6x + 3)(x - 4)$	$(3x) + (-24x) \neq -x$
$(6x + 4)(x - 3)$	$(4x) + (-18x) \neq -x$
$(6x + 6)(x - 2)$	$(6x) + (-12x) \neq -x$

(The other combination leads to a positive middle term.)

Before abandoning the $6x$ and x combination, interchange these factors:

$(x + 1)(6x - 12) \qquad (6x) + (-12x) \neq -x$

(Other combinations lead to a positive middle term.)

From this it is clear that $6x$ and x are *not* the proper pair of factors for $6x^2$. So try the pair $2x$ and $3x$.

$(2x + 1)(3x - 12) \qquad (3x) + (-24x) \neq -x$
$(2x + 2)(3x - 6) \qquad (6x) + (-12x) \neq -x$

(Other combinations lead to a positive middle term.)

Finally, interchange the factors $2x$ and $3x$:

$(3x + 1)(2x - 12) \qquad (2x) + (-36x) \neq -x$
$(3x + 2)(2x - 6) \qquad (4x) + (-18x) \neq -x$
$(3x + 4)(2x - 3) \qquad (8x) + (-9x) = -x$

Hence the factored form of $6x^2 - x - 12$ is $(3x + 4)(2x - 3)$.

Example 4 above was purposely labored to demonstrate thoroughly the technique of factoring a trinomial. Actually, experience with factoring tends to narrow the guesswork considerably. But when the correct combination is not found readily, it can be discouraging and a suspicion tends to arise that the trinomial is not factorable. A quick means to check whether or not a trinomial is indeed factorable, with integral coefficients, is to apply the following:

Trinomials of the form $ax^2 + bx + c$ (in which the numerical coefficients, denoted by a, b, and c, may be either positive or negative) *can be factored if* $(b^2 - 4ac)$ *is a perfect square.*

EXAMPLES

1. Will $3x^2 + 2x - 4$ factor?
   If $3x^2 + 2x - 4 = ax^2 + bx + c$, then $a = 3$, $b = 2$, and $c = -4$, and
   $(b^2 - 4ac) = (2)^2 - 4(3)(-4) = 4 - (-48) = +52$
   Since $+52$ is *not* a perfect square, $3x^2 + 2x - 4$ will *not* factor into the product of two binomials with integral (or even rational) coefficients.
2. Will $3x^2 + 10x - 8$ factor?
   Here $a = 3$, $b = 10$, $c = -8$.
   $$b^2 - 4ac = (10)^2 - 4(3)(-8) = 100 + 96 = 196 = (14)^2$$
   Hence $3x^2 + 10x - 8$ will factor.
3. Will $2x^2 + 11xy + 12y^2$ factor?
   $$b^2 - 4ac = 121 - (4)(2)(12) = 121 - 96 = 25 = (5)^2$$
   Hence $2x^2 + 11xy + 12y^2$ will factor.

(Complete Exercise 6.2)

6. FACTORING BY GROUPING

Ideas developed in the previous sections of this chapter can be usefully extended by broadening the word "term" to include any algebraic expression. For instance, according to Program 6.1:

$$3x^2 - 12xy = 3x(x - 4y)$$

However, had the common term been a binomial, say $(3 + x)$, instead of $3x$, it too could have been factored out. To start with a simple illustration, suppose we have the following expression:

$$a(x + y) + b(x + y)$$

Clearly $(x + y)$ is a common factor in each of the two major terms of the sum, and as such it can be factored out:

$$a(x + y) + b(x + y) = (x + y)(a + b)$$

Further,

$$a(x + y) + b(x + y) = ax + ay + bx + by$$

Hence:

$$ax + ay + bx + by = (x + y)(a + b)$$

Thus we see that in some cases a polynomial (such as $ax + ay + bx + by$) containing no universally common term may still be factored by appropriate grouping.

EXAMPLES

1. Factor: $xy + 5y + bx + 5b$.
   If we factor y out of the first two terms and b out of the last two terms, we can express the given polynomial as
   $$\underbrace{xy + 5y} + \underbrace{bx + 5b}$$
   $$y(x + 5) + b(x + 5)$$

Now if we factor out the common $(x + 5)$ we will have transformed the given polynomial into an equivalent product:

$$y(x + 5) + b(x + 5) = (x + 5)(y + b)$$

2. Factor: $3m + 3n - am - an$.
   Factor 3 from the first two terms and $-a$ from the last two terms:
   $$3m + 3n - am - an = 3(m + n) - a(m + n)$$
   $$= (m + n)(3 - a)$$

3. Factor: $cx + y - x - cy$.
   Rearrange, and factor out c:
   $$cx + y - x - cy = cx - cy + y - x$$
   $$= c(x - y) + y - x$$
   Express $y - x$ as $-1(x - y)$; then:
   $$c(x - y) + (y - x) = c(x - y) + (-1)(x - y)$$
   $$= (x - y)(c - 1)$$

4. Factor: $c^2d - 3c^2 - 9d + 27$.
   Factor c^2 from the first two terms and -9 from the last two terms:
   $$c^2(d - 3) - 9(d - 3) = (d - 3)(c^2 - 9)$$
   But $(c^2 - 9)$ can be factored further (Program 6.5) as $(c - 3)(c + 3)$. Hence:
   $$c^2d - 3c^2 - 9d + 27 = (c - 3)(c + 3)(d - 3)$$

There are still other polynomial sums whose terms are other than monomials which are factorable. Generally, the more experience one has with factoring the different basic types, the more likely he is to recognize a factorable pattern in these more complicated sums.

EXAMPLES

1. Factor: $(x - b)^2 + 9(x - b) + 14$.
   It may help insight to let $a = (x - b)$, in which case
   $$(x - b)^2 + 9(x - b) + 14 = a^2 + 9a + 14$$
   which readily factors to $(a + 7)(a + 2)$. However, since $a = (x - b)$, then
   $$(a + 7) = [(x - b) + 7]$$
   $$(a + 2) = [(x - b) + 2]$$
   and
   $$(x - b)^2 + 9(x - b) + 14 = (x - b + 7)(x - b + 2).$$

2. Factor: $(y + 2)^2 - (x + 3)^2$.
   If we let $a = (y + 2)$ and $b = (x + 3)$, then
   $$(y + 2)^2 - (x + 3)^2 = a^2 - b^2 = (a - b)(a + b).$$
   Substituting back for a and b, we get:
   $$(y + 2)^2 - (x + 3)^2 = [(y + 2) - (x + 3)] \times [(y + 2) + (x + 3)]$$
   $$= [y + 2 - x - 3] \times [y + 2 + x + 3]$$
   $$= [y - x - 1][y + x + 5]$$

With experience, need for the substi ution "crutch" will diminish, and factoring can take place directly.

3. Factor: $2(x - 3)^2 + (x - 3) - 6$.

   $2(x - 3)^2 \qquad -6$

   $[2(x - 3) - 3][(x - 3) + 2]$

   $-3(x - 3)$
   $4(x - 3)$ $\Big\} -3(x - 3) + 4(x - 3) = (x - 3)$

   Simplifying:
   $$2(x - 3)^2 + (x - 3) - 6 = [2(x - 3) - 3] \times [(x - 3) + 2]$$
   $$= [2x - 6 - 3] \times [x - 3 + 2]$$
   $$= [2x - 9][x - 1]$$

(Complete Exercise 6.3)

7. SIMPLIFYING ALGEBRAIC FRACTIONS

The subject of fractions in arithmetic was studied in Chapter 2. For our immediate purposes, three basic concepts developed there bear repeating:

1. The value of the fraction is unchanged when the numerator and denominator are multiplied or divided by the same number, not zero.
2. Reducing to lower terms is a form of division of numerator and denominator by a common divisor or factor.
3. A fraction is in lowest terms when there exists for both the numerator and the denominator no integral divisor (common factor) except 1.

In algebra, arithmetic terms such as *fraction, numerator, denominator, lowest terms*, etc., are extended to include algebraic expressions of many kinds. For example, the following parallels Program 2.2.

(6.8) *To write an algebraic fraction in lowest terms equivalent to a given fraction:*

Step 1. Factor the numerator of the given fraction completely.

Step 2. Factor the denominator of the given fraction completely.

Step 3. Divide out all common factors that exist in both the numerator and denominator of the given fraction.

EXAMPLES

1. Express $\dfrac{x^2 + 2x - 8}{x^2 + x - 12}$ in lowest terms.

   Step 1. $x^2 + 2x - 8 = (x - 2)(x + 4)$
   Step 2. $x^2 + x - 12 = (x + 4)(x - 3)$

$$\text{Step 3. } \frac{x^2 + 2x - 8}{x^2 + x - 12} = \frac{(x - 2)\cancel{(x + 4)}}{\cancel{(x + 4)}(x - 3)} = \frac{x - 2}{x - 3}$$

2. Express $\dfrac{3x^2 - 9ax + 6a^2}{6x^2 - 6a^2}$ in lowest terms.

Step 1. $3x^2 - 9ax + 6a^2 = 3(x - a)(x - 2a)$
Step 2. $6x^2 - 6a^2 = (2)(3)(x - a)(x + a)$

$$\text{Step 3. } \frac{3x^2 - 9ax + 2a^2}{6x^2 - 6a^2} = \frac{\cancel{3}\cancel{(x - a)}(x - 2a)}{(2)(\cancel{3})\cancel{(x - a)}(x + a)} = \frac{x - 2a}{2(x + a)}$$

3. Express $\dfrac{cp - 2a - ap + 2c}{p^2 + 4p + 4}$ in lowest terms.

Step 1. $cp - 2a - ap + 2c = (c - a)(p + 2)$
Step 2. $p^2 + 4p + 4 = (p + 2)(p + 2)$

$$\text{Step 3. } \frac{cp - 2a - ap + 2c}{p^2 + 4p + 4} = \frac{(c - a)\cancel{(p + 2)}}{(p + 2)\cancel{(p + 2)}} = \frac{(c - a)}{(p + 2)}$$

8. THE NEGATIVE OF AN ALGEBRAIC EXPRESSION

There is no difficulty in recognizing the difference between the expression for a negative and a positive integer, such as -3 and $+3$, nor is there usually any difficulty in recognizing that $-(c - a)$ is the negative of $(c - a)$. But when two such binomials are expressed in expanded form without parentheses, often the fact that one is the negative of the other is not so readily grasped. Thus:

$$-(c - a) = (-1)(c - a) = (-1)(c) + (-1)(-a)$$
$$= (-c + a) = (a - c)$$

Therefore it can be said that $(a - c)$ is the negative of $(c - a)$ and vice versa.

EXAMPLES

1. Write the negative of $(x - y)$.
   The negative of $(x - y)$ is
   $-(x - y) = -x + y = (y - x)$
2. Is $(a - b)$ the negative of $(b - a)$?
   The negative of $(a - b)$ is $-(a - b)$, which in expanded form is $-a + b$, and which in turn can be expressed as $b - a$. Hence:
   $(b - a) = -(a - b)$ and $(a - b) = -(b - a)$
3. Write the negative of $x^2 - x - 6$.
   The negative of $x^2 - x - 6$ is
   $-(x^2 - x - 6) = -x^2 + x + 6 = 6 + x - x^2$

Notice the effect upon the factors of these two expressions:

$x^2 - x - 6 = (x - 3)(x + 2)$
$6 + x - x^2 = (3 - x)(2 + x)$
Hence,
$(x - 3)(x + 2) = (-1)(3 - x)(2 + x)$
and
$(3 - x)(2 + x) = (-1)(x - 3)(x + 2)$

4. Express $\dfrac{x^2 + x - 12}{15 - 2x - x^2}$ in lowest terms.

Step 1. $x^2 + x - 12 = (x - 3)(x + 4)$
Step 2. $15 - 2x - x^2 = (5 + x)(3 - x)$

$$\text{Step 3. } \frac{\overset{-1}{\cancel{(x - 3)}}(x + 4)}{(5 + x)\cancel{(3 - x)}} = \frac{(-1)(x + 4)}{5 + x} = \frac{-(x + 4)}{(5 + x)} \quad \left[\text{or } -\frac{x + 4}{5 + x}\right]$$

[Note: Because we shall be dealing frequently with negative signs in fractions, it might be well to recall from Section 4, Chapter 4, that a fraction may be thought of as possessing three signs: the sign of the fraction, the sign of the numerator, and the sign of the denominator; and that any two of the three signs may be reversed without changing the value of the fraction.]

(Complete Exercise 6.4)

9. MULTIPLICATION WITH ALGEBRAIC FRACTIONS

The product of two or more algebraic fractions, as with fractions in arithmetic, is computed by multiplying numerators together and denominators together. As is true in arithmetic, effort can be saved by dividing out common factors from both the numerators and denominators before undertaking the multiplication steps. When this is done thoroughly, the resulting product fraction is automatically in lowest terms.

(6.9) *To compute the product of two algebraic fractions:*
Step 1. Factor completely the numerators and denominators of the given fractions.
Step 2. Divide out any factor common to any numerator and any denominator.
Step 3. Write as the desired product a fraction whose numerator consists of the remaining numerator factors and whose denominator consists of the remaining denominator factors.

EXAMPLES

1. Compute the product:

$$\frac{x^2 + x - 6}{2x^2 - 7x - 15} \cdot \frac{2x^2 - 3x - 9}{x^2 - 5x + 6}.$$

Step 1.

$$\left.\begin{aligned} x^2 + x - 6 &= (x + 3)(x - 2) \\ 2x^2 - 3x - 9 &= (2x + 3)(x - 3) \end{aligned}\right\} \text{Numerators}$$

$$\left.\begin{aligned} 2x^2 - 7x - 15 &= (2x + 3)(x - 5) \\ x^2 - 5x + 6 &= (x - 2)(x - 3) \end{aligned}\right\} \text{Denominators}$$

Step 2. $\dfrac{(x + 3)\cancel{(x - 2)}}{\cancel{(2x + 3)}(x - 5)} \cdot \dfrac{\cancel{(2x + 3)}\cancel{(x - 3)}}{\cancel{(x - 2)}\cancel{(x - 3)}}$

Step 3. $\dfrac{x + 3}{x - 5}$ is the desired product in lowest terms.

2. What is the product of

$$\frac{2x^2 - 18}{x^2 + 6x - 7} \cdot \frac{x^2 - 1}{8x^2 + 4x - 24}?$$

The computation:

$$\frac{\cancel{2}(x + 3)(x - 3)}{(x + 7)\cancel{(x - 1)}} \cdot \frac{\cancel{(x - 1)}(x + 1)}{\underset{2}{\cancel{4}}(x + 2)(2x - 3)}$$

$$= \frac{(x + 3)(x - 3)(x + 1)}{2(x + 7)(x + 2)(2x - 3)}$$

[Note: The computed product *can* be expanded if necessary or desired. Usually, however, it is left in factor form.]

3. Multiply: $\dfrac{ac - bc + ad - bd}{c^2 - d^2} \cdot \dfrac{d^2 + 2cd + c^2}{b^2 - a^2}$.

$$\frac{\overset{-1}{\cancel{(a - b)}}(c + d)}{(c - d)\cancel{(c + d)}} \cdot \frac{\cancel{(d + c)}(d + c)}{\cancel{(b - a)}(b + a)} = -\frac{(c + d)^2}{(c - d)(b + a)}$$

10. DIVISION WITH ALGEBRAIC FRACTIONS

In Section 8 of Chapter 2, two techniques were explained for computing quotients with fractions. When it comes to computing quotients with algebraic fractions, either method may be used, but the Inverted Divisor Method (Program 2.14) is usually the simpler.

(6.10) *To compute the quotient of two algebraic fractions:*

Step 1. Invert the divisor fraction.
Step 2. Multiply the dividend fraction with the inverted divisor fraction of Step 1.

EXAMPLES

1. Divide: $\dfrac{6a^5}{25b^4} \div \dfrac{a}{5b^2}$.

Step 1. The divisor $\dfrac{a}{5b^2}$ inverted is $\dfrac{5b^2}{a}$.

Step 2. $\dfrac{6a^5}{25b^4} \cdot \dfrac{5b^2}{a} = \dfrac{6\overset{a^4}{\cancel{a^5}} \cdot \cancel{5} \cdot \cancel{b^2}}{\underset{5}{\cancel{25}} \cdot \underset{b^2}{\cancel{b^4}} \cdot \cancel{a}}$

$$= \frac{6a^4}{5b^2}$$

2. Divide: $\dfrac{2a + 6}{a^2 - 6a + 9} \div \dfrac{a + 3}{a^2 - 5a + 6}$.

Step 1. Invert $\dfrac{a + 3}{a^2 - 5a + 6}$:

$$\frac{a^2 - 5a + 6}{a + 3} = \frac{(a - 3)(a - 2)}{(a + 3)}$$

Step 2.

$$\frac{2\cancel{(a + 3)}}{(a - 3)\cancel{(a - 3)}} \cdot \frac{\cancel{(a - 3)}(a - 2)}{\cancel{(a + 3)}} = \frac{2(a - 2)}{a - 3}$$

3. Divide: $\dfrac{c^2 - 2cd + d^2}{b^2 + 4} \div \dfrac{db - cb + 2d - 2c}{b^4 - 16}$.

Step 1. Invert $\dfrac{db - cb + 2d - 2c}{b^4 - 16}$:

$$\frac{b^4 - 16}{db - cb + 2d - 2c} = \frac{(b^2 + 4)(b - 2)(b + 2)}{(d - c)(b + 2)}$$

Step 2.

$$\frac{(c - d)\overset{-1}{\cancel{(c - d)}}}{\cancel{b^2 + 4}} \cdot \frac{\cancel{(b^2 + 4)}(b - 2)\cancel{(b + 2)}}{\cancel{(d - c)}\cancel{(b + 2)}}$$

$$= (-1)(c - d)(b - 2)$$
$$= [(-1)(c - d)][(b - 2)]$$
$$= (d - c)(b - 2)$$

(Complete Exercise 6.5)

11. LEAST COMMON DENOMINATOR AND EQUIVALENT FRACTIONS

In Chapter 2 it was noted that in order to add or subtract with fractions having different denominators, it is necessary to replace each fraction with an equivalent fraction having a denominator in common with the other fractions. It was also pointed out that *any* common denominator will do, though the least common denominator (LCD) of all the fractions under consideration usually makes for a simpler computation. This is also true when it comes to computing sums and differences with algebraic fractions. Remember, the LCD for a group of fractions is the smallest multiple that is *exactly* divisible by *all* of the denominators.

(6.11) *To compute the least common denominator* (LCD) *of several algebraic fractions:*

Step 1. *Factor each denominator completely.*
Step 2. *Form a product of the factors of the denominators such that each factor appears in it the number of times it appears in the denominator which contains it most. This is the least common denominator (usually left in factor form).*

EXAMPLES

1. Compute the LCD:

$$\frac{x+2}{x^2+2x-3}, \frac{7}{2x^2-8x+6}, \frac{x^2-2}{9x^2-54x+81}, \frac{2-x}{3x^2-15x+18}.$$

Step 1. $x^2+2x-3 = (x+3)(x-1)$
$2x^2-8x+6 = (2)(x-3)(x-1)$
$9x^2-54x+81 = (3)(3)(x-3)(x-3)$
$3x^2-15x+18 = (3)(x-2)(x-3)$

Step 2. $(x+3)(x-1)(2)(3)(3)(x-3)^2(x-2)$
LCD $= (2)(3)(3)(x+3)(x-1)(x-3)^2(x-2)$
or $18(x+3)(x-1)(x-3)^2(x-2)$

x^2+2x-3	$x+3$	$x-1$						
$2x^2-8x+6$		$x-1$	2	$x-3$				
$9x^2-54x+81$				$x-3$	3	3	$x-3$	
$3x^2-15x+18$				$x-3$	3			$x-2$
LCD	$x+3$	$x-1$	2	$x-3$	3	3	$x-3$	$x-2$

[Note: Sometimes a chart such as the one below is helpful in constructing an LCD. The guiding rule is: Put like factors, and those that are negatives of one another, in the same column wherever possible. The "bottom line" is the LCD.]

2. Compute the LCD:

$$\frac{x+5}{x^2-4}, \frac{3}{x^2+x-6}, \frac{8-x}{6-x-x^2}.$$

Step 1. $x^2-4 = (x-2)(x+2)$
$x^2+x-6 = (x-2)(x+3)$
$6-x-x^2 = (2-x)(3+x)$

Step 2. $(x-2)(x+2)(x+3) =$ LCD

[Note: $(2-x)$ divides the already present factor $(x-2)$ exactly (-1) times; therefore there is no need to introduce the factor $(2-x)$ among the factors of the LCD.]

After the least common denominator has been determined, the next step in computing sums and differences with algebraic fractions having different denominators (see Section 12 following) is to replace each fraction by an *equivalent fraction* whose denominator is the LCD. This parallels the arithmetic technique of expressing a fraction as an equivalent fraction having a given denominator (Program 2.3).

(6.12) *To write an algebraic fraction with a specified denominator equivalent to a given fraction:*
Step 1. *Divide the specified denominator by the denominator of the given fraction.*
Step 2. *Multiply numerator and denominator of the given fraction by the quotient of Step 1.*

EXAMPLES

1. Write a fraction equivalent to $\frac{3x+2}{x-7}$ having a denominator of $(x-7)(x+2)$.

Step 1. Divide $(x-7)(x+2)$ by $(x-7)$:

$$\frac{(x-7)(x+2)}{(x-7)} = (x+2)$$

Step 2. Multiply numerator and denominator of the given fraction, $\frac{3x+2}{x-7}$, by $(x+2)$:

$$\frac{(x+2)(3x+2)}{(x+2)(x-7)}$$

2. Express equivalents of each of the three fractions:

$$(1)\ \frac{x+2}{x^2+x-6}, \quad (2)\ \frac{3x}{20-6x-2x^2}, \quad (3)\ \frac{x-7}{x^2+8x+15}$$

each having as denominator the LCD of the three fractions.

(a) Factor the three denominators:
$x^2+x-6 = (x-2)(x+3)$ (*1*)
$20-6x-2x^2 = 2(2-x)(5+x)$ (*2*)
$x^2+8x+15 = (x+3)(x+5)$ (*3*)

(b) Find the LCD of the denominators:
$(2)(x-2)(x+3)(x+5)$

(c) Divide the LCD by each denominator:

$$\frac{(2)\cancel{(x-2)}\cancel{(x+3)}(x+5)}{\cancel{(x-2)}\cancel{(x+3)}} = 2(x+5) \qquad (1)$$

$$\frac{\cancel{(2)}\overset{-1}{\cancel{(x-2)}}(x+3)\cancel{(x+5)}}{\cancel{(2)}\cancel{(2-x)}\cancel{(5+x)}} = (-1)(x+3) \text{ or } -(x+3) \qquad (2)$$

$$\frac{(2)(x-2)\cancel{(x+3)}\cancel{(x+5)}}{\cancel{(x+3)}\cancel{(x+5)}} = 2(x-2) \qquad (3)$$

(d) Multiply the numerator and denominator of each fraction by its respective quotient in (c):

$$\frac{x+2}{(x-2)(x+3)} \cdot \frac{(2)(x+5)}{(2)(x+5)} \qquad (1)$$

$$= \frac{(2)(x+2)(x+5)}{(2)(x-2)(x+3)(x+5)}$$

$$\frac{3x}{(2)(2-x)(5+x)}\cdot\frac{(-1)(x+3)}{(-1)(x+3)} \quad (2)$$

$$= \frac{(-1)(3x)(x+3)}{(-1)(2)(2-x)(x+3)(5+x)}$$

$$= \frac{-3x(x+3)}{2(x-2)(x+3)(x+5)}$$

$$\uparrow \{(-1)(2-x)\}$$

$$\frac{x-7}{(x+3)(x+5)}\cdot\frac{(2)(x-2)}{(2)(x-2)} \quad (3)$$

$$= \frac{(2)(x-7)(x-2)}{(2)(x-2)(x+3)(x+5)}$$

12. ADDITION AND SUBTRACTION WITH ALGEBRAIC FRACTIONS

In Chapter 2 we noted that sums and differences could be computed with fractions by adding or subtracting numerators, *provided the denominators were the same.* When denominators were not the same, it was necessary to replace each fraction by an equivalent fraction whose denominator was the same as that of the other fractions. With algebraic fractions, similar requirements prevail.

(6.13) *To compute the sum (difference) with algebraic fractions:*

Step 1. If the denominators are different, find the least common denominator.

Step 2. Express each fraction as an equivalent fraction whose denominator is the LCD *of Step 1.*

Step 3. Write the sum (difference) of the numerators as the numerator of the result and the LCD *as its denominator.*

Step 4. (Optional, but usual.) Simplify the numerator of Step 3 and reduce to lowest terms.

EXAMPLES

1. Add: $\dfrac{x-23}{x^2-x-20}+\dfrac{x-3}{x-5}$.

Step 1. $x^2-x-20=(x-5)(x+4)$

$x-5=(x-5)$

$\text{LCD}=(x-5)(x+4)$

Step 2. $\dfrac{x-23}{(x-5)(x+4)}+\dfrac{(x-3)(x+4)}{(x-5)(x+4)}$

Step 3. $\dfrac{(x-23)+(x-3)(x+4)}{(x-5)(x+4)}$

Step 4. $\dfrac{x-23+x^2+x-12}{(x-5)(x+4)}$

$$= \frac{x^2+2x-35}{(x-5)(x+4)}$$

$$= \frac{(x+7)\cancel{(x-5)}}{\cancel{(x-5)}(x+4)} = \frac{(x+7)}{(x+4)}$$

Hence,

$$\frac{x-23}{x^2-x-20}+\frac{x-3}{x-5}=\frac{x+7}{x+4}.$$

2. Simplify: $\dfrac{7x-8}{x^2-9}-(3x+2)+\dfrac{3x^2}{x-3}$.

Step 1.

$x^2-9=(x-3)(x+3)$

$x-3=(x-3)$

$\text{LCD}=(x-3)(x+3)$ or (x^2-9)

Step 2.

$$\frac{7x-8}{(x-3)(x+3)}-\frac{(3x+2)(x^2-9)}{(x-3)(x+3)}+\frac{3x^2(x+3)}{(x-3)(x+3)}$$

Step 3.

$$\frac{(7x-8)-[(3x+2)(x^2-9)]+3x^2(x+3)}{(x-3)(x+3)}$$

$$= \frac{(7x-8)-(3x^3+2x^2-27x-18)+(3x^3+9x^2)}{(x-3)(x+3)}$$

$$= \frac{7x-8-3x^3-2x^2+27x+18+3x^3+9x^2}{(x-3)(x+3)}$$

$$= \frac{7x^2+34x+10}{(x-3)(x+3)}$$

[Note: $7x^2+34x+10$ is not factorable with integral coefficients because $(b^2-4ac)=1{,}156-280=876\neq$ perfect square: therefore the result is in lowest terms.]

3. Add: $\dfrac{3x}{2x-1}+\dfrac{2x}{1-2x}$.

Step 1. $\text{LCD}=2x-1$

[Note: $(2x-1)\div(1-2x)=-1$.]

Step 2. $\dfrac{3x}{2x-1}+\dfrac{-2x}{2x-1}$

Step 3. $\dfrac{3x-2x}{2x-1}=\dfrac{x}{2x-1}$

4. Simplify: $1-\dfrac{2}{3-\dfrac{1}{2+\dfrac{3}{x-1}}}$

(a) $1 - \cfrac{2}{3 - \cfrac{1}{2 + \cfrac{3}{x-1}}}$ $\left[= \frac{2x - 2 + 3}{x - 1} = \frac{2x + 1}{x - 1}\right]$

(b) $1 - \cfrac{2}{3 - \cfrac{1}{\cfrac{2x+1}{x-1}}}$ $\left[= \frac{x - 1}{2x + 1}\right]$

(c) $1 - \cfrac{2}{3 - \cfrac{x-1}{2x+1}}$ $\left[= \frac{6x + 3 - x + 1}{2x + 1} = \frac{5x + 4}{2x + 1}\right]$

(d) $1 - \cfrac{2}{\cfrac{5x+4}{2x+1}}$ $\left[= \frac{2(2x + 1)}{5x + 4} = \frac{4x + 2}{5x + 4}\right]$

(e) $1 - \frac{4x + 2}{5x + 4} = \frac{5x + 4 - 4x - 2}{5x + 4} = \frac{x + 2}{5x + 4}$

(Complete Exercise 6.6)

13. FRACTIONAL EQUATIONS

There are two types of fractional equations:

Type A: Equations in which at least one of the numerical coefficients is in fraction form, and the variable does not occur in any denominator.

Example: $\frac{2}{3}x + 4 = \frac{5}{2}x$.

Type B: Equations in which the variable occurs in the denominator of at least one term.

Example: $\frac{3}{x - 2} = 1 + \frac{2}{2x + 1}$.

In either case, the first step in solving a fractional equation is to multiply each member of the equation by the LCD of its terms. For Type A equations the result of this step (known as the *derived equation*) will always be an equivalent equation of the given equation. Equivalent equations, it will be recalled, have identical solution sets. For Type B equations the derived equation may or may not be equivalent to the given equation. In the latter cases it is essential that the solutions obtained from the derived equation be substituted back in the given equation to check for validity.

(6.14) *To solve a fractional equation:*

Step 1. Find the LCD *of the terms of the equation.*

Step 2. Multiply both members of the equation by the LCD *of Step 1 to obtain a derived equation without fractions.*

Step 3. Solve the derived equation of Step 2.

Step 4. Substitute the solution obtained in Step 3 for the variable in the given equation to verify its validity.

EXAMPLES

1. Solve for x: $\frac{5}{2}x - 7 = \frac{3}{4}x$ (Type A).

Step 1. LCD of 2, 4 is 4.

Step 2. $\left(\overset{2}{\cancel{4}} \cdot \frac{5}{\cancel{2}}x\right) - (4 \cdot 7) = \left(\cancel{4} \cdot \frac{3}{\cancel{4}}x\right)$

$10x - 28 = 3x$

Step 3.
$$10x - 28 = 3x$$
$$10x - 3x = 28$$
$$7x = 28$$
$$x = 4$$

Step 4. Substitute 4 for x in the given equation.

$$\frac{5}{2}x - 7 = \frac{3}{4}x$$
$x = 4$: $\frac{5}{2}(4) - 7 \stackrel{?}{=} \frac{3}{4}(4)$
$$10 - 7 \stackrel{?}{=} 3$$
$$3 = 3$$

The given equation is satisfied; therefore 4 is the valid solution.

2. Solve for x: $\frac{2}{3x} = \frac{3}{4x} - \frac{1}{2}$ (Type B).

Step 1. LCD $= 12x$

Step 2. $\left(\overset{4}{\cancel{12x}} \cdot \frac{2}{\cancel{3x}}\right) = \left(\overset{3}{\cancel{12x}} \cdot \frac{3}{\cancel{4x}}\right) - \left(\overset{6x}{\cancel{12x}} \cdot \frac{1}{\cancel{2}}\right)$

$8 = 9 - 6x$

Step 3.
$$8 = 9 - 6x$$
$$6x = 1$$
$$x = \tfrac{1}{6}$$

Step 4. $\frac{2}{3x} = \frac{3}{4x} - \frac{1}{2}$

$x = \frac{1}{6}$: $\frac{2}{3(\frac{1}{6})} \stackrel{?}{=} \frac{3}{4(\frac{1}{6})} - \frac{1}{2}$
$$\frac{2}{\frac{1}{2}} \stackrel{?}{=} \frac{3}{\frac{2}{3}} - \frac{1}{2}$$
$$4 \stackrel{?}{=} \frac{9}{2} - \frac{1}{2}$$
$$4 = 4$$

3. Solve for x: $\frac{8}{3x + 1} + 2 = \frac{2x}{x - 1}$ (Type B).

Step 1. LCD $= (3x+1)(x-1)$

Step 2.

$$\left[\cancel{(3x+1)}(x-1)\cdot\frac{8}{\cancel{3x+1}}\right] + [(3x+1)(x-1)\cdot 2]$$

$$= \left[(3x+1)\cancel{(x-1)}\cdot\frac{2x}{\cancel{(x-1)}}\right]$$

$$8(x-1) + 2(3x+1)(x-1) = 2x(3x+1)$$

Step 3.

$$8x - 8 + 2(3x^2 - 2x - 1) = 6x^2 + 2x$$
$$8x - 8 + \cancel{6x^2} - 4x - 2 = \cancel{6x^2} + 2x$$
$$8x - 4x - 10 = 2x$$
$$2x = 10$$
$$x = 5$$

Step 4. $\dfrac{8}{3x+1} + 2 = \dfrac{2x}{x-1}$

$x = 5$: $\dfrac{8}{3(5)+1} + 2 \stackrel{?}{=} \dfrac{2(5)}{(5)-1}$

$$\tfrac{8}{16} + 2 \stackrel{?}{=} \tfrac{10}{4}$$
$$2\tfrac{1}{2} = 2\tfrac{1}{2}$$

4. Solve for x: $\dfrac{3x}{x-2} = 4 + \dfrac{6}{x-2}$ (Type B).

Step 1. LCD $= (x-2)$

Step 2.

$$\left[\cancel{(x-2)}\cdot\frac{3x}{\cancel{(x-2)}}\right] = [(x-2)\cdot 4] + \left[\cancel{(x-2)}\cdot\frac{6}{\cancel{x-2}}\right]$$

$$3x = 4(x-2) + 6$$

Step 3.

$$3x = 4x - 8 + 6$$
$$-x = -2$$
$$x = 2$$

Step 4. Substitute 2 for x in the given equation:

$$\frac{3x}{x-2} = 4 + \frac{6}{x-2}$$

$x = 2$: $\dfrac{3(2)}{(2)-2} \stackrel{?}{=} 4 + \dfrac{6}{(2)-2}$

$$\frac{6}{0} \stackrel{?}{=} 4 + \frac{6}{0}$$

$\frac{6}{0}$ is undefined; therefore 2 is not a valid solution of the given equation. The equation has no solution.

5. Solve the system:

$$\begin{cases} \dfrac{1}{x} + \dfrac{1}{y} = 5 \\[2mm] \dfrac{2}{x} - \dfrac{1}{y} = 1 \end{cases}$$

LCD $= xy$:

$$\frac{1}{x} + \frac{1}{y} = 5 \longrightarrow y + x = 5xy$$

$$\frac{2}{x} - \frac{1}{y} = 1 \longrightarrow 2y - x = xy$$

Program 6.14 is of no help in systems of equations such as this because both variables appear in one of the terms.

Instead, solve for $\frac{1}{x}$ or $\frac{1}{y}$ first, using Program 5.6:

① $\dfrac{1}{x} + \dfrac{1}{y} = 5$

② $\dfrac{2}{x} - \dfrac{1}{y} = 1$

Add ① and ②: $\dfrac{3}{x} = 6$

Solve for x:
$$3 = 6x$$
$$\tfrac{1}{2} = x$$

Substitute $\frac{1}{2}$ for x in one of the equations and solve for y:

① $\dfrac{1}{x} + \dfrac{1}{y} = 5$

$x = \frac{1}{2}$: $\dfrac{1}{\frac{1}{2}} + \dfrac{1}{y} = 5$

$$2 + \frac{1}{y} = 5$$
$$\frac{1}{y} = 3$$
$$y = \tfrac{1}{3}$$

Verification:

② $\dfrac{2}{x} - \dfrac{1}{y} = 1$

$\left.\begin{matrix} x = \frac{1}{2} \\ y = \frac{1}{3} \end{matrix}\right\}$ $\dfrac{2}{\frac{1}{2}} - \dfrac{1}{\frac{1}{3}} \stackrel{?}{=} 1$

$$4 - 3 \stackrel{?}{=} 1$$
$$1 = 1$$

Solution: $(\frac{1}{2}, \frac{1}{3})$

(Complete Exercise 6.7)

Name.. Date..............................

EXERCISE 6.1

Write the products directly.

1. $(-3)(a-6) =$

2. $a(x-y) =$

3. $4c(a^2-c^2) =$

4. $5x(a-b+c) =$

5. $3xy(x-y+z) =$

6. $-(4x-3y+2z) =$

7. $-p^2(p^3+p^2-3p) =$

8. $-ab(ab-ab^2+a^2b) =$

9. $3m(n^2+2mn-m^2) =$

10. $-5(4a^3-3a^2+2a+1) =$

Find the common term, and factor

11. $5x-10y =$

12. $6x-2x^2 =$

13. $3a^2-12a =$

14. $27a^2p^2+3ap =$

15. $34pq-51p =$

16. $6a-36a^2 =$

17. $14a^2xy-2ax^2y =$

18. $x^3-6x^2 =$

19. $2a^2+a^2-3a =$

20. $g^5+3g^4-2g^2 =$

21. $2mn-4m+6n =$

22. $2x^2-4xy+12y^2 =$

23. $2x^3y+4x^2y^2+16xy^3 =$

24. $a^6-2a^4+3a^2 =$

25. $a^3b^2c^4+a^2b^3c^3+a^4b^2c^3 =$

26. $3a^2-18a^2b+27ab^2 =$

27. $9x^2y-8xy^2+12xy =$

28. $2a^5-4a^4+8a^3+10a =$

Expand, using Program 6.2.

29. $(x+y)^2 =$

30. $(a+b)^2 =$

31. $(m+n)^2 =$

32. $(x-y)^2 =$

33. $(a-b)^2 =$

34. $(m-n)^2 =$

35. $(x+3)^2 =$

36. $(b-4)^2 =$

37. $(3x+1)^2 =$

38. $(2y-1)^2 =$

39. $(3a - 2b)^2 =$

40. $(2x + 3y)^2 =$

41. $(8x + 7y)^2 =$

42. $(pm + qn)^2 =$

43. $(ax - by)^2 =$

44. $(\frac{1}{2}x - \frac{3}{4}y)^2 =$

45. $(12)^2 = (10 + 2)^2 =$

46. $(15)^2 = (10 + \quad)^2 =$

47. $(25)^2 =$

48. $(61)^2 =$

In the parentheses supply the term necessary to make the polynomial a perfect square trinomial.

49. $x^2 - (\quad)xy + y^2$

50. $9x^2 + (\quad)x + 16$

51. $25a^2 + (\quad)ab + 81b^2$

52. $25x^2 - (\quad)x + 144$

53. $36p^2 + (\quad)pt + 49t^2$

54. $16a^4 + (\quad)a^2 + 9$

55. $a^2x^2 + (\quad)x + c^2$

56. $4a^2b^2 - (\quad)d + 25c^2d^2$

Factor the trinomials completely when possible; if a trinomial cannot be factored with integral coefficients, write "prime" next to it.

57. $x^2 + 2xy + y^2 =$

58. $a^2 - 2ab + b^2 =$

59. $x^2 + 4x + 4 =$

60. $2x^2 - 12x + 18 =$

61. $16a^2 + 8a - 1 =$

62. $9x^2y^2 + 6xy + 1 =$

63. $18x^2 + 24xy + 8y^2 =$

64. $4a^2 - 12a + 9 =$

65. $25x^4 + 80x^2y^2 + 64y^4 =$

66. $16a^2 + 14a + 9 =$

67. $25x^2 + 110xy + 121y^2 =$

68. $9 - 12x + 4x^2 =$

69. $405x^2 + 360xy + 80y^2 =$

70. $m^2x^2 - 2mx + 1 =$

Name.. Date.............................. **EXERCISE 6.2**

Expand directly.

1. $(m - n)(m + n) =$

2. $(s + t)(s - t) =$

3. $(x - 3)(x + 3) =$

4. $(y + 7)(y - 7) =$

5. $(2x - y)(2x + y) =$

6. $(3x - 2y)(3x + 2y) =$

7. $(8x - 9y)(8x + 9y) =$

8. $(ab - c)(ab + c) =$

9. $(15)(25) = (20 - 5)(20 + 5) =$

10. $(37)(43) =$

Factor.

11. $a^2 - b^2 =$

12. $a^2 - 9b^2 =$

13. $16a^2 - 1 =$

14. $9x^2y^2 - 64z^2 =$

15. $25p^2 - 4q^2 =$

16. $121m^2 - 144 =$

17. $a^2x^2 - b^2y^2 =$

18. $a^4 - b^4 =$

19. $12m^2 - 27n^2 =$

20. $36a^2b^2 - a^2 =$

21. $a^5b^2 - a^3 =$

22. $x^4 - x^2y^2 =$

Expand, using Program 6.6.

23. $(x + 3)(x + 2) =$

24. $(5a + 1)(a + 7) =$

25. $(3x + 4)(x + 3) =$

26. $(2x - 5)(3x - 2) =$

27. $(4y - 7)(2y - 5) =$

28. $(3x + 1)(x - 5) =$

29. $(2x - 7)(x + 4) =$

30. $(5a - 4)(3a + 1) =$

31. $(2m + 3)(2m + 5) =$

32. $(8p - 3)(5p + 2) =$

33. $(3x - 7)(3x - 7) =$

34. $(3x - 7)(5x + 2) =$

35. $(2a - 3)(2a + 3) =$

36. $(3xy - 7)(2xy + 4) =$

37. $(7x + 2y)(7x + 2y) =$

38. $(ax + b)(cx + d) =$

39. $(kx - c)(mx + d) =$

40. $(2px + 3)(x - p) =$

Factor; for those trinomials which you cannot factor, apply the $b^2 - 4ac$ test and write its value instead of the factors.

41. $x^2 + 5x + 6 =$

42. $y^2 - 7y + 12 =$

43. $a^2 - 10a + 16 =$

44. $a^2 + 12a + 36 =$

45. $y^2 + 4y - 5 =$

46. $x^2 + 2x - 15 =$

47. $x^2 - 2x + 3 =$

48. $18x^2 + 33x + 14 =$

49. $3x^2 + 29x + 40 =$

50. $6a^2 + 11a - 10 =$

51. $4x^2 - 16xy + 15y^2 =$

52. $3a^2 + 14a - 5 =$

53. $2x^2 - 3x - 9 =$

54. $100 - 20y + y^2 =$

55. $4 - 12b + 9b^2 =$

56. $6b^2 - 19b + 10 =$

57. $x^2 - 3x + 2 =$

58. $12a^2 - 35ab + 18b^2 =$

59. $m^3 + 10m^2n + 24mn^2 =$

60. $9x^2 - 28xy + 3y^2 =$

61. $18a^4 + 21a^2 - 4a =$

62. $a^4 - 6a^2 + 8 =$

63. $30x^2 + 4xy - 2y^2 =$

64. $15r^2 + 20r + 8 =$

65. $6a^3 + 5a - 13a^2 =$

66. $4apx^2 - 17apx + 18ap =$

67. $a^4 - 4a^2b^2 + 3b^4 =$

68. $a - 7a^2 - 18 =$

Name.. Date.............................. **EXERCISE 6.3**

Factor by grouping.

1. $ax + bx + 3a + 3b =$

2. $cx + cy - 4x - 4y =$

3. $dx - c + x - dc =$

4. $cn - 4c - bn + 4b =$

5. $gx - 3gh + ghx - 3g =$

6. $ab + 6a + 2b + 12 =$

7. $a^2x + a^2d - x - d =$

8. $xy - hx + x^2 - hy =$

9. $2at + bt - bs - 2as =$

10. $2ay - 6ax - 3bx + by =$

11. $5ac - ad - 2bd + 10bc =$

12. $bc + 2c - 4b - 8 =$

13. $kx^2 + ky^2 - mx^2 - my^2 =$

14. $a^2x^2 - 3x^2 - 4a^2 + 12 =$

Factor.

15. $(x + 1)^2 + 2(x + 1) + 1 =$

16. $(a - b)^2 + 4(a - b) + 4 =$

17. $(x + 5)^2 - (x + 2)^2 =$

18. $4a^2(b + c) - b^2(b + c) =$

19. $2(x + y)^2 - (x + y) - 3 =$

20. $4(m + n)^2 + 20(m + n) + 25 =$

21. $2(a^2 + 2ab + b^2) - 3(a + b) - 9 =$

22. $(x - y)(x + y)^2 + 6(x + y)(x - y) + 9(x - y) =$

Factor completely.

23. $8a^2 - 2ab - 3b^2 =$

24. $y^2 - 2xy^2 + 2x - 1 =$

25. $a + 3b - a^2 + 9b^2 =$

26. $a^4 - a^2 - 12 =$

27. $100a^2 - 10{,}000b^2 =$

28. $ab - a - b + 1 =$

29. $25x^2y^2 - x^2 =$

30. $2(x + y)^2 + 12(x + y) + 10 =$

31. $2b^2 - 10 - 5a^2 + a^2b^2 =$

32. $a^4 - 4a^2 + 4 =$

33. $(9a^2 + 30a + 25) - (a - 4)^2 =$

34. $2x^4 - 10x^2 - 12 =$

35. $16p^4 - 81q^4 =$

36. $(m + n)^2 - m^2 =$

37. $a^2 + 6ab + 9b^2 - 4 =$

38. $x^6 - x^4 - 16x^2 + 16 =$

39. $2x^4 + 6x^3 - 8x^2 - 24x =$

40. $(x^2 + 4x - 6)^2 - 36 =$

Name.. Date.............................. **EXERCISE 6.4**

Express in lowest terms.

1. $\dfrac{18a^2}{30a^2 - 12a} =$

2. $\dfrac{3x - 9}{2x - 6} =$

3. $\dfrac{ab - 4b}{5a - 20} =$

4. $\dfrac{ab - ac}{b^2 - c^2} =$

5. $\dfrac{xy}{x^2y^2 - 3xy} =$

6. $\dfrac{x^4 - 16}{x^3 + 4x} =$

7. $\dfrac{x^2 - 5x + 6}{x^2 - 4x + 4} =$

8. $\dfrac{x^2 + 6x + 9}{x^2 - 9} =$

9. $\dfrac{6x^2 - 11x - 10}{3x^2 - 19x - 14} =$

10. $\dfrac{2x^2 - x - 10}{x^2 - 2x - 8} =$

11. $\dfrac{ax - ay - 2x + 2y}{a^2 - 7a + 10} =$

12. $\dfrac{x^3 - 5x^2 + 6x}{2x^3 - 8x} =$

13. $\dfrac{4x^3y - 12x^2y + 9xy}{2axy - 3ay - 2bxy + 3by} =$

14. $\dfrac{4a^2 - b^2}{(2a - b)(2a^2 + ab)} =$

15. $\dfrac{(x + y)^2 - 3(x + y) - 10}{x^2 + xy + 2x} =$

16. $\dfrac{(a - b)^2 - 9}{bc - ac - 3c} =$

Make the equalities valid by inserting the proper sign (+ or −) in the parentheses.

17. $\frac{x}{x-y} = (\ \ )\frac{x}{y-x} = (\ \ )\frac{-x}{y-x} = (\ \ )\frac{-x}{x-y}$

18. $-\frac{a}{3x-4} = (\ \ )\frac{-a}{3x-4} = (\ \ )\frac{-a}{4-3x}$

19. $\frac{m-n}{a-b} = (\ \ )\frac{n-m}{b-a} = (\ \ )\frac{n-m}{a-b} = (\ \ )\frac{m-n}{b-a}$

20. $\frac{1}{(x-y)(s-t)} = (\ \ )\frac{1}{(y-x)(s-t)}$

$= (\ \ )\frac{1}{(y-x)(t-s)}$

21. $\frac{a-5}{(a-4)(a-2)} = (\ \ )\frac{a-5}{(a-2)(a-4)} = (\ \ )\frac{5-a}{(2-a)(4-a)} = (\ \ )\frac{5-a}{(2-a)(a-4)} = (\ \ )\frac{-(a-5)}{(4-a)(2-a)}$

Express in lowest terms.

22. $\frac{a-ax}{x-1} =$

23. $\frac{4-2t}{3t-6} =$

24. $-\frac{(a-1)}{p-ap} =$

25. $\frac{3-a}{a^2-9} =$

26. $-\frac{4x^2-9}{9x^2-4x^4} =$

27. $\frac{x^2-y^2}{(y-x)^2} =$

28. $\frac{3x^2+2x-8}{4+x-3x^2} =$

29. $\frac{xy-ay-ab+bx}{ab-bx+ay-xy} =$

30. $-\frac{6x^2-23x+21}{12-5x-2x^2} =$

31. $\frac{(x-2)^2}{(2-x)^2} =$

32. $\frac{(x-2)(x-3)(x-4)}{(2-x)(3-x)(4-x)} =$

33. $\frac{(x+2)(x^2-9)}{6+x-x^2} =$

34. $\frac{12x^2y-4x^3-9xy^2}{4x^3+4x^2y-15xy^2} =$

35. $\frac{(x^2-8x+16)-4}{1-(x^2-2x+1)} =$

Name.. Date.............................. **EXERCISE 6.5**

Multiply, and express products in lowest terms.

1. $\dfrac{24}{3x-6}\cdot\dfrac{x^2-2}{4}=$

2. $\dfrac{a^2-b^2}{a^2}\cdot\dfrac{ab}{a+b}=$

3. $\dfrac{9xy}{x+4}\cdot\dfrac{2x^2+5x-12}{3x^2y^2-6xy}=$

4. $\dfrac{4x^2-y^2}{y-2x}\cdot\dfrac{3y}{xy+2x^2}=$

5. $\dfrac{6a-18}{9a^2+6a-24}\cdot\dfrac{12a-16}{8a-12}=$

6. $\dfrac{a+b}{2a+b}\cdot\dfrac{4a^2-b^2}{a^2-b^2}=$

7. $\dfrac{3ay^2-9ay}{10y^2+5y}\cdot\dfrac{2y^3+y^2}{a^2y-3a^2}=$

8. $\dfrac{x^2-8x+15}{x^2+4x-21}\cdot\dfrac{x^2-6x-16}{x^2+9x+14}=$

9. $\dfrac{x^2-2x-3}{x^2-1}\cdot\dfrac{x^2+6x+9}{x^2-9}=$

10. $\dfrac{xy-ay-bx+ab}{2x-a}\cdot\dfrac{a-2x}{a-x}=$

11. $\dfrac{2x-y}{x+y}\cdot\dfrac{x-y}{y-2x}\cdot\dfrac{x+y}{y-x}=$

12. $\dfrac{x^2-2xy+y^2-16}{2x-y}\cdot\dfrac{2y-4x}{2x-8-2y}=$

Divide, and express quotients in lowest terms.

13. $\dfrac{x-3y}{x^2-6x+8}\div\dfrac{3x-9y}{x^2-16}=$

14. $\dfrac{3c-bc}{5x-ax}\div\dfrac{3a-ab}{5m-am}=$

15. $\dfrac{3x + 6}{x^2 + 4x} \div \dfrac{3x^2 - 9}{x^2 + 2x} =$

16. $\dfrac{x^2 - 9}{3x - 3y} \div \dfrac{x^2 + 9 - 6x}{y^2 - x^2} =$

17. $\dfrac{2m + 4}{m^2 - 7m - 18} \div \dfrac{4m - 16}{m^2 - 81}$

18. $\dfrac{x^2 - 16}{x^2 - 9} \div \dfrac{x - 4x^2}{9x + 27} =$

19. $(a^2 - b^2) \div \dfrac{a^2 - 4ab + 3b^2}{a + b} =$

20. $\dfrac{as - bs + bt - at}{3m^3 - 3mn^2} \div \dfrac{as - at}{n^2 - 2mn + m^2} =$

21. $\dfrac{x^2 - 5x + 6}{x - 2} \div (x - 2) =$

22. $\dfrac{a^2 - 18 + 3a}{30 - a^2 + a} \div \dfrac{15 - 2a - a^2}{-(a^4 - 36a^2)} =$

23. $\dfrac{\dfrac{x - 5}{x^2 - 25}}{2x + 3} =$

24. $\dfrac{\dfrac{x^2 - y^2}{(x + y)^2}}{\dfrac{3x + 3y}{x - y}} =$

Simplify.

25. $\dfrac{\dfrac{x^2 + x - 6}{x^2 + 3x - 28} \cdot \dfrac{x^2 - 49}{x^2 - 10x + 16}}{\dfrac{x^2 - 4x - 21}{x^2 - 11x + 24}} =$

Name.. Date.............................. **EXERCISE 6.6**

Simplify.

1. $\frac{x + 3y}{18} + \frac{x - 2y}{24} =$

2. $\frac{s - t}{14s} - \frac{s - t}{21t} =$

3. $\frac{3a - 6b}{12a^2b} + \frac{3a - 5b}{18ab^2} =$

4. $\frac{1}{f_1} + \frac{1}{f_2} =$

5. $\frac{5}{r} - \frac{6}{t} + \frac{8}{r} =$

6. $a - 3 + \frac{5}{a} =$

7. $2x - \frac{6x^2}{3x - 2} =$

8. $\frac{5}{m} - \frac{6}{n} + 3 =$

9. $x - 2 + \frac{4x + 3}{x - 7} =$

10. $x + 3 - \frac{2x - 5}{7 - x} =$

11. $\frac{3x - 5}{x + 3} + \frac{2x - 1}{x - 4} =$

12. $\frac{1}{p + q} - \frac{1}{p - q} =$

13. $\frac{3}{a - b} - \frac{4}{b - a} =$

14. $\frac{x - 2}{4x} - \frac{3x + 5}{6x} =$

15. $\frac{3x}{x^2 - 4} - \frac{2}{x + 2} =$

16. $\frac{8y}{y^2 - 9} + \frac{4}{3 - y} =$

17. $\frac{3a - 4}{a^2 - a - 20} - \frac{3}{a - 5} =$

18. $\frac{x + 2}{x^2 - 9} + \frac{3x - 1}{x^2 + x - 12} =$

19. $\dfrac{3c - d}{2c - d} + \dfrac{3c^2}{d^2 - 4c^2} =$

20. $\dfrac{x - 4}{x - 2} + \dfrac{3 + 5x}{2 - x} =$

21. $\dfrac{1}{x^2 - 5x + 6} - \dfrac{4}{4 - x^2} =$

22. $\dfrac{2x}{x^2 + xy} - \dfrac{3}{xy + y^2} =$

23. $2 - \dfrac{x}{x - 2} + \dfrac{3(2x - 9)}{x^2 - 5x + 6} =$

24. $2 - \dfrac{3 - 2x}{2x - 3} + 2x =$

25. $\dfrac{x + 2}{2x^2 - 3x - 9} - \dfrac{x - 2}{3x^2 - 11x + 6} =$

26. $\dfrac{2}{x^2 - 9} - \dfrac{3}{x^2 - 1} + \dfrac{1}{x^2 + 2x - 3} =$

27. $\dfrac{8}{x^2 - x - 2} + \dfrac{3 + x}{x^2 - 4} + \dfrac{4 - 2x}{x^2 + 3x + 2} =$

28. $\dfrac{\dfrac{x - x^2}{x^2 - 1}}{\dfrac{x}{1 + x} - x} =$

29. $\dfrac{1 - \dfrac{3}{3a + 2}}{a + \dfrac{2}{3a - \dfrac{a + 2}{a}}} =$

Name.. Date................................ **EXERCISE 6.7**

Solve for x.

1. $\frac{x}{3} - 3 = \frac{x}{6}$

2. $3 = \frac{1}{4x}$

3. $\frac{x+2}{4} - \frac{x-3}{9} = \frac{2-x}{12}$

4. $\frac{7}{3x+1} = \frac{3}{2x-1}$

5. $\frac{x-5}{x+4} = \frac{x-1}{x-4}$

6. $\frac{a+x}{a-x} = \frac{a+b}{a-b}$

7. $\frac{6}{x+2} - \frac{5}{x} = \frac{5-4x}{x^2+2x}$

8. $\frac{3}{x+2} + \frac{2x}{4-x^2} = \frac{2}{x-2}$

9. $\frac{2x-3}{3x-1} = \frac{2x+7}{3x+7}$

10. $\frac{9}{x+1} - \frac{2x+3}{x-2} = \frac{7x-2x^2}{x^2-x-2}$

11. $\frac{x+a}{x-b} - \frac{x-a}{x+b} - \frac{a^2-b^2}{x^2-b^2} = 0$

12. $\frac{5x}{x^2+x-6} - \frac{3x}{x^2+2x-8} = \frac{2}{x+3}$

Solve.

13. $\begin{cases} \dfrac{2}{a} + \dfrac{3}{b} = 2 \\ \dfrac{4}{a} - \dfrac{3}{b} = 1 \end{cases}$

14. $\begin{cases} \dfrac{1}{2x} + \dfrac{3}{4y} = 4 \\ \dfrac{5}{3x} - \dfrac{2}{3y} = 7 \end{cases}$

15. $\begin{cases} \dfrac{3}{m} - \dfrac{2}{n} = 4 \\ \dfrac{5}{m} - \dfrac{3}{n} = 5 \end{cases}$

Solve.

16. A salesman figures that by driving at a certain speed he can make the next stop in 4 hours. By driving 12 mph faster, he can get there an hour earlier. How far is the next stop?

....

17. When the wind velocity is 30 mph it takes a certain airplane as long to travel 800 miles against the wind as 950 miles with the wind. How fast would the plane be flying in still air?

....

18. The sum of the reciprocals of two numbers is $\frac{5}{6}$ and the difference between these two reciprocals is $\frac{1}{6}$. What are the numbers?

....

19. A farmer has a tractor that can plow a certain field in 21 hours, and his neighbor has one that can do the job in 14 hours. How long would it take the two of them, working together, to plow the field?

....

Write the products directly.

1. $3a(2b + 3a)$

2. $5x(x - 2y + 3z)$

3. $2m(3a - 4b + c)$

4. $3t^2(st - 2t + 3)$

5. $(m - n)^2$

6. $(a + b)^2$

7. $(2a + b)^2$

8. $(m - 3n)^2$

9. $(3x - 4y)^2$

10. $(a^2 + b^2)^2$

Supply the term to make the polynomial a perfect square trinomial.

11. $a^2 + (\ )ab + 4b^2$

12. $x^2 - (\ )x + 36$

13. $9m^2 - (\ ) + 25$

14. $4a^2b^2 + (\ ) + 49c^2$

Factor.

15. $2a^3 + 2a^2b - 6ab$

16. $-p^2q^2 + p^3q - p^2q^3$

17. $x^2 - 4xy + 4y^2$

18. $a^2 + 6ab - 9b^2$

19. $4m^2 + 12mn + 9n^2$

20. $4x^2 - 20xy + 25y^2$

21. $18 - 24y + 8y^2$

22. $12mn + m^2 + 36n^2$

Expand.

23. $(2x + 3y)(2x - 3y)$

24. $(4x - 1)(4x + 1)$

25. $(x - 7)(x - 3)$

26. $(2x + 1)(3x + 2)$

27. $(4p - q)(3p + 2q)$

28. $(3x + 2y)(4x - 5y)$

29. $(mn + 2)(3mn + 4)$

30. $(kx - 2y)(3kx - y)$

Factor.

31. $9x^2 - 49y^2$

32. $a^4b^4 - c^4$

33. $9a^2 + 3a - 2$

34. $2m^2 + 9m - 18$

35. $8x^2 - 2x - 15$

36. $12x^2 + 16x - 3$

37. $81a^2 + 42a - 8$

38. $36m^2 + 35m + 6$

39. $6a^3 + a^2b - 2ab^2$

40. $36x^2 - 36xy^2 + 8y^2$

Factor by grouping.

41. $ax + ay - bx - by$

42. $mn - an - bm - ab$

43. $2ab + 3b + 3c + 2ac$

44. $3xy - 5y - 15z + 9xz$

45. $2abx - 3acx - 4aby + 6acy$

46. $abxy - 3bxy + ax^2y - 3x^2y$

Factor completely.

47. $a^2 - 2b^2x - b^2 + 2a^2x$

48. $ax^2 - ay^2 + by^2 - bx^2$

49. $x - y - x^2 + y^2$

50. $2a^3 + 5a^2 - 4a - 10$

51. $x^4 - 13x^2 + 36$

52. $16a^4 - 8a^2 + 1$

Express in lowest terms.

53. $\dfrac{2a^2 - 17a + 21}{a^2 - 12a + 35}$

54. $\dfrac{6x^2 - 5x - 4}{6x^2 - 23x + 20}$

55. $\dfrac{(3 - a)(a^2 + 9)}{a^4 - 81}$

56. $\dfrac{xy + bx - ay - ab}{ab - bx + ay - xy}$

57. $\dfrac{4b + 2a - 8 - ab}{ab - 3b + 6 - 2a}$

58. $\dfrac{x^3 + 2x^2 - 9x - 18}{x + 6 - x^2}$

Simplify

59. $\dfrac{2a + b}{4a^2 - b} \cdot \dfrac{a^3 - 2a^2b}{a^2 - 2ab}$

60. $\dfrac{2x + 3}{35x^2 + 19x + 2} \cdot \dfrac{49x^2 - 1}{4x^2 + 12x + 19}$

61. $\dfrac{2a^2 + a - 3}{9 - 4a^2} \cdot \dfrac{2a - 3}{a^2 - 1}$

62. $\dfrac{x^2 + 5x + 6}{x^2 - 1} \cdot \dfrac{x^2 - 2x - 3}{x^2 - 9}$

63. $\dfrac{m^2 - 5m + 4}{m^2 - 6m + 8} \cdot \dfrac{m^2 - 5m + 6}{m^2 - 4m + 3}$

64. $\dfrac{1 + x}{1 + x^2} \cdot \dfrac{1 - 2x + x^2}{1 - x^2} \cdot \dfrac{x}{x - 1}$

65. $\dfrac{8a^3b^2}{7xy^2} \div \dfrac{a^4b^2 - 3a^2b^3}{2x^3y - x^2y^2}$

66. $\dfrac{4a^2 - 28a + 49}{12a^2 - 17a + 6} \div \dfrac{4a^2 - 49}{12a^2 - a - 6}$

67. $\dfrac{2a^2 - 13a + 15}{3a^2 - 17a + 10} \div \dfrac{4a^2 - 9}{3ap - 2p}$

68. $\dfrac{3 + 4a + a^2}{a^2 + a - 6} \div \dfrac{2 + 3a + a^2}{a^2p - 2ap}$

69. $\dfrac{9a^2 + 6ab + b^2 - 4}{2a^2 - 5ab + 3b^2} \div \dfrac{10 + 5b + 15a}{a^2 - 7ab + 6b^2}$

70. $\dfrac{ab + a + b + 1}{mn + 2n + 2m + 4} \div \dfrac{a^2 - 1}{n^2 + 4n + 4}$

71. $\dfrac{2}{a - 2} - \dfrac{3}{5a - 10}$

72. $\dfrac{2}{x^2 - y^2} - \dfrac{5}{xy - y^2} + \dfrac{3}{x + y}$

73. $\dfrac{3}{x - 2} - \dfrac{x}{x^2 - 4} - \dfrac{2}{x + 2}$

74. $\dfrac{2x + 6}{x^2 - x - 12} - \dfrac{3(x + 5)}{x^2 + 3x - 10}$

75. $\dfrac{x}{2y^2 - xy} + \dfrac{1}{x - 2y} + \dfrac{y}{2y - x}$

76. $\dfrac{a^2 + 1}{a^3 - a^2 - a - 1} - \dfrac{a^2 - 1}{a^3 + a^2 - a - 1}$

77. $\dfrac{3a^2}{a^4 - 4} + \dfrac{5a^2 - 3}{2a^4 + a^2 - 6}$

78. $\dfrac{2 - 2x}{x^3 + 3x^2 + 3x - 1} + \dfrac{2}{1 - x} + \dfrac{x + 1}{x^2 - 2x + 1}$

Solve for the variable.

79. $\dfrac{3}{2x - 1} = \dfrac{4}{3x}$

80. $\dfrac{5}{4k + 2} = \dfrac{1}{k + 1}$

81. $\dfrac{3x - 2}{2x + 3} = \dfrac{9x - 5}{6x + 1}$

82. $\dfrac{x - 4}{1 + 2x} = \dfrac{2x^2 - 7x - 4}{4x^2 + 5x}$

83. $\dfrac{y + 6}{6} - \dfrac{24}{9y + 36} = \dfrac{3y + 24}{18}$

84. $\dfrac{2}{a - 4} - \dfrac{1}{a - 1} = \dfrac{a + 3}{a^2 - 5a + 4}$

85. $\dfrac{7x}{x - 3} + \dfrac{5x}{x + 1} - \dfrac{12x^2 - 12}{x^2 - 2x - 3} = 0$

86. $\dfrac{2n - 3}{2n^2 - 3n - 2} + \dfrac{n + 1}{2n - 4} = \dfrac{n}{2n + 1}$

Solve the system.

87. $\begin{cases} \dfrac{4}{x} + \dfrac{3}{y} = 3 \\ \dfrac{2}{x} - \dfrac{6}{y} = -1 \end{cases}$

88. $\begin{cases} \dfrac{2}{x} + \dfrac{1}{y} = 0 \\ \dfrac{3}{x} + \dfrac{2}{y} = 1 \end{cases}$

REVIEW

Solve for the variable; the replacement set is the set of integers.

1. $x - 2 < 5$
2. $y + 3 > 4$
3. $-3 > m + 3$
4. $x - 3 \geq 0$
5. $4x \leq -16$
6. $20 \geq 5a$

Solve for the variable; no restriction is placed on the domain of the variable. Use a number line.

7. $t \leq 5$
8. $m > 0$
9. $3x > -15$
10. $2k + 3 < -1$
11. $0 \leq 3x + 4 + 5x$
12. $2a - 4 \leq 3a + 5$

Solve by addition-subtraction, or by substitution.

13. $\begin{cases} 2x + 4y = -1 \\ 4x + 5y = 4 \end{cases}$
14. $\begin{cases} x - y = -2 \\ 3x - y = 1 \end{cases}$
15. $\begin{cases} 2x - 3y = 5 \\ 6y - 4x = -10 \end{cases}$
16. $\begin{cases} -6x + 4y = 3 \\ 3x - 2y = 1 \end{cases}$

Evaluate.

17. $\begin{vmatrix} 3 & 1 \\ 4 & -2 \end{vmatrix}$
18. $\begin{vmatrix} 2 & 0 \\ 6 & 0 \end{vmatrix}$
19. $\begin{vmatrix} -3 & -3 \\ -5 & -5 \end{vmatrix}$
20. $\begin{vmatrix} 7 & 5 \\ -2 & -9 \end{vmatrix}$

Solve, using determinants.

21. Problem 13
22. Problem 14
23. Problem 15
24. Problem 16

7

EXPONENTS AND RADICALS

1. LAWS OF EXPONENTS

A power was defined in Chapter 4 as a product of equal factors. In the usual expression of a power, the exponent indicates the number of times the equal factor (base) occurs in the product:

$$\underbrace{a \cdot a \cdot a \cdot a}_{\text{4 factors of } a} = a^{4} \begin{matrix}\leftarrow \text{exponent} \\ \leftarrow \text{base}\end{matrix} \Big\} \text{ power}$$

Also in Chapter 4 meaning was given to 0 as an exponent: *Any power having a nonzero base and an exponent of 0 is equal to 1.* Thus we have, up to now, interpretations for any exponent that is a nonnegative integer.

In this chapter we extend the interpretation of exponents to include negative numbers and numbers expressed as fractions. Before doing so, however, account must be taken of the basic *Laws of Exponents*, which are consequences of the properties of the numbers to which they apply. The first two Laws (I and II) we have already met in the form of Programs 4.10 and 4.14.

Laws of Exponents *If a and b denote rational numbers, and m and n nonnegative integers, then:*

I $a^m \times a^n = a^{m+n}$ (Program 4.10)

II $a^m \div a^n = a^{m-n}$ if $m > n$, $a \neq 0$ (Program 4.14a)

$= \dfrac{1}{a^{n-m}}$ if $m < n$, $a \neq 0$ Program 4.14b)

$= 1$ if $m = n$, $a \neq 0$ (Definition)

III $(a^n)^m = a^{m \times n}$

IV $(ab)^n = a^n b^n$

V $\left(\dfrac{a}{b}\right)^n = \dfrac{a^n}{b^n}$ $(b \neq 0)$

EXAMPLES

III $(a^n)^m = a^{m \times n}$

1. $(b^5)^3 = b^{3 \times 5} = b^{15}$.
   This can be verified by Program 4.10:
   $(b^5)^3 = b^5 \cdot b^5 \cdot b^5 = b^{5+5+5} = b^{15}$
2. $(x^3)^5 = x^{5 \times 3} = x^{15}$.
3. $(x)^4 = x^{4 \times 1} = x^4$.
4. $(2^2)^3 = 2^{3 \times 2} = 2^6$.

IV $(ab)^n = a^n b^n$

1. $(bc^2)^3 = b^{3 \times 1} c^{3 \times 2} = b^3 c^6$.
   This can be verified by Program 4.10:
   $$\begin{aligned}(bc^2)^3 &= bc^2 \cdot bc^2 \cdot bc^2 = b \cdot c^2 \cdot b \cdot c^2 \cdot b \cdot c^2 \\ &= b \cdot b \cdot b \cdot c^2 \cdot c^2 \cdot c^2 \\ &= b^{1+1+1} c^{2+2+2} \\ &= b^3 c^6\end{aligned}$$
2. $(x^2y^3)^4 = x^{4 \times 2} y^{4 \times 3} = x^8 y^{12}$.
3. $(3xy^2)^3 = 3^{3 \times 1} x^{3 \times 1} y^{3 \times 2} = 3^3 x^3 y^6 = 27x^3y^6$.
4. $(2 \cdot 5)^2 = 2^{2 \times 1} \cdot 5^{2 \times 1} = 2^2 \cdot 5^2$.

V $\left(\dfrac{a}{b}\right)^n = \dfrac{a^n}{b^n}$ $(b \neq 0)$

1. $\left(\dfrac{p^2}{q^4}\right)^3 = \dfrac{p^{3 \times 2}}{q^{3 \times 4}} = \dfrac{p^6}{q^{12}}$ $(q \neq 0)$.
   This can be verified by Program 4.10:
   $$\begin{aligned}\left(\frac{p^2}{q^4}\right)^3 &= \left(\frac{p^2}{q^4}\right)\left(\frac{p^2}{q^4}\right)\left(\frac{p^2}{q^4}\right) \\ &= \frac{p^2 \cdot p^2 \cdot p^2}{q^4 \cdot q^4 \cdot q^4} = \frac{p^{2+2+2}}{q^{4+4+4}} = \frac{p^6}{q^{12}}.\end{aligned}$$
2. $\left(\dfrac{2x}{y^2}\right)^4 = \dfrac{(2^{4 \times 1})(x^{4 \times 1})}{y^{4 \times 2}} = \dfrac{2^4x^4}{y^8} = \dfrac{16x^4}{y^8}$ $(y \neq 0)$.
3. $\left(\dfrac{3}{4}\right)^4 = \dfrac{3^4}{4^4}$.

2. NEGATIVE EXPONENTS

Acceptance of the definition of the zero exponent allows us to give a reasonable definition to negative exponents, one that is perfectly consistent with the already-stated Laws of Exponents:

Any power consisting of a nonzero base, a, and negative exponent, $-m$, is equivalent to $\frac{1}{a^m}$.

To state this another way:

a^{-m} *and* a^m *are reciprocals of one another.*

Thus,

$$a^m \times a^{-m} = a^{m+(-m)} = a^0 = 1$$

is consistent with Program 4.10 (Exponent Law I) and meets the requirement that the product of a number and its reciprocal is 1.

EXAMPLES

1 $a^{-4} = \frac{1}{a^4}$.

[a^{-4} is usually read "a to the minus fourth."]

2. $8^{-1} = \frac{1}{8^1} = \frac{1}{8}$.

3. $\frac{1}{5} = 5^{-1}$.

This definition of a negative exponent has an important relationship with fraction notation, which may be generalized as follows: any nonzero factor appearing in the numerator of a fraction may be placed in the denominator, and any nonzero factor appearing in the denominator of a fraction may be placed in the numerator without changing the value of the fraction *provided the sign of the exponent is reversed.*

EXAMPLES

1. $a^3b^{-2} = a^3\left(\frac{1}{b^2}\right) = \frac{a^3}{b^2}$.

2. $\frac{1}{a^{-2}} = \frac{1}{\frac{1}{a^2}} = 1 \div \frac{1}{a^2} = 1 \times a^2 = a^2$.

3. $\frac{a^{-2}}{b^{-3}} = \frac{b^3}{a^2}$.

4. $\frac{(a+b)^{-2}}{(p+q)} = \frac{1}{(a+b)^2(p+q)} \quad \left[\text{or } \frac{(p+q)^{-1}}{(a+b)^2}\right]$.

5. $\left(\frac{2}{3}\right)^{-1} = \frac{2^{-1}}{3^{-1}} = \frac{3^1}{2^1} = \frac{3}{2}$.

It should also be noted that in introducing negative exponents, the three parts of Exponent Law II may be collapsed into one:

$$a^m \div a^n = a^{m-n} \quad (a \neq 0)$$

When $m > n$, the quotient will have a positive exponent; when $m < n$ the quotient will have a negative exponent; when $m = n$, the exponent will be 0, which means that the quotient is 1.

(Complete Exercise 7.1)

3. ROOTS

The inverse process of "raising to a power" is that of "extracting a root." When asked to find the product of a number of equal factors we say:

$2^2 = 2 \cdot 2 = 4$ (or "two-squared $= 4$")

$2^3 = 2 \cdot 2 \cdot 2 = 8$ (or "two-cubed $= 8$")

$2^4 = 2 \cdot 2 \cdot 2 \cdot 2 = 16$ (or "two-to-the-fourth $= 16$")

When the circumstances are reversed and we ask "What number squared is 4?" or "What number cubed is 8?" and so on, we are seeking a *root*; thus:

2 is a *square root* of 4
2 is a *cube root* of 8
2 is a *fourth root* of 16

Powers of rational numbers are always rational numbers, but roots of rational numbers are not always rational numbers. For example, the square roots of 9 are the rational numbers 3 and -3, since

$$(3)^2 = 9 \quad \text{and} \quad (-3)^2 = 9$$

On the other hand, the square roots of 5, or 2, or 6, are not rational numbers, but *irrational numbers.* Included among the irrational numbers are all roots of positive rational numbers and odd roots of negative rational numbers that are themselves not rational numbers. (Even roots of negative rationals belong to still another set, the complex numbers, as discussed in Chapter 8, Section 3.)

Together the rational and irrational numbers make up a set called the *real numbers* (Fig. 7.1).

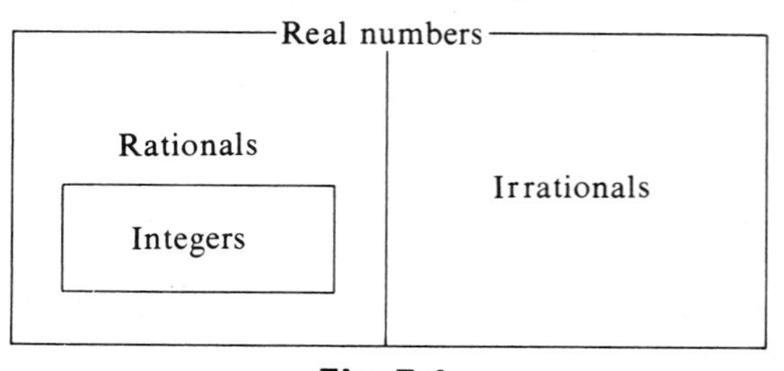

Fig. 7-1

Standard numerals for irrational numbers that are the roots of rational numbers are usually expressed with a *radical* symbol, $\sqrt{}$. A numeral appearing in the crook of the radical to denote the root is called the *index*. The numeral appearing within the radical is called the *radicand*. Thus:

index ↴

Cube root of 7 = $\sqrt[3]{7}$ ←radicand

radical ↑

Square root of 5 = $\sqrt[2]{5}$ or $\sqrt{5}$. (The index is usually omitted from the square-root radical.)

Fourth root of 5 = $\sqrt[4]{5}$.

nth root of a = $\sqrt[n]{a}$.

In general, the nth root of a number is one of n equal factors whose product is that number:

$$\sqrt{3} \times \sqrt{3} = 3$$

$$\sqrt[3]{4} \times \sqrt[3]{4} \times \sqrt[3]{4} = 4$$

Some numbers have more than one real-number root—e.g., the square roots of 4 are 2 and -2, since $(2)(2) = 4$ and $(-2)(-2) = 4$. This leads to the convention of the *principal root*:

The principal nth root of a positive number is the positive root.

The principal nth root of 0 is 0.

The principal nth root of a negative number is the negative root when n is odd.

(The principal nth root of a negative number when n is even does not exist among the real numbers. See Section 3, Chapter 8, for an explanation.)

EXAMPLES

1. The principal square root of 9 is $\sqrt{9}$ or 3; $-\sqrt{9}$ or -3 is also a square root of 9, but not the principal square root.
2. The principal cube root of 8 is $\sqrt[3]{8}$ or 2.
3. The principal cube root of -8 is $\sqrt[3]{-8}$ or -2.
4. The principal square root of -9 does not exist among the real numbers; $(-3)^2 = +9$ and $(+3)^2 = +9$, and any real number other than 3 when squared equals a number other than 9.

4. FRACTIONAL EXPONENTS

Up to here we have interpretations for exponents of 0 and the positive and negative integers. Now we give a useful and consistent interpretation to exponents in the form of a unit fraction, e.g., $a^{1/3}$. If we were to "cube" this expression, according to Exponent Law III the resulting power would have an exponent of 1:

$$(a^{1/3})^3 = a^{3 \times 1/3} = a^1.$$

It is reasonable, then, to interpret $a^{1/3}$ as the "cube root of a," or

$$a^{1/3} = \sqrt[3]{a}$$

Similarly:

$$a^{1/2} = \sqrt{a}$$

$$a^{1/7} = \sqrt[7]{a}$$

$$x^{1/17} = \sqrt[17]{x}$$

$$8^{1/3} = \sqrt[3]{8} = 2$$

$$9^{1/2} = \sqrt{9} = 3$$

In general,

$$a^{1/n} = \sqrt[n]{a}$$

where $\sqrt[n]{a}$ represents the *principal nth root* of a, and a is nonnegative when n represents an even integer. The latter condition rules out expressions such as $\sqrt{-2}$ and $\sqrt[4]{-12}$ which have no meaning among the real numbers.

Now that we have an interpretation for exponents of the form $1/n$, all we need is one for exponents of the form m/n and our discussion of exponential interpretation is complete.

Consider the product:

$$a^p \cdot a^p \cdot a^p = a^{p+p+p} = a^{3p}$$

If we let p represent an exponent of the form $1/n$, say $1/4$, then:

$$a^{1/4} \cdot a^{1/4} \cdot a^{1/4} = a^{1/4+1/4+1/4} = a^{3/4} \quad \text{(Exp. Law I)}$$

Stating the product another way, as the cube of the factor $a^{1/4}$, we have:

$$a^{1/4} \cdot a^{1/4} \cdot a^{1/4} = (a^{1/4})^3$$

Consequently,

$$a^{3/4} = (a^{1/4})^3$$

Exponent Law III confirms this:

$$(a^{1/4})^3 = a^{3(1/4)} = a^{3/4}$$

Consider now the related example:

$$\sqrt[4]{a \cdot a \cdot a} = \sqrt[4]{a^3} = (a^3)^{1/4}$$

To be consistent with Exponent Law III:

$$(a^3)^{1/4} = a^{(1/4) \times 3} = a^{3/4}$$

Thus we see that

$$a^{3/4} = \sqrt[4]{a^3} = (\sqrt[4]{a})^3$$

which illustrates the generalization:

The numerator of a fractional exponent may be interpreted as the power to which the base is to be raised, and the denominator of the exponent as the root to be extracted.

EXAMPLES

1. Simplify $8^{2/3}$.

$$8^{2/3} = \begin{cases} \sqrt[3]{8^2} = \sqrt[3]{64} = 4 \\ (\sqrt[3]{8})^2 = (2)^2 = 4 \end{cases}$$

2. Simplify $9^{3/2}$.

$$9^{3/2} = \begin{cases} \sqrt{9^3} = \sqrt{729} = 27 \\ (\sqrt{9})^3 = (3)^3 = 27 \end{cases}$$

[Note: When the radicand involves a rational root, it is usually simpler computationally to extract the root first and then apply the power.]

5. SUMMARY

In the previous sections we have built up plausible and consistent interpretations of exponents from the original positive integers to, now, the whole set of rational numbers. In summary:

If a and b denote real numbers, and m and n rational numbers, then:

I $a^m \cdot a^n = a^{m+n}$

II $a^m \div a^n = a^{m-n} \quad (a \neq 0)$

III $(a^n)^m = a^{mn}$

IV $(ab)^n = a^n b^n$

V $\left(\dfrac{a}{b}\right)^n = \dfrac{a^n}{b^n} \quad (b \neq 0)$

To *simplify an exponential expression* means to carry out all operations possible by the above Laws of Exponents (I to V), to eliminate all negative exponents from the final expression, and to expand all numerical coefficients having integral exponents.

EXAMPLES

1. $(x^{3/4})^8 = x^{8(3/4)} = x^6$

2. $x^{1/2}x^{2/3} = x^{1/2+2/3} = x^{3/6+4/6} = x^{7/6}$
$= x^{1+1/6} = x^1 x^{1/6} = x\sqrt[6]{x}$

3. $(9x^{-3})^{2/3} = 9^{2/3}x^{(2/3)(-3)} = (3^2)^{2/3}(x^{-2})$
$= 3^{4/3}x^{-2} = \dfrac{3\sqrt[3]{3}}{x^2}$

4. $(-\frac{1}{27})^{-2/3} = [(-\frac{1}{3})^3]^{-2/3} = (-\frac{1}{3})^{(-2/3)(3)}$
$= (-\frac{1}{3})^{-2} = \dfrac{1}{(-\frac{1}{3})^2} = \dfrac{1}{\frac{1}{9}} = 9$

5. $\dfrac{9x^{-2}y^3}{3x^4y^{-5}} = 3^2 \cdot x^{-2} \cdot y^3 \cdot 3^{-1} \cdot x^{-4} \cdot y^5$
$= 3^{2-1}x^{-2-4}y^{3+5}$
$= 3^1x^{-6}y^8 = \dfrac{3y^8}{x^6}$

6. $(a^2 - b^2)\sqrt{(a-b)^{-2}} = (a^2 - b^2)[(a-b)^{-2}]^{1/2}$
$= (a^2 - b^2)(a - b)^{-1}$
$= \dfrac{(a^2 - b^2)}{(a - b)} = \dfrac{\cancel{(a-b)}(a+b)}{\cancel{(a-b)}}$
$= a + b$

7. $\dfrac{(3x^2y)^{1/3}}{9x^{-2/3}y^2} = \dfrac{3^{1/3}x^{2/3}y^{1/3}}{3^2x^{-2/3}y^2}$
$= 3^{1/3} \cdot x^{2/3} \cdot y^{1/3} \cdot 3^{-2} \cdot x^{2/3} \cdot y^{-2}$
$= 3^{-5/3}x^{4/3}y^{-5/3}$
$= \dfrac{x \cdot x^{1/3}}{3 \cdot 3^{2/3} \cdot y \cdot y^{2/3}} = \dfrac{x\sqrt[3]{x}}{3y\sqrt[3]{9y^2}}$

(Complete Exercise 7.2)

6. RENAMING RADICAL EXPRESSIONS

There is frequent occasion in the computational phases of algebra to rename a radical expression, i.e., to re-express it equivalently. Strict adherence to the Laws of Exponents is essential.

(7.1) *To reduce the order (index) of a radical expression:*

Step 1. Write the terms of the radical expression exponentially with all numerical coefficients in prime factor form.

Step 2. Reduce the fractional exponents in Step 1 to lowest terms.

Step 3. Write the reduced expression of Step 2 in radical form.

EXAMPLES

1. Reduce the order of $\sqrt[4]{9a^2b^2}$.
   Step 1. $(3^2a^2b^2)^{1/4} = 3^{2/4}a^{2/4}b^{2/4}$
   Step 2. $3^{2/4}a^{2/4}b^{2/4} = 3^{1/2}a^{1/2}b^{1/2} = (3ab)^{1/2}$
   Step 3. $\sqrt{3ab}$
2. Express with a lower-order radical: $\sqrt[6]{25x^4b^{12}}$.
   Step 1. $(5^2x^4b^{12})^{1/6} = 5^{2/6}x^{4/6}b^{12/6}$
   Step 2. $5^{2/6}x^{4/6}b^{12/6} = 5^{1/3}x^{2/3}b^2$
   Step 3. $b^2\sqrt[3]{5x^2}$
3. Reduce to a lower order: $\sqrt[4]{25x^2y^3}$.
   Step 1. $(5^2x^2y^3)^{1/4} = 5^{2/4}x^{2/4}y^{3/4}$
   Step 2. $5^{2/4}x^{2/4}y^{3/4} = 5^{1/2}x^{1/2}y^{3/4}$
   Step 3. Except to express it as a product of radicals, $(\sqrt{5x})(\sqrt[4]{y^3})$, the original $\sqrt[4]{25x^2y^3}$ is already at its lowest order.

(7.2) *To raise the order (index) of a radical expression to a given order:*

Step 1. Write the terms of the radical expression exponentially with all numerical coefficients in prime factor form.
Step 2. Replace each exponent with an equivalent fractional exponent whose denominator is the desired order.
Step 3. Write the expression of Step 2 in radical form.

EXAMPLES

1. Express $\sqrt[3]{9xy^2}$ with an index of 6.
   Step 1. $\sqrt[3]{9xy^2} = (3^2xy^2)^{1/3} = 3^{2/3}\cdot x^{1/3}\cdot y^{2/3}$
   Step 2. $3^{2/3}x^{1/3}y^{2/3} = 3^{4/6}x^{2/6}y^{4/6}$
   Step 3. $\sqrt[6]{3^4x^2y^4} = \sqrt[6]{81x^2y^4}$
2. Express $\sqrt[4]{7a^2b^3}$ with an index of 12.
   Step 1. $\sqrt[4]{7a^2b^3} = 7^{1/4}a^{2/4}b^{3/4}$
   Step 2. $7^{1/4}a^{2/4}b^{3/4} = 7^{3/12}a^{6/12}b^{9/12}$
   Step 3. $\sqrt[12]{7^3a^6b^9} = \sqrt[12]{343a^6b^9}$

(7.3) *To remove a factor from the radicand:*

Step 1. Factor the radicand so that the exponent of one or more of its factors is a multiple of the index.
Step 2. Replace any of the factors of the radicand with its root as a factor outside the radical sign.

EXAMPLES

1. Simplify: $\sqrt{x^3}$.
   Step 1. $\sqrt{x^3} = \sqrt{x^2\cdot x}$
   Step 2. $x\sqrt{x}$
2. Simplify: $\sqrt{18a^5}$.
   Step 1. $\sqrt{18a^5} = \sqrt{2\cdot 3^2\cdot a^4\cdot a}$
   Step 2. $3a^2\sqrt{2a}$
3. Simplify: $\sqrt[3]{24a^5b^8}$.
   Step 1. $\sqrt[3]{24a^5b^8} = \sqrt[3]{3\cdot 2^3\cdot a^3\cdot a^2\cdot b^6\cdot b^2}$
   Step 2. $2ab^2\sqrt[3]{3a^2b^2}$

(7.4) *To introduce a factor into the radicand:*

Step 1. Express the factor to be introduced with a fractional exponent whose denominator is the index of the radical.
Step 2. Insert under the radical the factor raised to the power of the numerator of the fractional exponent of Step 1.

EXAMPLES

1. Express under one radical: $b^2\sqrt[6]{a^5}$.
   Step 1. $b^2 = b^{12/6}$
   Step 2. $\sqrt[6]{a^5b^{12}}$
2. Express $3\sqrt{5}$ under one radical:
   Step 1. $3 = 3^{2/2}$
   Step 2. $\sqrt{3^2\cdot 5} = \sqrt{9\cdot 5} = \sqrt{45}$
3. Express under one radical: $\sqrt{x}\cdot\sqrt[4]{a}$.
   Step 1. $\sqrt{x} = x^{1/2} = x^{2/4}$
   Step 2. $\sqrt[4]{ax^2}$

7. RATIONALIZING THE DENOMINATOR

To *rationalize* the denominator of a fraction means to express the fraction equivalently as one having a denominator whose exponents are positive integers and whose coefficients are rational numbers. In terms of radical symbolism, a denominator is said to be rational if it contains no radicals, the exponents are positive integers, and the numerical coefficients are rational.

The procedure for rationalizing denominators is based upon the fact that a fraction remains unchanged in value if its numerator and denominator are both multiplied by the same nonzero number. The desired multiplier is one that yields a denominator whose terms have only positive integral exponents and rational coefficients.

[Note: The equivalent fraction with a rational denominator may or may not have a rational numerator.]

(7.5) *To rationalize a denominator:*

Step 1. Express the denominator exponentially.
Step 2. Determine a factor which when multiplied with the given denominator yields a product containing only positive integral exponents and rational coefficients.
Step 3. Multiply numerator and denominator by the factor of Step 2.

EXAMPLES

1. Express equivalently with a rational denominator: $\frac{5}{\sqrt[3]{a}}$.

Step 1. $\sqrt[3]{a} = a^{1/3}$

Step 2. The desired factor is $a^{2/3}$, since $a^{1/3} \cdot a^{2/3} = a^{1/3+2/3} = a^{3/3} = a^1$

Step 3. $\frac{5}{a^{1/3}} \cdot \frac{a^{2/3}}{a^{2/3}} = \frac{5a^{2/3}}{a^1} = \frac{5\sqrt[3]{a^2}}{a}$

2. Rationalize the denominator: $\frac{p}{\sqrt[3]{x^2y}}$.

Step 1. $\sqrt[3]{x^2y} = x^{2/3}y^{1/3}$

Step 2. The desired factor is $x^{1/3}y^{2/3}$, since $(x^{2/3}y^{1/3})(x^{1/3}y^{2/3}) = x^{2/3+1/3}y^{1/3+2/3} = x^1y^1$

Step 3. $\frac{p}{x^{2/3}y^{1/3}} \cdot \frac{x^{1/3}y^{2/3}}{x^{1/3}y^{2/3}} = \frac{p\sqrt[3]{xy^2}}{xy}$

3. Express $\sqrt{\frac{3}{5}}$ with a rational denominator.

$\left[\text{Note: } \sqrt{\frac{3}{5}} = (\frac{3}{5})^{1/2} = \frac{3^{1/2}}{5^{1/2}}.\right]$

Step 1. Express the denominator $\sqrt{5}$ exponentially: $5^{1/2}$

Step 2. The desired factor is $5^{1/2}$, since $5^{1/2} \cdot 5^{1/2} = 5^{1/2+1/2} = 5^1$

Step 3. $\frac{3^{1/2}}{5^{1/2}} \cdot \frac{5^{1/2}}{5^{1/2}} = \frac{3^{1/2} \cdot 5^{1/2}}{5^1}$

$$= \frac{(3 \cdot 5)^{1/2}}{5} = \frac{(15)^{1/2}}{5} = \frac{\sqrt{15}}{5}$$

At times a denominator of a radical expression is a binomial containing one or two square-root radicals. The denominator can be rationalized by a method based upon the fact that the product of the sum and difference of two numbers is the difference of their squares (Program 6.4).

(7.6) *To rationalize a binomial denominator involving square roots:*

Step 1. Multiply both numerator and denominator by the denominator with the sign of the second term reversed.

Step 2. Simplify if possible.

[Note: This program, as stated, holds only for square roots; it does not hold when the index of the root is other than 2.]

EXAMPLES

1. Express $\frac{1}{\sqrt{5} - \sqrt{3}}$ equivalently with a rational denominator.

Step 1. $\frac{\sqrt{5} + \sqrt{3}}{\sqrt{5} + \sqrt{3}} \cdot \frac{1}{\sqrt{5} - \sqrt{3}} = \frac{\sqrt{5} + \sqrt{3}}{(\sqrt{5})^2 - (\sqrt{3})^2}$

Step 2. $\frac{\sqrt{5} + \sqrt{3}}{(\sqrt{5})^2 - (\sqrt{3})^2} = \frac{\sqrt{5} + \sqrt{3}}{5 - 3}$

$$= \frac{\sqrt{5} + \sqrt{3}}{2}$$

2. Rationalize the denominator: $\frac{8}{\sqrt{7} + \sqrt{3}}$.

$$\frac{\sqrt{7} - \sqrt{3}}{\sqrt{7} - \sqrt{3}} \cdot \frac{8}{\sqrt{7} + \sqrt{3}} = \frac{8(\sqrt{7} - \sqrt{3})}{7 - 3}$$

$$= \frac{\overset{2}{\cancel{8}}(\sqrt{7} - \sqrt{3})}{\cancel{4}}$$

$$= 2(\sqrt{7} - \sqrt{3})$$

$$= 2\sqrt{7} - 2\sqrt{3}$$

3. Rationalize the denominator: $\frac{y}{1 - \sqrt{y}}$.

$$\frac{1 + \sqrt{y}}{1 + \sqrt{y}} \cdot \frac{y}{1 - \sqrt{y}} = \frac{y(1 + \sqrt{y})}{1 - y}$$

(Complete Exercise 7.3)

8. ADDITION AND SUBTRACTION WITH RADICAL EXPRESSIONS

There is no essential need to state separate programs for computing with numbers in radical notation. A knowledge of the basic Laws of Exponents and of algebraic addition, subtraction, multiplication, and division is sufficient to handle most problems involving radicals. However, the programs given in this and the next two sections can speed up computation with radical expressions. It would be well for the student to verify the procedures by computing solutions to some of the examples exponentially.

Radical expressions are said to be *like* when they have the same index and radicand. Thus $3\sqrt[3]{4}$, $6\sqrt[3]{4}$, and $x\sqrt[3]{4}$ are like because each possesses the same index (3) and the same radicand (4); on the other hand $2\sqrt{6}$ and $2\sqrt{5}$ are not like because they differ in radicand; $3\sqrt[3]{2}$ and $3\sqrt{2}$ are not like because they differ in index.

Similar to computing algebraic sums and differences, addition and subtraction with like radical expressions amount to addition and subtraction of coefficients of like radicals:

$$3\sqrt{5} + 4\sqrt{5} = (3 + 4)\sqrt{5} = 7\sqrt{5}$$

The sum or difference of unlike radical expressions is expressed (indicated) simply as the sum or difference of unlike terms:

$$\begin{array}{r} 2\sqrt{6} \\ +3\sqrt[3]{6} \\ \hline 2\sqrt{6} + 3\sqrt[3]{6} \end{array} \qquad \begin{array}{r} 4\sqrt{7} \\ -3\sqrt{5} \\ \hline 4\sqrt{7} - 3\sqrt{5} \end{array}$$

(7.7) *To compute a sum (difference) with radical expressions:*

Step 1. *Simplify each term involving a radical expression.*

Step 2. *Add (subtract) like radical expressions by adding (subtracting) their coefficients; indicate the sum (difference) of the unlike radical expressions.*

EXAMPLES

1. Add: $5\sqrt{27} + 7\sqrt{12} + 3\sqrt{3}$.

Step 1. $5\sqrt{27} = 5\sqrt{9 \cdot 3} = 5 \cdot \sqrt{9} \cdot \sqrt{3}$
$= 5 \cdot 3 \cdot \sqrt{3} = 15\sqrt{3}$
$7\sqrt{12} = 7\sqrt{4 \cdot 3} = 7 \cdot \sqrt{4} \cdot \sqrt{3}$
$= 7 \cdot 2 \cdot \sqrt{3} = 14\sqrt{3}$
$3\sqrt{3} = 3\sqrt{3}$

Step 2. $15\sqrt{3} + 14\sqrt{3} + 3\sqrt{3} = (15 + 14 + 3)\sqrt{3}$
$= 32\sqrt{3}$

2. Subtract: $9\sqrt{2} - \dfrac{6}{\sqrt{2}}$.

Step 1. $9\sqrt{2} = 9\sqrt{2}$
$$\frac{6}{\sqrt{2}} = \frac{6 \cdot \sqrt{2}}{\sqrt{2} \cdot \sqrt{2}} = \frac{6\sqrt{2}}{2} = 3\sqrt{2}$$

Step 2. $9\sqrt{2} - 3\sqrt{2} = (9 - 3)\sqrt{2} = 6\sqrt{2}$

3. Simplify: $3\sqrt{125} - 4\sqrt[3]{5} - 15\sqrt{\tfrac{1}{5}} + 3\sqrt[3]{625}$.

Step 1. $3\sqrt{125} = 3\sqrt{25 \cdot 5}$
$= 3 \cdot 5 \cdot \sqrt{5} = 15\sqrt{5}$
$4\sqrt[3]{5} = 4\sqrt[3]{5}$
$$15\sqrt{\tfrac{1}{5}} = 15 \cdot \frac{\sqrt{1}}{\sqrt{5}} = \frac{15}{\sqrt{5}}$$
$$= \frac{15\sqrt{5}}{\sqrt{5} \cdot \sqrt{5}} = \frac{15\sqrt{5}}{5} = 3\sqrt{5}$$
$3\sqrt[3]{625} = 3\sqrt[3]{125 \cdot 5}$
$= 3 \cdot 5 \cdot \sqrt[3]{5} = 15\sqrt[3]{5}$

Step 2. $15\sqrt{5} - 4\sqrt[3]{5} - 3\sqrt{5} + 15\sqrt[3]{5}$
$= (15\sqrt{5} - 3\sqrt{5}) - (4\sqrt[3]{5} - 15\sqrt[3]{5})$
$= 12\sqrt{5} - (-11\sqrt[3]{5})$
$= 12\sqrt{5} + 11\sqrt[3]{5}$

9. MULTIPLICATION WITH RADICAL EXPRESSIONS

Two alternatives are possible for computing products with exponential expressions. Products may be computed by adding exponents *if their bases are alike*, or by multiplying their bases together *if their exponents are alike*. Translating this into terms of radical expressions, we get the following:

1. Products with radical expressions may be computed by adding index reciprocals if the radicands are alike.
2. Products with radical expressions may be computed by multiplying their radicands if their indexes are alike.

These two cases and the case in which the factors differ in both index and radicand can all be handled by the following general program.

(7.8) *To compute a product with radical expressions:*

Step 1. *If all of the radical factors are not of the same order (index), express each with an equivalent radical expression of a common order.*

Step 2. *Write the product of the coefficients as the coefficient of the product and the product of the radicands as the radicand of the product (expressed under the radical of common order).*

Step 3. *Simplify if possible.*

EXAMPLES

1. Multiply: $2\sqrt{3x} \cdot 4\sqrt{6xy} \cdot \sqrt{2py}$.

Step 1. All factors are of a common order.

Step 2. Multiply the coefficients together:
$2 \cdot 4 \cdot 1 = 8$
Multiply the radicands together:
$3x \cdot 6xy \cdot 2py = 36x^2y^2p$
Express as a product:
$8\sqrt{36x^2y^2p}$

Step 3. Simplify:
$8\sqrt{36x^2y^2p} = 8 \cdot 6 \cdot x \cdot y \cdot \sqrt{p} = 48xy\sqrt{p}$

2. Multiply: $4x\sqrt[3]{2x^2y} \cdot y\sqrt{2xy}$.

Step 1. A common order is 6.
$\sqrt[3]{2x^2y} = \sqrt[6]{(2x^2y)^2} = \sqrt[6]{4x^4y^2}$
$\sqrt{2xy} = \sqrt[6]{(2xy)^3} = \sqrt[6]{8x^3y^3}$

Step 2. $4x \cdot y = 4xy$
$\sqrt[6]{4x^4y^2} \cdot \sqrt[6]{8x^3y^3} = \sqrt[6]{4x^4y^2 \cdot 8x^3y^3}$
$= \sqrt[6]{4 \cdot 8 \cdot x^{4+3} \cdot y^{2+3}}$
$= \sqrt[6]{32x^7y^5}$

Step 3. $4xy\sqrt[6]{32x^7y^5} = 4x^2y\sqrt[6]{32xy^5}$

3. Multiply: $x\sqrt[3]{5xy^2}\cdot y\sqrt{2y}\cdot 3\sqrt[4]{8x^2y}$.
Step 1. A common index is 12.

$$\sqrt[3]{5xy^2} = \sqrt[12]{(5xy^2)^4} = \sqrt[12]{5^4x^4y^8}$$
$$\sqrt{2y} = \sqrt[12]{(2y)^6} = \sqrt[12]{2^6y^6}$$
$$\sqrt[4]{8x^2y} = \sqrt[12]{(2^3x^2y)^3} = \sqrt[12]{2^9x^6y^3}$$

Step 2. $x\cdot y\cdot 3 = 3xy$

$$\sqrt[12]{5^4x^4y^8}\cdot\sqrt[12]{2^6y^6}\cdot\sqrt[12]{2^9x^6y^3}$$
$$= \sqrt[12]{5^4\cdot x^4\cdot y^8\cdot 2^6\cdot y^6\cdot 2^9\cdot x^6\cdot y^3}$$
$$= \sqrt[12]{5^4\cdot 2^{6+9}\cdot x^{4+6}\cdot y^{8+6+3}}$$
$$= \sqrt[12]{5^4\cdot 2^{15}\cdot x^{10}\cdot y^{17}}$$

Step 3. $3xy\sqrt[12]{5^4\cdot 2^{15}\cdot x^{10}\cdot y^{17}}$

$$= 3xy\cdot 2\cdot y\cdot\sqrt[12]{5^4\cdot 2^3\cdot x^{10}\cdot y^5}$$
$$= 6xy^2\sqrt[12]{5{,}000x^{10}y^5}$$

4. Multiply: $(3\sqrt{x} + 2\sqrt{y})(2\sqrt{x} - 5\sqrt{y})$.
We may compute the product either by the standard method:

$$\begin{array}{l} 3\sqrt{x} + 2\sqrt{y} \\ 2\sqrt{x} - 5\sqrt{y} \\ \hline 6\sqrt{x^2} + 4\sqrt{xy} \\ \qquad - 15\sqrt{xy} - 10\sqrt{y^2} \\ \hline 6\sqrt{x^2} - 11\sqrt{xy} - 10\sqrt{y^2} = 6x - 11\sqrt{xy} - 10y \end{array}$$

or directly:

$6\sqrt{x^2}$ $\quad -10\sqrt{y^2}$

$(3\sqrt{x} + 2\sqrt{y})(2\sqrt{x} - 5\sqrt{y})$

$(4\sqrt{xy})$
$(-15\sqrt{xy})$ $\Big\}$ $4\sqrt{xy} - 15\sqrt{xy}$
$= -11\sqrt{xy}$

$$6\sqrt{x^2} - 11\sqrt{xy} - 10\sqrt{y^2} = 6x - 11\sqrt{xy} - 10y$$

10. DIVISION WITH RADICAL EXPRESSIONS

Computing a quotient with radical expressions is most readily performed by expressing the terms of the division as a fraction in which the dividend is the numerator and the divisor is the denominator. If the divisor contains a radical expression, the denominator is ordinarily rationalized; the resulting equivalent fraction when simplified is the desired quotient.

(7.9) *To compute a quotient with radical expressions:*
Step 1. Express the dividend as the numerator and the divisor as the denominator of a fraction.
Step 2. Rationalize the denominator.
Step 3. Simplify the rationalized fraction.

EXAMPLES

1. Divide $12\sqrt{5}$ by $\sqrt{3}$.

Step 1. $12\sqrt{5} \div \sqrt{3} = \dfrac{12\sqrt{5}}{\sqrt{3}}$

Step 2. $\dfrac{\sqrt{3}}{\sqrt{3}}\cdot\dfrac{12\sqrt{5}}{\sqrt{3}} = \dfrac{12\sqrt{15}}{3}$

Step 3. $\dfrac{12\sqrt{15}}{3} = 4\sqrt{15}$ (quotient)

2. Divide $(4x + \sqrt{6})$ by $3\sqrt{2}$.

Step 1. $(4x + \sqrt{6}) \div 3\sqrt{2} = \dfrac{4x + \sqrt{6}}{3\sqrt{2}}$

Step 2. $\dfrac{\sqrt{2}}{\sqrt{2}}\cdot\dfrac{4x + \sqrt{6}}{3\sqrt{2}} = \dfrac{4x\sqrt{2} + \sqrt{12}}{3\cdot 2}$

Step 3. $\dfrac{4x\sqrt{2} + \sqrt{12}}{6} = \dfrac{4x\sqrt{2}}{6} + \dfrac{2\sqrt{3}}{6}$

$$= \frac{2x\sqrt{2}}{3} + \frac{\sqrt{3}}{3}$$

(or $\frac{2}{3}x\sqrt{2} + \frac{1}{3}\sqrt{3}$)

3. Divide $(4\sqrt{2} + 6\sqrt[3]{3})$ by $\sqrt[3]{2}$.

Step 1. $(4\sqrt{2} + 6\sqrt[3]{3}) \div \sqrt[3]{2} = \dfrac{4\sqrt{2} + 6\sqrt[3]{3}}{\sqrt[3]{2}}$

Step 2. $\dfrac{4\sqrt{2} + 6\sqrt[3]{3}}{\sqrt[3]{2}}\cdot\dfrac{\sqrt[3]{2^2}}{\sqrt[3]{2^2}}$

$$= \frac{4\sqrt{2}\sqrt[3]{2^2} + 6\sqrt[3]{3\cdot 2^2}}{\sqrt[3]{2^3}}$$

Step 3. $\dfrac{4\sqrt{2}\sqrt[3]{2^2} + 6\sqrt[3]{3\cdot 2^2}}{\sqrt[3]{2^3}}$

$$= \frac{4\sqrt[6]{2^3}\sqrt[6]{2^4} + 6\sqrt[3]{12}}{2}$$
$$= \frac{4\sqrt[6]{2^7} + 6\sqrt[3]{12}}{2}$$
$$= \frac{4\cdot 2\cdot\sqrt[6]{2} + 6\sqrt[3]{12}}{2}$$
$$= 4\sqrt[6]{2} + 3\sqrt[3]{12}$$

4. Divide $2\sqrt{3} - 3\sqrt{2}$ by $\sqrt{2} - \sqrt{3}$.
Step 1. $(2\sqrt{3} - 3\sqrt{2}) \div (\sqrt{2} - \sqrt{3})$

$$= \frac{2\sqrt{3} - 3\sqrt{2}}{\sqrt{2} - \sqrt{3}} = -\frac{2\sqrt{3} - 3\sqrt{2}}{\sqrt{3} - \sqrt{2}}\text{*}$$

Step 2. $-\dfrac{2\sqrt{3} - 3\sqrt{2}}{\sqrt{3} - \sqrt{2}}\cdot\dfrac{\sqrt{3} + \sqrt{2}}{\sqrt{3} + \sqrt{2}}$

$$= -\frac{2\sqrt{9} - \sqrt{6} - 3\sqrt{4}}{\sqrt{9} - \sqrt{4}}$$

* Expressing the fraction negatively is not essential, but it does facilitate the multiplication in Step 2.

Step 3. $-\dfrac{2\sqrt{9} - \sqrt{6} - 3\sqrt{4}}{\sqrt{9} - \sqrt{4}} = -\dfrac{6 - \sqrt{6} - 6}{3 - 2}$

$$= -\frac{-\sqrt{6}}{1} = \sqrt{6}$$

5. Divide $4\sqrt{5} - 2$ by $3\sqrt{2} - 2\sqrt{5}$.

Step 1. $(4\sqrt{5} - 2) \div (3\sqrt{2} - 2\sqrt{5})$

$$= \frac{4\sqrt{5} - 2}{3\sqrt{2} - 2\sqrt{5}}$$

Step 2. $\dfrac{3\sqrt{2} + 2\sqrt{5}}{3\sqrt{2} + 2\sqrt{5}} \cdot \dfrac{4\sqrt{5} - 2}{3\sqrt{2} - 2\sqrt{5}}$

$$= \frac{12\sqrt{10} - 6\sqrt{2} + 8\sqrt{25} - 4\sqrt{5}}{(3\sqrt{2})^2 - (2\sqrt{5})^2}$$

Step 3. $= \dfrac{12\sqrt{10} - 6\sqrt{2} + (8 \cdot 5) - 4\sqrt{5}}{18 - 20}$

$$= \frac{12\sqrt{10} - 6\sqrt{2} + 40 - 4\sqrt{5}}{-2}$$

$$= -6\sqrt{10} + 3\sqrt{2} - 20 + 2\sqrt{5}$$

(Complete Exercise 7.4)

Name.. Date.............................. **EXERCISE 7.1**

Simplify without using negative exponents. Assume variable exponents represent positive integers.

1. $2^3 =$

2. $2^5 =$

3. $-3^2 =$

4. $-3^3 =$

5. $(-3)^3 =$

6. $(-2)^4 =$

7. $(-\frac{2}{3})^2$

8. $(|-5|)^3 =$

9. $a^3 \cdot a^5 =$

10. $a^3 \cdot a^2 \cdot a =$

11. $x^3 \cdot x^2 \cdot x^0 =$

12. $(a^2y^3)(a^2y) =$

13. $x^4 \div x =$

14. $y^6 \div y^6 =$

15. $\dfrac{a^2b^3}{a^3b} =$

16. $\dfrac{x^3y^3}{x^3y} =$

17. $\dfrac{a^2b^2c^4}{a^3bc^3} =$

18. $(x^2)^3 =$

19. $-(b^2)^2 =$

20. $(-m^3)^4 =$

21. $2^2 \cdot 2^4 =$

22. $(3^2)^3 =$

23. $(2)^2(3)^0 =$

24. $-(a^2b^3)^3 =$

25. $(-a^5b^4)^3 =$

26. $(2xy^2)^3 =$

27. $(3a^2bc)^4 =$

28. $(-2x^2yz^3)^3 =$

29. $\left(\dfrac{x^3}{y^5}\right)^4 =$

30. $\left(\dfrac{3a^2}{4b^3}\right)^0 =$

31. $\left(-\dfrac{3}{v^2}\right)^4 =$

32. $(a^{2m})^p =$

33. $\left(\dfrac{3^n}{2}\right)^3 =$

34. $\left(-\dfrac{x^2}{y^n}\right)^4 =$

35. $x^{4n} \div x^n =$

36. $x^{5n} \div x =$

37. $x^{2n} \div x^{3n} =$

38. $a^{2+n} \cdot a^n =$

39. $(3a + x)^0 =$

40. $r^{xy} \cdot r^{x+1} =$

41. $3(x^n)^2 =$

42. $(m^a)^{p+2} =$

43. $(a^2b^3)^r =$

44. $\left(-\dfrac{a^2b^0}{2b}\right)^7 =$

Express with all negative exponents.

45. $a^4 =$

46. $a^2b^3 =$

47. $\dfrac{1}{y^2} =$

48. $\dfrac{2}{p^3} =$

49. $x^0y^3 =$

50. $\dfrac{1}{16} =$

51. $\dfrac{x^{-1}}{b} =$

52. $-c^3 =$

53. $-(x)^2(yz)^3 =$

54. $(a - b)^2 =$

55. $(a^2 - b^2) =$

56. $(a^2 - b^2)^{-2} =$

Simplify the expressions so that only positive exponents are involved; carry out all expansions, additions and subtractions.

57. $(3x)^{-3} =$

58. $\left(\dfrac{2}{x}\right)^{-2} =$

59. $(x^{-2})^{-2} =$

60. $(xy^{-1})^{-2} =$

61. $(3x^0y)^{-4} =$

62. $(2^3x^{-3})^{-2} =$

63. $\left(-\dfrac{a^{-1}}{b}\right)^{-2} =$

64. $\left(-\dfrac{3}{2}\right)^{-3} =$

65. $\left(\dfrac{ax^4}{b^3y}\right)^3\left(\dfrac{by^3}{a^2x^2}\right)^2 =$

66. $x^{-1} - y^{-1} =$

67. $(x^{-2})^{-2} \cdot (y^{-3})^2 =$

68. $2x^{-1} - y^{-2} =$

69. $x^{-3} \div y^{-2} =$

70. $a^{-2} \div \left(\dfrac{b}{a}\right)^{-3} =$

71. $\dfrac{a}{b^{-1}} + \dfrac{b}{a^{-1}} + ab =$

72. $(a^{-1} + b^{-1})^{-1} =$

73. $6x^{-2} + y =$

74. $(a^{-1} + 2b^{-1})^{-2} =$

Name.. Date.............................. **EXERCISE 7.2**

Write the principal real-number root for each.

1. $\sqrt{4} =$

2. $\sqrt{36} =$

3. $\sqrt[3]{8} =$

4. $\sqrt[3]{-8} =$

5. $\sqrt{.01} =$

6. $\sqrt[3]{-\frac{1}{8}} =$

7. $\sqrt{-36} =$

8. $\sqrt{.16} =$

9. $\sqrt[3]{-.001} =$

10. $\sqrt[3]{-\frac{8}{27}} =$

11. $\sqrt[6]{-64} =$

12. $\sqrt[5]{-32} =$

Express in exponential form.

13. $\sqrt{p} =$

14. $\sqrt[3]{s} =$

15. $\sqrt[10]{t} =$

16. $\sqrt[3]{p^2} =$

17. $\sqrt[5]{x^2y^2} =$

18. $\sqrt[n]{y^2} =$

19. $\dfrac{1}{\sqrt[r]{p^s}} =$

20. $\sqrt[k]{p^{3m}} =$

21. $\sqrt{c-6} =$

22. $\sqrt[4]{\dfrac{c^3}{b}} =$

23. $\sqrt[4]{x^2y^{1/2}} =$

24. $\dfrac{1}{\sqrt[n]{c+d^2}} =$

Express in radical form.

25. $a^{1/5} =$

26. $b^{1/3}c^{2/3} =$

27. $a^{5/8} =$

28. $3x^{1/2} =$

29. $(3x)^{1/2} =$

30. $(8x)^{1/3} =$

31. $(b-c)^{1/2} =$

32. $\dfrac{1}{(a+2)^{-1/3}} =$

33. $[(x+y)^2]^{1/3} =$

34. $(s^2t^{1/4})^{1/n} =$

35. $\dfrac{x^{-1/2}}{y^{1/2}} =$

36. $\left(p - \dfrac{pr}{r}\right)^{-1/3} =$

Simplify.

37. $x^{2/3} \div x^{1/2} =$

38. $m^{2/5} \div m^{2/3} =$

39. $\dfrac{a^{7/2}}{a^{1/2}} =$

40. $\dfrac{b^{5/2}}{\sqrt[3]{b^{10}}} =$

41. $\dfrac{a^{.8}}{a^{.2}} =$

42. $\dfrac{a^{.12}}{a^{.6}} =$

43. $\dfrac{m^{2/3}n^{1/3}}{m^0 n^{1/4}} =$

44. $\left(\dfrac{4x^{-1}y}{x^{2/3}y^{-1}}\right)^{-3} =$

45. $\left(-\dfrac{8}{125}\right)^{2/3} =$

46. $\left(\dfrac{1}{16}\right)^{-3/4} =$

47. $\sqrt{(x+y)^{-2}} =$

48. $\sqrt{x^{-1}y^{-1}} =$

49. $(-2^{-6})^{2/3} =$

50. $-(2^6)^{-2/3} =$

51. $\dfrac{(2x^2y^3)^{1/4}}{16x^{-2}y} =$

52. $\dfrac{x^{(2m+3)/c}}{x^{(m-4)/c}} =$

53. $\sqrt[5]{s^{1/2}} =$

54. $(2^0 + 3^0 + 4^0)^{-1/2} =$

55. $\sqrt[4]{a^{-3/4}} =$

56. $\left(\dfrac{a^{m+2}n^{m-3}}{a^2n^{-3}}\right)^{-1/2} =$

57. $\left(\dfrac{a^{2/3}b^{1/2}}{3a^{3/4}}\right)^4 =$

58. $\sqrt[4]{\sqrt[3]{\sqrt{x}}} =$

Name.. Date.................... **EXERCISE 7.3**

Reduce the order.

1. $\sqrt[4]{9a^2} =$ **2.** $\sqrt[4]{64a^2b^2} =$ **3.** $\sqrt[6]{4m^2} =$

4. $\sqrt[9]{27m^6}$ **5.** $\sqrt[12]{36s^8} =$ **6.** $\sqrt[6]{\dfrac{27x^3}{64y^3}}$

Supply the radicand.

7. $\sqrt{3} = \sqrt[4]{\quad}$ **8.** $\sqrt{2xy} = \sqrt[6]{\quad}$ **9.** $\sqrt[3]{4xy} = \sqrt[6]{\quad}$

10. $\sqrt{16} = \sqrt[3]{\quad}$ **11.** $\sqrt[4]{13x^2y^3z} = \sqrt[8]{\quad}$ **12.** $\sqrt[3]{(a-b)^2} = \sqrt[6]{\quad}$

Express with all terms under the radical.

13. $3\sqrt{2} =$ **14.** $5\sqrt{3} =$ **15.** $2\sqrt[3]{6} =$

16. $m\sqrt{3} =$ **17.** $a\sqrt{5a} =$ **18.** $4a\sqrt[3]{2a} =$

19. $4x^2y\sqrt{2xy} =$ **20.** $3mn^2\sqrt[3]{2m^2n} =$ **21.** $(a-b)\sqrt[6]{(a-b)} =$

Simplify

22. $\sqrt{75} =$ **23.** $\sqrt[3]{32} =$ **24.** $\sqrt[3]{81} =$

25. $\sqrt{8x} =$ **26.** $\sqrt{50x^3} =$ **27.** $\sqrt[3]{27x^4y^2} =$

28. $\sqrt{18x^2y^3z^4} =$ **29.** $\sqrt[3]{kx^5y^4} =$ **30.** $3\sqrt{12x^3} =$

31. $\sqrt{\dfrac{x}{16}} =$ **32.** $\sqrt{\dfrac{5s^5}{t^6}} =$ **33.** $4\sqrt[3]{\dfrac{m^2}{8}} =$

34. $3 + a\sqrt{45} =$ **35.** $a + b\sqrt{(a+b)^3} =$ **36.** $\sqrt[3]{(a-5)^4} =$

Express in simplest terms with rational denominators.

37. $\dfrac{1}{\sqrt{3}} =$

38. $\dfrac{2a}{\sqrt{2a}} =$

39. $\dfrac{3x}{\sqrt{5}} =$

40. $\dfrac{7a}{2\sqrt{3}} =$

41. $\dfrac{3}{\sqrt[3]{4}} =$

42. $\dfrac{2p}{\sqrt[3]{p^2}} =$

43. $\dfrac{12x}{\sqrt[3]{x}} =$

44. $\sqrt{\dfrac{6a}{5b}} =$

45. $\sqrt{7m^{-1}} =$

46. $\dfrac{1}{\sqrt{3} - \sqrt{2}}$

47. $\dfrac{2}{\sqrt{3} + \sqrt{5}}$

48. $\dfrac{\sqrt{2}}{\sqrt{2} - \sqrt{3}}$

49. $\dfrac{\sqrt{3}}{3 - \sqrt{3}}$

50. $\dfrac{x}{1 - \sqrt{x}}$

51. $\dfrac{\sqrt{a}}{\sqrt{a} - \sqrt{b}}$

52. $\dfrac{\sqrt{2} + \sqrt{3}}{\sqrt{2} - \sqrt{3}}$

53. $\dfrac{\sqrt{5} - \sqrt{3}}{\sqrt{5} + \sqrt{3}}$

54. $\dfrac{\sqrt{3}}{2\sqrt{3} - 3\sqrt{2}}$

55. $\dfrac{\sqrt{a} - \sqrt{b}}{\sqrt{a} + \sqrt{b}}$

56. $\dfrac{2\sqrt{2} - \sqrt{3}}{3\sqrt{2} + \sqrt{3}}$

57. $\sqrt{\dfrac{a}{3} - \dfrac{b}{4}}$

Name.. Date.............................. **EXERCISE 7.4**

Simplify.

1. $4\sqrt{a} + 3\sqrt{a} - 2\sqrt{a} =$

2. $\sqrt{18} + \sqrt{50} - \sqrt{72} =$

3. $\sqrt[3]{81} - 2\sqrt[3]{3} + 3\sqrt[3]{24} =$

4. $3\sqrt{45} - 2\sqrt{25} + \sqrt{5} =$

5. $\sqrt{\frac{1}{2}} + 2\sqrt{\frac{1}{8}} =$

6. $\sqrt[3]{2} + \sqrt[3]{-54} + \sqrt[3]{16} =$

7. $\sqrt[3]{6} - 3\sqrt[3]{48} + 2\sqrt[3]{\frac{1}{36}} =$

8. $\sqrt{8} - \sqrt[3]{2} + \sqrt{18} =$

9. $\sqrt{\frac{25}{12}} - \sqrt{\frac{9}{8}} =$

10. $2\sqrt{24} + 3\sqrt{150} - 3\sqrt{96} =$

11. $3\sqrt{72} - 7\sqrt{18} + 2\sqrt[3]{54} =$

12. $8\sqrt{3} - \dfrac{2}{\sqrt{3}} + \dfrac{7}{2\sqrt{3}} =$

13. $\sqrt{\frac{3}{5}} - \sqrt{\frac{16}{15}} + \sqrt{\frac{20}{3}} =$

14. $\sqrt{6x^2} - \sqrt{24x^2} + \sqrt{9x^0} =$

15. $\sqrt[3]{(a-b)^2} + \sqrt[3]{(a-b)^5} =$

16. $\sqrt{a^2b} - \sqrt{9b} - a\sqrt{b} =$

Multiply, and simplify.

17. $\sqrt[3]{2a} \cdot \sqrt[3]{4a^2} =$

18. $(4x\sqrt{3x})^2 =$

19. $\sqrt{3a^3} \cdot \sqrt{6a^3} =$

20. $\sqrt[3]{4} \cdot \sqrt{6} =$

21. $\sqrt{6}(\sqrt{2} - 3\sqrt{3} + 2\sqrt{6}) =$

22. $18(2\sqrt{12x} - \sqrt{10x} + 2\sqrt{98x}) =$

23. $(3x\sqrt[3]{2xy^2})(x\sqrt{2x}) =$

24. $(2\sqrt{6} + 3\sqrt{2})^2 =$

25. $(\sqrt{a} - \sqrt{b})^2 =$

26. $(\sqrt{2} - \sqrt{x})(3\sqrt{2} + 4\sqrt{x}) =$

Divide, and simplify quotients.

27. $8 \div 2\sqrt{2} =$

28. $3 \div \sqrt[3]{9} =$

29. $(\sqrt{6} - 2\sqrt{15}) \div \sqrt{3} =$

30. $a \div (1 + \sqrt{a}) =$

31. $(3\sqrt{2} + \sqrt{3} + 2\sqrt{6}) \div \sqrt{6} =$

32. $(2\sqrt{5} - 7\sqrt{3}) \div (\sqrt{5} - 2\sqrt{3}) =$

33. $(\sqrt{6} - 3\sqrt{2}) \div (4\sqrt{2} - 3\sqrt{6}) =$

34. $(7 - \sqrt{x}) \div (\sqrt{x} + 4) =$

35. $(\sqrt{x} - y) \div (y - \sqrt{x}) =$

36. $\dfrac{\sqrt{x} - \sqrt{a}}{\sqrt{x} + \sqrt{y}} =$

Solve.

37. If $x = 2 - \sqrt{2}$, then $x^2 + x - 1 =$ ....

38. If $x = \sqrt{3} + 2$, then $3x^2 + 2x - 4 =$ ....

Simplify. Use only positive exponents. Carry out all expansions, additions and subtractions. Assume variable exponents represent positive integers.

1. $(-3)^4$

2. $(\frac{1}{2})^3$

3. $m^2 \cdot m^3 \cdot m$

4. $a \cdot a^3 \cdot a^0$

5. $t^4 \div t^4$

6. $m^6 \div m^5$

7. $(x^3)^2$

8. $(a^4)^4$

9. $(3ab^2)^3$

10. $(7m^2)^3$

11. $\left(\frac{3m^3}{2p}\right)^0$

12. $\left(-\frac{3m^0}{2}\right)^1$

13. $\left(-\frac{a^3}{b^2}\right)^8$

14. $\left(-\frac{x}{y^3}\right)^3$

15. $y^{3+n} \cdot y^n$

16. $a^{2+n} \cdot a^2$

17. $(d^a)^{n+2}$

18. $(m^k)^{2+n}$

19. $a^5 \cdot a^{-3} \cdot a^{-2}$

20. $x^{-3} \cdot x^{-2} \cdot x^6$

21. $\left(\frac{a^{-3}}{c^{-4}}\right)^3$

22. $\left(\frac{c^{-1}}{d^3}\right)^2$

23. $(x^2y^3)^3\left(\frac{a^2x^3}{y^2}\right)^2$

24. $\left(\frac{x^2y}{x}\right)^3(xy)^2$

25. $3x^{-2} + 3y^{-2}$

26. $\frac{3}{x^{-2}} + \frac{4}{y^{-2}}$

27. $\left(\frac{a}{b} - \frac{b}{a}\right)^{-1}$

28. $\left(\frac{1}{x} + \frac{1}{y}\right)^{-2}$

29. $\frac{x^{-1}y}{x^{-1} + y}$

30. $\frac{a^{-1}b^{-1}}{b^{-1} + a}$

Write the principal real-number root for each.

31. $\sqrt[3]{64}$

32. $\sqrt[5]{32}$

33. $\sqrt[3]{-125}$

34. $\sqrt[3]{-\frac{1}{8}}$

35. $\sqrt[4]{.0001}$

36. $\sqrt[3]{-.001}$

Express in exponential form.

37. $\sqrt[3]{x}$

38. $\sqrt[5]{-m}$

39. $\sqrt[5]{a^2b^3}$

40. $\sqrt[3]{x^3y^6}$

41. $\sqrt[k]{x^2y}$

42. $\sqrt[n]{a^2b^3}$

Express in simplest radical form.

43. $m^{3/4}$

44. $a^{2/3}$

45. $(8a)^{1/3}$

46. $(4a^3)^{1/3}$

47. $\frac{1}{(m+n)^{-1/2}}$

48. $\frac{1}{(a-b)^{-1/3}}$

49. $\frac{x^{1/2}}{y^{-1/2}}$

50. $\frac{a^{-1/3}}{b^{1/3}}$

Simplify.

51. $a^{3/4} \div a^{1/2}$

52. $x^{2/3} \div x^{-1}$

53. $\frac{x^{2/3}}{x^{1/3}}$

54. $\frac{m^{3/4}}{m^{3/2}}$

55. $\frac{x^{1/5}y^{3/10}}{xy^{-7/10}}$

56. $\frac{a^{3/4}b^{1/2}}{a^0b^{1/4}}$

57. $\sqrt{(m-n)^{-2}}$

58. $\sqrt[3]{(a+b^{-1}}$

59. $\sqrt{\sqrt[3]{x^2}}$

60. $((a^{1/2})^{1/3})^{-1}$

Reduce the order.

61. $\sqrt[4]{16x^2}$

62. $\sqrt[4]{49a^2b^2}$

63. $\sqrt[6]{8x^3}$

64. $\sqrt[6]{25m^4}$

65. $\sqrt[12]{16a^4b^8}$

66. $\sqrt[12]{27x^6y^9}$

Simplify.

67. $\sqrt{63x^3}$

68. $\sqrt{36a^3b}$

69. $\sqrt[3]{ab^4c^2}$

70. $x\sqrt{3xy^3}$

71. $\sqrt{\frac{2a^2}{b^6}}$

72. $\sqrt[3]{(x+y)^4}$

Express in simplest terms, with rational denominators.

73. $\frac{3a}{\sqrt{5}}$

74. $\frac{3x}{\sqrt{3x}}$

75. $\frac{y}{3\sqrt{3}}$

76. $\sqrt{\frac{2x}{3y}}$

77. $\sqrt{8b^{-1}}$

78. $\sqrt[3]{x^{-2}}$

79. $\frac{1}{2-\sqrt{2}}$

80. $\frac{3}{\sqrt{3}-1}$

81. $\frac{\sqrt{2}}{\sqrt{3}-\sqrt{2}}$

82. $\frac{\sqrt{5}}{\sqrt{5}+\sqrt{2}}$

83. $\frac{m}{\sqrt{m}-1}$

84. $\frac{\sqrt{k}}{\sqrt{k}+1}$

85. $\frac{\sqrt{3}-\sqrt{4}}{\sqrt{3}+\sqrt{4}}$

86. $\frac{\sqrt{a}+\sqrt{b}}{\sqrt{a}-\sqrt{b}}$

87. $\frac{3\sqrt{x}+2\sqrt{y}}{\sqrt{x}-\sqrt{y}}$

88. $\frac{2\sqrt{3}+3\sqrt{2}}{2\sqrt{3}-\sqrt{2}}$

Simplify.

89. $\sqrt{2} - \sqrt{18} + \sqrt{32}$

90. $\sqrt{72} + \sqrt{90} - \sqrt{50}$

91. $6\sqrt{3} + \sqrt{27} - \sqrt{108}$

92. $3\sqrt{24} - \sqrt{54} - \sqrt{150}$

93. $\sqrt[3]{16} + \sqrt[3]{54} + \sqrt[3]{128}$

94. $2\sqrt[3]{5} + \sqrt[3]{135} - \sqrt[3]{40}$
95. $\sqrt{5a^3} + a\sqrt{125a} - \sqrt{20a}$
96. $\sqrt{x^2y} - \sqrt{16x} - x\sqrt{y}$
97. $\dfrac{2}{\sqrt{3}} + \dfrac{\sqrt{3}}{2} - 3\sqrt{3}$
98. $\dfrac{3}{\sqrt{5}} - \dfrac{2}{\sqrt{5}} + \sqrt{\dfrac{60}{3}}$

Multiply, and simplify.

99. $(3x\sqrt{2x})^2$
100. $(2ab\sqrt{3a})^3$
101. $\sqrt{4x^3} \cdot \sqrt{3x^2}$
102. $\sqrt{2m} \cdot \sqrt{8m^3}$
103. $(\sqrt{3} - \sqrt{2})^2$
104. $(\sqrt{5} + \sqrt{3})^2$
105. $(\sqrt{3} - \sqrt{x})(\sqrt{3} + 2\sqrt{x})$
106. $(\sqrt{2} + \sqrt{3y})(2\sqrt{2} - 3\sqrt{3y})$
107. $(2\sqrt{x} - 3\sqrt{y})^2$
108. $(\sqrt{a} - 3\sqrt{b})^2$

Divide, and simplify the quotient.

109. $10 \div 3\sqrt{2}$
110. $6 \div \sqrt[3]{4}$
111. $\sqrt{84} \div 3\sqrt{7}$
112. $3\sqrt[3]{56} \div \sqrt[3]{14}$
113. $(2\sqrt{3} + 3\sqrt[3]{2}) \div \sqrt[3]{3}$
114. $(3\sqrt[3]{4} - 2\sqrt[3]{3} + 2\sqrt{6}) \div \sqrt[3]{6}$
115. $4 \div (2 - \sqrt{2})$
116. $1 \div (1 + \sqrt{2})$
117. $(\sqrt{3} - 1) \div (\sqrt{3} + 1)$
118. $(2 + \sqrt{3}) \div (2 - \sqrt{3})$
119. $(2\sqrt{5} - \sqrt{2}) \div (3\sqrt{5} + 2\sqrt{2})$
120. $(4\sqrt{5} + \sqrt{3}) \div (2\sqrt{3} - \sqrt{5})$

REVIEW

Factor.

1. $a^2 + 6ad + 9d^2$
2. $16x^2 - 8xy + y^2$
3. $25 - 9a^2b^2$
4. $4m^2 - 81n^2$
5. $x^2 - 10x + 21$
6. $x^2 - 12x + 35$
7. $5y^2 - 40y + 60$
8. $2ax^2 - 4ax - 6a$
9. $6a^2 - 11a - 2$
10. $6p^2 - 23p + 20$
11. $mp - np - nq + mq$
12. $b + ab + ac + c$
13. $ab - cb - ad + cd$
14. $-2yz + 3xz + 3x - 2y$
15. $2ab + 4bc + 4cd + 2ad$
16. $6bc - 2ac - 3ad + 9bd$
17. $x^2 - 2yz - y^2 - z^2$
18. $(x + 3)^2 + (x + 3)(x + 2)$
19. $p + q + (p + q)^2$
20. $c^2 + 2cd + d^2 + 2c + 2d$

Solve by equation.

21. A man invested \$6000 at one rate of interest and \$4000 at half that rate. If his annual return from the two investments is \$720, what must be the two rates?
22. Four consecutive odd numbers have 112 as their sum. Name the smallest of the four numbers.
23. What amounts of 20% and 35% solutions of an acid should be mixed together to produce 50 cc of a 26% solution?
24. A truck makes a round-trip each day between two cities in 5 hrs of actual travel time. On one half of the round-trip, the driver can average 45 mph, but on the other half only 30 mph. How long does he travel at the faster speed?
25. A man is five times as old as his daughter, but 15 years from now he will only be twice her age. How old is each at present?

Name.. Date.............................. ....

ACHIEVEMENT TEST NO. 3 (Chapters 6 and 7)

Place answers, in simplest form, in the blanks provided. (4 points each 17 through 20, others 3 points each, total 100.)

Expand:

1. $-ab(a^2b - ab^2 + 3ab - c) =$ ....

2. $(ax - by)^2 =$ ....

3. $(pq - y)(pq + y) =$ ....

4. $(x + y)(a - b) =$ ....

Factor:

5. $10x^2 - 6x^4 + 12x^3 - 2x^5 =$ ....

6. $16x^2 - 56xy + 49y^2 =$ ....

7. $16x^4y^4 - 1 =$ ....

8. $9a^2 + 3 - 28a =$ ....

9. $ax^2 - ay^2 - bx^2 + by^2 =$ ....

10. $(x^2 - 5x + 6) - (x - 3)^2 =$ ....

Reduce to lowest terms:

11. $\dfrac{(b - a)^2}{a^2 - b^2} =$ ....

12. $\dfrac{xy - 3y + 2x - 6}{x^2 - x - 6} =$ ....

Perform the operations and simplify:

13. $\dfrac{6x^2 - 19x + 15}{5 + 2x - 3x^2} \cdot \dfrac{x^2 - 1}{4x^2 - 9} =$ ....

14. $\dfrac{x^2 + x - 12}{x^2 - 7x + 12} \div \dfrac{2x^2 + 3x - 20}{x^2 - 16}$ ....

15. $\dfrac{4a}{a^2 - 4} + \dfrac{3}{2 - a} + \dfrac{a}{2 + a} =$ ....

16. $\dfrac{2}{x^2 - 4} - \dfrac{3}{x^2 - 1} + \dfrac{1}{x^2 - x - 2} =$ ....

Solve:

17. $\dfrac{5}{1-x} + \dfrac{1}{3x-4} = \dfrac{3}{8-6x}$ ....

18. $\dfrac{6}{x-2} + \dfrac{4}{2-x} = \dfrac{x}{x-2}$ ....

19. $\begin{cases} \dfrac{1}{2x} - 1 = \dfrac{2}{y} \\ \dfrac{5}{x} + \dfrac{1}{y} = 3 \end{cases}$ ....

20. The sum of the reciprocals of two numbers is $2\frac{1}{4}$; the difference of the reciprocals is $\frac{3}{4}$. What are the two numbers?

....

Simplify:

21. $\left(\dfrac{a^{-2}}{b^{-3}}\right)^3 =$ ....

22. $\dfrac{x}{y^{-1}} + \dfrac{y}{x^{-1}} - xy =$ ....

23. $(x^{-1} + 2y^{-1})^2 =$ ....

24. $\left(-\dfrac{27}{125}\right)^{-2/3} =$ ....

25. $\left(\dfrac{x^{2/3}y^{1/2}}{3x^{3/4}}\right)^0 =$ ....

26. $\dfrac{x^{(2a-3)/c}}{x^{(a-4)/c}} =$ ....

27. $\sqrt[4]{\sqrt[3]{\sqrt{a}}} =$ ....

28. $\dfrac{6}{\sqrt[3]{2}} =$ ....

29. $\dfrac{2x}{\sqrt[3]{x^2}} =$ ....

30. $\dfrac{\sqrt{2}}{1-\sqrt{2}} =$ ....

31. $5\sqrt{8x} + 2\sqrt{72x} + 3\sqrt{18x} =$ ....

32. $(\sqrt{a} - b)^2 =$ ....

8

QUADRATIC EQUATIONS

1. DEFINITIONS

An equation in a single variable in which the highest power of the variable is two is called a *quadratic*, or *second-degree*, equation in a single variable. Examples of such equations are:

(a) $x^2 = 9$

(b) $x^2 + 1 = 0$

(c) $x^2 + 3x = 0$

(d) $x^2 = 2x$

(e) $x^2 - 13x + 42 = 0$

Quadratic equations which involve only constant terms and second-degree expressions of the variable are called *pure quadratic* equations. Equations (a) and (b) above are examples.

Whereas a conditional linear (or first-degree) equation may have one solution, a conditional quadratic equation may have two solutions. In general, a conditional equation in a single variable may have as many solutions as the number of its degree, provided the domain of the variable is sufficiently extensive.

2. SOLVING PURE QUADRATIC EQUATIONS

Although a pure quadratic equation may be solved by the more general methods of Sections 4 and 6 of this chapter, a simpler procedure may also be used, as outlined in Program 8.1.

[Note: In solving equations, whether quadratic or higher ordered, we are interested in all roots of numbers, not just principal roots.]

(8.1) *To solve a pure quadratic equation*:

Step 1. Transform the equation so that all terms containing the variable are in one member and all other terms are in the other member, and simplify.

Step 2. Divide both members by the numerical coefficient of the variable term.

Step 3. Reduce the order of the variable by taking the square root of both members; the resulting numbers are the solutions.

Step 4. Verify each solution by substituting it in the given quadratic equation.

EXAMPLES

1. Solve: $4x^2 - 12 = x^2$.

Step 1. $4x^2 - 12 = x^2$

$4x^2 - x^2 = 12$

$3x^2 = 12$

Step 2. $\dfrac{3x^2}{3} = \dfrac{12}{3}$

$x^2 = 4$

Step 3. $x = \pm\sqrt{4} = \pm 2$

(i.e., $+2$ and -2)

Step 4. $x = +2$: $4x^2 - 12 = x^2$

$4(2)^2 - 12 \stackrel{?}{=} (2)^2$

$4(4) - 12 \stackrel{?}{=} 4$

$16 - 12 \stackrel{?}{=} 4$

$4 = 4$

$x = -2$: $4x^2 - 12 = x^2$

$4(-2)^2 - 12 \stackrel{?}{=} (-2)^2$

$4(4) - 12 \stackrel{?}{=} 4$

$16 - 12 \stackrel{?}{=} 4$

$4 = 4$

Solution set for $4x^2 - 12 = x^2$: $\{+2, -2\}$

2. Solve: $12x^2 - 5 = 3x^2 + 2$.

Step 1. $12x^2 - 5 = 3x^2 + 2$

$12x^2 - 3x^2 = 2 + 5$

$9x^2 = 7$

Step 2. $\dfrac{9x^2}{9} = \dfrac{7}{9}$

$x^2 = \frac{7}{9}$

Step 3. $x = \pm\sqrt{\frac{7}{9}} = \pm\dfrac{\sqrt{7}}{3}$ (or $\pm\frac{1}{3}\sqrt{7}$)

Step 4. $x = +\dfrac{\sqrt{7}}{3}$: $12x^2 - 5 = 3x^2 + 2$

$$12\left(\frac{\sqrt{7}}{3}\right)^2 - 5 \stackrel{?}{=} 3\left(\frac{\sqrt{7}}{3}\right)^2 + 2$$

$$12(\tfrac{7}{9}) - 5 \stackrel{?}{=} 3(\tfrac{7}{9}) + 2$$
$$\tfrac{28}{3} - 5 \stackrel{?}{=} \tfrac{7}{3} + 2$$
$$9\tfrac{1}{3} - 5 \stackrel{?}{=} 2\tfrac{1}{3} + 2$$
$$4\tfrac{1}{3} = 4\tfrac{1}{3}$$

$x = -\dfrac{\sqrt{7}}{3}$: $12x^2 - 5 = 3x^2 + 2$

$$12\left(-\frac{\sqrt{7}}{3}\right)^2 - 5 \stackrel{?}{=} 3\left(-\frac{\sqrt{7}}{3}\right)^2 + 2$$

$$12(+\tfrac{7}{9}) - 5 \stackrel{?}{=} 3(+\tfrac{7}{9}) + 2$$
$$\tfrac{28}{3} - 5 \stackrel{?}{=} \tfrac{7}{3} + 2$$
$$9\tfrac{1}{3} - 5 \stackrel{?}{=} 2\tfrac{1}{3} + 2$$
$$4\tfrac{1}{3} = 4\tfrac{1}{3}$$

Solutions: $\{\frac{1}{3}\sqrt{7}, -\frac{1}{3}\sqrt{7}\}$

3. Solve: $7x^2 = 18 + 3x^2$.

Step 1. $7x^2 = 18 + 3x^2$
$7x^2 - 3x^2 = 18$
$4x^2 = 18$

Step 2. $\dfrac{4x^2}{4} = \dfrac{18}{4}$

$x^2 = \frac{9}{2}$

Step 3. $x = \pm\sqrt{\frac{9}{2}}$

$$= \pm\frac{3}{\sqrt{2}} = \left(\pm\frac{3\sqrt{2}}{\sqrt{2}\cdot\sqrt{2}}\right) = \pm\frac{3\sqrt{2}}{2}$$

Step 4. $x = \pm\dfrac{3\sqrt{2}}{2}$: $7x^2 = 18 + 3x^2$

$$7\left(\pm\frac{3\sqrt{2}}{2}\right)^2 \stackrel{?}{=} 18 + 3\left(\pm\frac{3\sqrt{2}}{2}\right)^2$$

$$7(\tfrac{9}{4}\cdot 2) \stackrel{?}{=} 18 + 3(\tfrac{9}{4}\cdot 2)$$
$$7(\tfrac{9}{2}) \stackrel{?}{=} 18 + 3(\tfrac{9}{2})$$
$$\tfrac{63}{2} \stackrel{?}{=} 18 + \tfrac{27}{2}$$
$$\tfrac{63}{2} \stackrel{?}{=} \tfrac{36}{2} + \tfrac{27}{2}$$
$$\tfrac{63}{2} = \tfrac{63}{2}$$

Solutions: $\{\frac{3}{2}\sqrt{2}, -\frac{3}{2}\sqrt{2}\}$

3. IMAGINARY AND COMPLEX NUMBERS

The real numbers, i.e., the rationals and irrationals, are not sufficient to satisfy some quadratic equations. For example:

$$x^2 + 1 = 0$$
$$x^2 = -1$$

No real number when squared is equal to -1. Consequently, the equation above has no solution among the real numbers. There is a larger set of numbers, however, called the *complex numbers*, that includes the real numbers as a subset and which contains numbers that do satisfy $x^2 + 1 = 0$. Basic to the set of complex numbers is a number called the *imaginary unit*,* symbolized by the letter i and having the property $i^2 = -1$. Thus:

$$\sqrt{-1} = \sqrt{i^2} = i$$

Included in the set of complex numbers are all the real numbers and all those numbers that are the even roots of negative numbers. (Recall our bypassing the latter in Section 3 of Chapter 7.) For example:

$$\sqrt{-1} = i$$
$$\sqrt{-3} = \sqrt{(-1)(3)} = \sqrt{-1}\cdot\sqrt{3} = i\cdot\sqrt{3} \text{ or } \sqrt{3}i$$
$$\sqrt{-4} = \sqrt{(-1)(4)} = \sqrt{4}\cdot\sqrt{-1} = 2i$$
$$\sqrt{-12} = \sqrt{(-1)(4)(3)} = \sqrt{4}\cdot\sqrt{3}\cdot\sqrt{-1} = 2\sqrt{3}i$$

Numbers such as those represented above, of the type bi where b denotes a real number and i the imaginary unit, are called *pure imaginary* numbers. They are part of a larger set called the *imaginary* numbers, which are the sums and differences of real and pure imaginary numbers. For example:

$2 + 3i$, $5 - i$, $6 + \sqrt{2}i$ or $6 + i\sqrt{2}$,
$\sqrt{7} - 4i$, $\sqrt{3} - i\sqrt{13}$

Every complex number may be represented by a numeral of the form $a + bi$ in which a and b denote real numbers and i the imaginary unit. Thus, when $b = 0$, we have a real number; when $a = 0$, we have a pure imaginary number. For example:

Real ($b = 0$):	$3 + 0i = 3$
Imaginary ($b \neq 0$):	$5 + 2i$
Pure imaginary ($a = 0, b \neq 0$):	$0 + 3i = 3i$

* At one time, the square root of a negative number was inconceivable to mathematicians and was passed off as being "imaginary." Unfortunately, the term has endured.

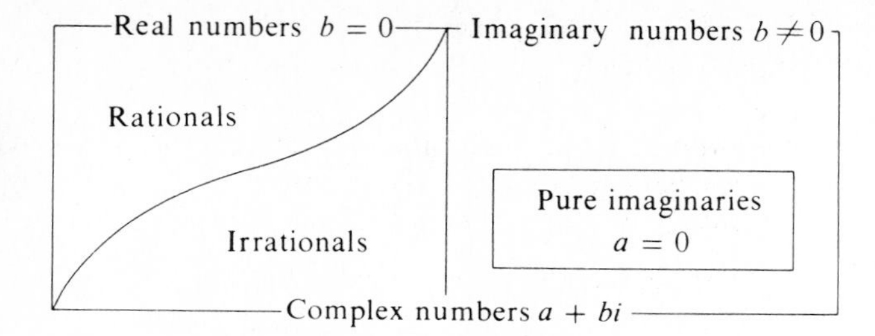

Fig. 8.1

Figure 8.1 illustrates the major subsets of the set of complex numbers.

There is an important cyclic property about the imaginary unit, i, worth noting:

$$i = \sqrt{-1}$$

$$i^2 = i \cdot i = (\sqrt{-1})(\sqrt{-1}) = -1$$

$$(\text{not } \sqrt{(-1)(-1)} = \sqrt{+1} = +1)$$

$$i^3 = i^2 \cdot i = (-1)(\sqrt{-1}) = -\sqrt{-1} \quad (= -i)$$

$$i^4 = i^2 i^2 = (-1)(-1) = +1$$

$$i^5 = i^4 i^1 = (+1)(\sqrt{-1}) = \sqrt{-1} = i$$

$$i^6 = i^4 i^2 = (+1)(-1) = -1$$

etc.

In fact:

$$i = i^5 = i^9 = i^{13} \ldots$$

$$i^2 = i^6 = i^{10} = i^{14} \ldots$$

$$i^3 = i^7 = i^{11} \ldots$$

$$i^4 = i^8 = i^{12} \ldots$$

In order to avoid certain contradictions, it is best to use the i-notation rather than the radical notation in operating with imaginary numbers. Otherwise a result such as this is possible:

$$\sqrt{-4} \cdot \sqrt{-9} = \sqrt{(-4)(-9)}$$

$$= \sqrt{+36} = 6 \text{ (invalid)}$$

By expressing the terms using i, we obtain a product that is consistent with our definitions:

$$\left.\begin{array}{l}\sqrt{-4} = 2i \\ \sqrt{-9} = 3i\end{array}\right\} 2i \cdot 3i = 6i^2$$

$$= 6(-1) = -6$$

Now that we have available this broader set of numbers, unless otherwise specified, the domain of the variable for all equations hereafter is assumed to be the set of complex numbers.

EXAMPLES

1. Solve for x: $3x^2 + 31 = 4$.
   By Program 8.1:
   Step 1. $3x^2 + 31 = 4$
   $$3x^2 = 4 - 31 = -27$$
   Step 2. $\dfrac{3x^2}{3} = -\dfrac{27}{3}$
   $$x^2 = -9$$
   Step 3. $x = \pm\sqrt{-9} = \pm 3i$
   Step 4. $x = \pm 3i$: $3x^2 + 31 = 4$
   $$3(\pm 3i)^2 + 31 \stackrel{?}{=} 4$$
   $$3(9i^2) + 31 \stackrel{?}{=} 4$$
   $$3(9)(-1) + 31 \stackrel{?}{=} 4$$
   $$-27 + 31 \stackrel{?}{=} 4$$
   $$4 = 4$$
   Solution set for $3x^2 + 31 = 4$: $\{3i, -3i\}$
2. Solve for x: $2x^2 + 7 = 2$
   Step 1. $2x^2 + 7 = 2$
   $$2x^2 = 2 - 7 = -5$$
   Step 2. $\dfrac{2x^2}{2} = -\dfrac{5}{2}$
   $$x^2 = -\tfrac{5}{2}$$
   Step 3. $x = \pm\sqrt{\dfrac{-5}{2}} = \pm\dfrac{i\sqrt{5}}{\sqrt{2}}$
   $$= \pm\frac{i\sqrt{10}}{2}$$
   Step 4. $x = \pm\dfrac{i\sqrt{10}}{2}$: $2x^2 + 7 = 2$
   $$2\left(\pm\frac{i\sqrt{10}}{2}\right)^2 + 7 \stackrel{?}{=} 2$$
   $$2\left(\frac{10i^2}{4}\right) + 7 \stackrel{?}{=} 2$$
   $$2\left(\frac{-10}{4}\right) + 7 \stackrel{?}{=} 2$$
   $$-5 + 7 \stackrel{?}{=} 2$$
   $$2 = 2$$
   Solution set: $\{\tfrac{1}{2}i\sqrt{10}, -\tfrac{1}{2}i\sqrt{10}\}$

(Complete Exercise 8.1)

4. SOLVING QUADRATIC EQUATIONS BY FACTORING

Underlying the solution-by-factoring procedure is the generalization that if the product of two factors is zero, then one or both of the factors must be zero.

Thus, if $M \times N = 0$, then either $N = 0$, or $M = 0$, or both $M = 0$ and $N = 0$. Consequently, if we can transform a quadratic equation into a product of factors equal to zero, we may set each factor equal to zero and solve for the variable in each of these sub-equations. Any solution or variable replacement that makes one factor of the given equation equal to zero will invariably make the product of factors equal to zero, and thereby satisfy the given equation.

(8.2) *To solve a quadratic equation by factoring:*
Step 1. Transform the equation by transposing terms to make one member zero.
Step 2. Factor the member containing the variable.
Step 3. Set each factor of Step 2 equal to zero.
Step 4. Solve the equations of Step 3 separately.
Step 5. Verify each solution by substituting it in the given quadratic equation.

EXAMPLES

1. Solve: $x^2 = 5x - 6$.

Step 1. $x^2 = 5x - 6$
$x^2 - 5x + 6 = 0$
Step 2. $(x - 3)(x - 2) = 0$
Step 3. $(x - 3) = 0 \qquad (x - 2) = 0$
Step 4. $x - 3 = 0 \qquad x - 2 = 0$
$x = 3 \qquad x = 2$
Step 5. $x = 3$: $x^2 = 5x - 6$
$(3)^2 \stackrel{?}{=} 5(3) - 6$
$9 \stackrel{?}{=} 15 - 6$
$9 = 9$

$x = 2$: $x^2 = 5x - 6$
$(2)^2 \stackrel{?}{=} 5(2) - 6$
$4 \stackrel{?}{=} 10 - 6$
$4 = 4$

Solution set for $x^2 = 5x - 6$: $\{3, 2\}$

2. Solve: $x^2 + x = 20$.

Step 1. $x^2 + x = 20$
$x^2 + x - 20 = 0$
Step 2. $(x - 4)(x + 5) = 0$
Step 3. $(x - 4) = 0 \qquad (x + 5) = 0$
Step 4. $x - 4 = 0 \qquad x + 5 = 0$
$x = 4 \qquad x = -5$
Step 5. $x = 4$: $x^2 + x = 20$
$(4)^2 + (4) \stackrel{?}{=} 20$
$16 + (4) \stackrel{?}{=} 20$
$20 = 20$

$x = -5$: $x^2 + x = 20$
$(-5)^2 + (-5) \stackrel{?}{=} 20$
$25 - 5 \stackrel{?}{=} 20$
$20 = 20$

Solution set: $\{4, -5\}$

3. Solve: $5x^2 = 3x$.

Step 1. $5x^2 = 3x$
$5x^2 - 3x = 0$
Step 2. $(x)(5x - 3) = 0$
Step 3. $(x) = 0 \qquad (5x - 3) = 0$
Step 4. $x = 0 \qquad 5x - 3 = 0$
$5x = 3$
$x = \frac{3}{5}$
Step 5. $x = 0$: $5x^2 = 3x$
$5(0)^2 \stackrel{?}{=} 3(0)$
$0 = 0$

$x = \frac{3}{5}$: $5x^2 = 3x$
$5(\frac{3}{5})^2 \stackrel{?}{=} 3(\frac{3}{5})$
$5(\frac{9}{25}) \stackrel{?}{=} 3(\frac{3}{5})$
$\frac{9}{5} = \frac{9}{5}$

Solution set: $\{0, \frac{3}{5}\}$

4. Solve: $x^2 = 16$ (pure quadratic).

Step 1. $x^2 = 16$
$x^2 - 16 = 0$
Step 2. $(x - 4)(x + 4) = 0$
Step 3. $(x - 4) = 0 \qquad (x + 4) = 0$
Step 4. $x - 4 = 0 \qquad x + 4 = 0$
$x = 4 \qquad x = -4$
Step 5. $x = 4$: $x^2 - 16 = 0$
$(4)^2 - 16 \stackrel{?}{=} 0$
$16 - 16 \stackrel{?}{=} 0$
$0 = 0$

$x = -4$: $x^2 - 16 = 0$
$(-4)^2 - 16 \stackrel{?}{=} 0$
$16 - 16 \stackrel{?}{=} 0$
$0 = 0$

Solution set: $\{4, -4\}$

5. Solve: $\frac{1}{x + 3} + \frac{1}{x - 2} = \frac{1}{6}$.

$$\frac{1}{x + 3} + \frac{1}{x - 2} = \frac{1}{6}$$

$\text{LCD} = (6)(x + 3)(x - 2)$

$$\frac{(6)(x - 2)}{\text{LCD}} + \frac{(6)(x + 3)}{\text{LCD}} = \frac{(x + 3)(x - 2)}{\text{LCD}}$$

$(6)(x - 2) + (6)(x + 3) = (x + 3)(x - 2)$
$6x - 12 + 6x + 18 = x^2 + x - 6$
$-x^2 + 11x + 12 = 0$
$x^2 - 11x - 12 = 0$
$(x - 12)(x + 1) = 0$
$(x - 12) = 0 \qquad (x + 1) = 0$
$x - 12 = 0 \qquad x + 1 = 0$
$x = 12 \qquad x = -1$

$x = 12$: $\frac{1}{x + 3} + \frac{1}{x - 2} = \frac{1}{6}$

$$\frac{1}{12+3} + \frac{1}{12-2} \stackrel{?}{=} \frac{1}{6}$$

$$\tfrac{1}{15} + \tfrac{1}{10} \stackrel{?}{=} \tfrac{1}{6}$$

$$\tfrac{2}{30} + \tfrac{3}{30} \stackrel{?}{=} \tfrac{5}{30}$$

$$\tfrac{5}{30} = \tfrac{5}{30}$$

$x = -1$: $$\frac{1}{x+3} + \frac{1}{x-2} = \frac{1}{6}$$

$$\frac{1}{-1+3} + \frac{1}{-1-2} \stackrel{?}{=} \frac{1}{6}$$

$$\frac{1}{2} + \frac{1}{-3} \stackrel{?}{=} \frac{1}{6}$$

$$\tfrac{1}{2} - \tfrac{1}{3} \stackrel{?}{=} \tfrac{1}{6}$$

$$\tfrac{3}{6} - \tfrac{2}{6} \stackrel{?}{=} \tfrac{1}{6}$$

$$\tfrac{1}{6} = \tfrac{1}{6}$$

Solution set: $\{12, -1\}$

5. PROBLEMS

Some problems, when translated into algebraic terms, take the form of a quadratic equation. The *problem* may be satisfied either by both solutions of the equation, or by only one, or perhaps none. For instance, a negative solution of an equation may satisfy the equation but be absurd or impossible in terms of the problem situation—in which case that solution is rejected. (See Example 1.)

EXAMPLES

1. The area of a square building lot is 6,400 square feet. Find the length of one of its sides.
   Let x represent the length of the side.
   The area of the lot may then be expressed as $x \cdot x$ or x^2. Hence:
   $x^2 = 6{,}400$
   $x = +80, -80$
   Since the lot could not have a dimension of -80 feet, the negative solution is rejected.
2. The product of a number decreased by 2 and the number increased by 2 is 96. Find the number.
   Let x represent the number. Then,
   $x - 2$ represents the number decreased by 2, and
   $x + 2$ represents the number increased by 2.
   The equation:
   $(x-2)(x+2) = 96$
   $x^2 - 4 = 96$
   $x^2 = 100$
   $x = +10, -10$
   When $x = +10$: $x + 2 = 12$
   $x - 2 = 8$
   $(8) \times (12) = 96$
   When $x = -10$: $x + 2 = -8$
   $x - 2 = -12$
   $(-12) \times (-8) = 96$
   Thus both solutions of the equation satisfy the problem *provided* the domain of the variable includes positive and negative integers. If the circumstances of the problem were such as to limit the domain of the variable to the positive numbers, say, then only the $+10$ solution would be valid, and the -10 solution would be rejected as meaningless in the context of the problem.

(Complete Exercise 8.2)

6. SOLVING QUADRATIC EQUATIONS BY FORMULA

The best means for solving quadratic equations in most cases is the quadratic formula. A technique known as "completing the square" is important to the derivation of that formula.

(8.3) *To complete the square for an expression of the form $x^2 + bx$:*
Step 1. Halve the coefficient (b) of the linear or first-degree term and square it.
Step 2. Add the result of Step 1 to the original expression to form a perfect square trinomial.

EXAMPLES

1. Complete the square for $x^2 + 10x$.
   Step 1. $\frac{1}{2} \times 10 = 5 : 5^2 = 25$
   Step 2. $x^2 + 10x + 25 = (x+5)^2$
2. Complete the square for $x^2 - 6x$.
   Step 1. $\frac{1}{2} \times (-6) = -3 : (-3)^2 = 9$
   Step 2. $x^2 - 6x + 9 = (x-3)^2$
3. Complete the square for $x^2 + 3x$.
   Step 1. $\frac{1}{2} \times 3 = \frac{3}{2} : (\frac{3}{2})^2 = \frac{9}{4}$
   Step 2. $x^2 + 3x + \frac{9}{4} = (x + \frac{3}{2})^2$
4. Complete the square for $x^2 - \frac{3}{5}x$.
   Step 1. $\frac{1}{2} \times (-\frac{3}{5}) = -\frac{3}{10} : (-\frac{3}{10})^2 = \frac{9}{100}$
   Step 2. $x^2 - \frac{3}{5}x + \frac{9}{100} = (x - \frac{3}{10})^2$

Any quadratic equation in a single variable may be written in general form as

$$ax^2 + bx + c = 0$$

where a, b, and c denote real numbers and $a \neq 0$. (When $b = 0$, we have a pure quadratic equation: $ax^2 + c = 0$.)

To derive the quadratic formula we first transform the general quadratic equation so that all terms containing the variable are alone in one member. Then we divide through by the coefficient of the x^2 (quadratic) term:

$$ax^2 + bx = -c$$

$$x^2 + \frac{b}{a}x = -\frac{c}{a}$$

Next, we add to both members a term that will "complete the square" of the member containing the variables (Program 8.3), namely $\left(\frac{b}{2a}\right)^2$, and simplify:

$$x^2 + \frac{b}{a}x + \left(\frac{b}{2a}\right)^2 = -\frac{c}{a} + \left(\frac{b}{2a}\right)^2$$

$$x^2 + \frac{b}{a}x + \frac{b^2}{4a^2} = -\frac{4ac}{4a^2} + \frac{b^2}{4a^2}$$

$$\left(x + \frac{b}{2a}\right)^2 = \frac{b^2 - 4ac}{4a^2}$$

Finally, we reduce the order of both members,

$$x + \frac{b}{2a} = \pm\frac{\sqrt{b^2 - 4ac}}{2a}$$

and solve for x:

$$x + \frac{b}{2a} = +\frac{\sqrt{b^2 - 4ac}}{2a} \qquad x + \frac{b}{2a} = -\frac{\sqrt{b^2 - 4ac}}{2a}$$

$$x = \frac{-b + \sqrt{b^2 - 4ac}}{2a} \qquad x = \frac{-b - \sqrt{b^2 - 4ac}}{2a}$$

These two solutions of the general quadratic equation are frequently combined into a single expression.

$$x = \frac{-b \pm \sqrt{b^2 - 4ac}}{2a} \quad (a \neq p)$$

called the *quadratic formula*. Since a, b, and c denote numbers, the right member of the formula will produce two numbers that are the solutions of the quadratic equation.

(8.4) *To solve a quadratic equation by the quadratic formula*:

Step 1. Express the quadratic equation in the form $ax^2 + bx + c = 0$.

Step 2. Identify the numerical coefficients a, b, and c in the equation of Step 1.

Step 3. Substitute the values for a, b, and c of Step 2 in the quadratic formula:

$$x = \frac{-b \pm \sqrt{b^2 - 4ac}}{2a} \quad (a \neq 0)$$

Step 4. Simplify the results of Step 3, the solutions.

Step 5. Verify each solution by substituting it in the given quadratic equation.

EXAMPLES

1. Solve by the quadratic formula: $3x^2 - 10x + 8 = 0$.

Step 1. $\mathbf{a}x^2 + \mathbf{b}x + \mathbf{c} = 0$

$\mathbf{3}x^2 - \mathbf{10}x + \mathbf{8} = 0$

Step 2. $a = +3, b = -10, c = +8$

Step 3. $x = \frac{-b \pm \sqrt{b^2 - 4ac}}{2a}$

$= \frac{-(-10) \pm \sqrt{(-10)^2 - 4(3)(8)}}{2(3)}$

Step 4. $x = \frac{10 \pm \sqrt{100 - 96}}{6}$

$= \frac{10 \pm \sqrt{4}}{6} = \frac{10 \pm 2}{6} = \frac{12}{6}, \frac{8}{6}$

$= 2, \frac{4}{3}$

Step 5. $x = 2$: $3x^2 - 10x + 8 = 0$

$3(2)^2 - 10(2) + 8 \stackrel{?}{=} 0$

$12 - 20 + 8 \stackrel{?}{=} 0$

$0 = 0$

$x = \frac{4}{3}$: $3x^2 - 10x + 8 = 0$

$3(\frac{4}{3})^2 - 10(\frac{4}{3}) + 8 \stackrel{?}{=} 0$

$\frac{16}{3} - \frac{40}{3} + \frac{24}{3} \stackrel{?}{=} 0$

$0 = 0$

Solution set: $\{2, \frac{4}{3}\}$

2. Solve: $2x^2 + x = 1$.

Step 1. $2x^2 + x = 1$, in general form, is

$2x^2 + x - 1 = 0$

Step 2. $a = +2, b = +1, c = -1$

Step 3. $x = \frac{-b \pm \sqrt{b^2 - 4ac}}{2a}$

$= \frac{-1 \pm \sqrt{(1)^2 - 4(2)(-1)}}{2(2)}$

Step 4. $x = \frac{-1 \pm \sqrt{1 + 8}}{4}$

$= \frac{-1 \pm \sqrt{9}}{4} = \frac{-1 \pm 3}{4} = \frac{2}{4}, \frac{-4}{4}$

$= \frac{1}{2}, -1$

Step 5. $x = \frac{1}{2}$: $2x^2 + x = 1$

$2(\frac{1}{2})^2 + (\frac{1}{2}) \stackrel{?}{=} 1$

$2(\frac{1}{4}) + \frac{1}{2} \stackrel{?}{=} 1$

$\frac{1}{2} + \frac{1}{2} \stackrel{?}{=} 1$

$1 = 1$

$x = -1$: $2x^2 + x = 1$

$2(-1)^2 + (-1) \stackrel{?}{=} 1$

$2 - 1 \stackrel{?}{=} 1$

$1 = 1$

Solution set: $\{\frac{1}{2}, -1\}$

3. Solve: $x^2 - 3 = -4x$.
Step 1. $x^2 + 4x - 3 = 0$
Step 2. $a = +1, b = +4, c = -3$

Step 3. $x = \dfrac{-b \pm \sqrt{b^2 - 4ac}}{2a}$

$= \dfrac{-4 \pm \sqrt{16 - 4(1)(-3)}}{2(1)}$

Step 4. $x = \dfrac{-4 \pm \sqrt{28}}{2}$

$= \dfrac{-4 \pm 2\sqrt{7}}{2}$

$= -2 \pm \sqrt{7}$

Step 5. $x = -2 + \sqrt{7}$:

$$x^2 - 3 = -4x$$
$$(-2 + \sqrt{7})^2 - 3 \stackrel{?}{=} -4(-2 + \sqrt{7})$$
$$4 - 4\sqrt{7} + 7 - 3 \stackrel{?}{=} 8 - 4\sqrt{7}$$
$$8 - 4\sqrt{7} = 8 - 4\sqrt{7}$$

$x = -2 - \sqrt{7}$:

$$x^2 - 3 = -4x$$
$$(-2 - \sqrt{7})^2 - 3 \stackrel{?}{=} -4(-2 - \sqrt{7})$$
$$4 + 4\sqrt{7} + 7 - 3 \stackrel{?}{=} 8 + 4\sqrt{7}$$
$$8 + 4\sqrt{7} = 8 + 4\sqrt{7}$$

Solution set: $\{-2 + \sqrt{7}, -2 - \sqrt{7}\}$

4. Solve: $x^2 + 3 = 2x$.
Step 1. $x^2 + 3 = 2x$
$x^2 - 2x + 3 = 0$
Step 2. $a = +1, b = -2, c = +3$

Step 3. $x = \dfrac{-b \pm \sqrt{b^2 - 4ac}}{2a}$

$= \dfrac{-(-2) \pm \sqrt{(-2)^2 - 4(1)(3)}}{2(1)}$

$= \dfrac{2 \pm \sqrt{4 - 12}}{2}$

Step 4. $x = \dfrac{2 \pm \sqrt{-8}}{2} = \dfrac{2 \pm 2i\sqrt{2}}{2}$

$= 1 \pm i\sqrt{2}$

Step 5. $x = 1 \pm i\sqrt{2}$

$$x^2 + 3 = 2x$$
$$(1 \pm i\sqrt{2})^2 + 3 \stackrel{?}{=} 2(1 \pm i\sqrt{2})$$
$$(1 \pm 2i\sqrt{2} + 2i^2) + 3 \stackrel{?}{=} 2 \pm 2i\sqrt{2}$$
$$1 \pm 2i\sqrt{2} - 2 + 3 \stackrel{?}{=} 2 \pm 2i\sqrt{2}$$
$$2 \pm 2i\sqrt{2} = 2 \pm 2i\sqrt{2}$$

Solution set: $\{1 + i\sqrt{2}, 1 - i\sqrt{2}\}$

5. Solve: $4x^2 - 12x + 9 = 0$.
Step 1. $4x^2 - 12x + 9 = 0$
Step 2. $a = +4, b = -12, c = +9$

Step 3. $x = \dfrac{-b \pm \sqrt{b^2 - 4ac}}{2a}$

$= \dfrac{-(-12) \pm \sqrt{(-12)^2 - 4(4)(9)}}{2(4)}$

$= \dfrac{12 \pm \sqrt{144 - 144}}{8}$

Step 4. $x = \dfrac{12 \pm 0}{8} = \dfrac{12}{8}, \dfrac{12}{8}$

$= \dfrac{3}{2}, \dfrac{3}{2}$

Step 5. $x = \frac{3}{2}$: $4x^2 - 12x + 9 = 0$

$$4(\tfrac{3}{2})^2 - 12(\tfrac{3}{2}) + 9 \stackrel{?}{=} 0$$
$$4(\tfrac{9}{4}) - 12(\tfrac{3}{2}) + 9 \stackrel{?}{=} 0$$
$$9 - 18 + 9 \stackrel{?}{=} 0$$
$$0 = 0$$

Solution set: $\{\frac{3}{2}\}$

6. Solve the pure quadratic equation, $x^2 - 9 = 0$, by the quadratic formula.
Step 1. $x^2 - 9 = 0$
Step 2. $a = +1, b = 0, c = -9$

Step 3. $x = \dfrac{-b \pm \sqrt{b^2 - 4ac}}{2a}$

$= \dfrac{0 \pm \sqrt{(0)^2 - 4(1)(-9)}}{2(1)}$

$= \dfrac{0 \pm \sqrt{0 + 36}}{2}$

Step 4. $x = \dfrac{\pm\sqrt{36}}{2} = \dfrac{6}{2}, -\dfrac{6}{2}$

$= 3, -3$ (or ± 3)

Step 5. $x = \pm 3$: $x^2 - 9 = 0$

$$(\pm 3)^2 - 9 \stackrel{?}{=} 0$$
$$9 - 9 \stackrel{?}{=} 0$$
$$0 = 0$$

Solution set: $\{3, -3\}$

(Complete Exercise 8.3)

7. RELATIONS

In Chapter 5, Sections 8 and 9, we dealt with equations in two variables, e.g., $x + y = 5$. Here we return to them, and expand upon their interpretation. Recall that the equation

$$x + y = 5$$

has not one solution, but a set of solutions, all ordered pairs. Among them are (5, 0), (2, 3), (1, 4), (0, 5), (−1, 6), etc.

In mathematics, any set of ordered pairs is called a *relation.* The set of ordered pairs that is the solution set for the equation, $x + y = 5$, is a relation. So is a set of ordered pairs in which the first element of each pair is the time of day, and the second is the temperature at that time; e.g., {(4 am, 32°), (5 am, 34°), (6 am, 38°), . . . , (12 noon, 44°)}. Similarly, the set of whole numbers and their respective cubes, when expressed as ordered pairs, forms a relation: $\{(0, 0), (1, 1), (2, 8), (3, 27), (4, 64), \ldots\}$.

The set of all first components in a relation is called the *domain of the relation.* The set of all second components is called the *range of the relation.*

EXAMPLES

1. Relation: $\{(1, 1), (2, 4), (3, 9), (4, 16)\}$
   Domain of the relation: $\{1, 2, 3, 4\}$
   Range of the relation: $\{1, 4, 9, 16\}$
2. Relation: $\{(1, 2), (2, 2), (3, 2), (4, 2)\}$
   Domain of the relation: $\{1, 2, 3, 4\}$
   Range of the relation: $\{2\}$
3. Relation: {(1, Thurs.), (2, Fri.), . . . , (31, Sat.)}
   Domain of the relation: $\{1, 2, 3, \ldots, 31\}$
   Range of the relation: {Sun., Mon., . . . , Sat.}

An equation in two variables may be thought to define a relation. Thus, the equation above, $x + y = 5$, defines a relation, which includes $\{(5, 0), (2, 3), (1, 4), (-1, 6), \ldots\}$. When an equation in two variables defines a relation, the variable that represents the elements in the domain (first components) of the relation is called the *independent variable.* The variable that represents the elements in the range is called the *dependent variable.*

EXAMPLES

1. $y = 2x$: $\{(1, 2), (2, 4), (3, 6), (4, 8), \ldots\}$
   Independent variable: x
   Dependent variable: y
   [Note: An alternative way to express the above is by *set-builder notation*: $\{(x, y)|y = 2x\}$. That expression is read: "the set of ordered pairs, x, y, such that $y = 2x$."
2. $\{(m, n)|m = n^2\}$ ("The set of ordered pairs, m, n, such that $m = n^2$.")
   Independent variable: n
   Dependent variable: m
   Some ordered pairs in the relation:

   $(0, 0), (\frac{1}{2}, \frac{1}{4}), (-1, 1), (6, 36)$

The following program outlines the procedure for generating the ordered pairs of a relation specified by an equation in two variables.

(8.5) *To determine the set of ordered pairs defined by an equation in two variables:*
Step 1. *Solve the equation for the dependent variable.*
Step 2. *Substitute for the independent variable in the equation each element of the domain, and compute the corresponding value of the dependent variable occurring in the range.*
Step 3. *The set of ordered pairs made up of numbers of Step 2 is the relation defined by the equation.*

EXAMPLES

1. Determine five ordered pairs of the relation defined by $2x + y = 4$, x and y being integers only. (That is, $\{(x, y)|2x + y = 4, x, y \text{ integers}\}$.)
   Step 1. Solve the equation, $2x + y = 4$, for the dependent variable, y:

   $2x + y = 4$

   $y = 4 - 2x$

   Step 2. Substitute five elements from the domain (set of integers) for the independent variable (x) and compute the corresponding values for the dependent variable (y), which must be members of the range (set of integers also).

x	$y = 4 - 2x$	y
-1	$y = 4 - 2(-1)$	6
0	$y = 4 - 2(0)$	4
1	$y = 4 - 2(1)$	2
2	$y = 4 - 2(2)$	0
3	$y = 4 - 2(3)$	-2

   Step 3. The ordered pairs, $(-1, 6), (0, 4), (1, 2), (2, 0), (3, -2)$, are members of the relation $\{(x, y)|2x + y = 4, x, y \text{ integers}\}$.
2. Identify the relation defined by $2x - y = 3$, x an integer between 2 and 7. (Or, $\{(x, y)|2x - y = 3, x \text{ an integer and } 2 < x < 7\}$.)
   Step 1. Solve $2x - y = 3$ for y:

   $y = 2x - 3$

   Step 2.

x	3	4	5	6
y	3	5	7	9

   Step 3. $\{(x, y)|2x - y = 3, x \text{ an integer and } 2 < x < 7\} = \{(3, 3), (4, 5), (5, 7), (6, 9)\}$.
3. Graph the relation $\{(x, y)|x + y = 3, x \text{ an integer and } -3 < x \leq 2\}$.

Step 1. $x + y = 3$

$y = 3 - x$

Step 2.

x	-2	-1	0	1	2
y	5	4	3	2	1

Step 3.

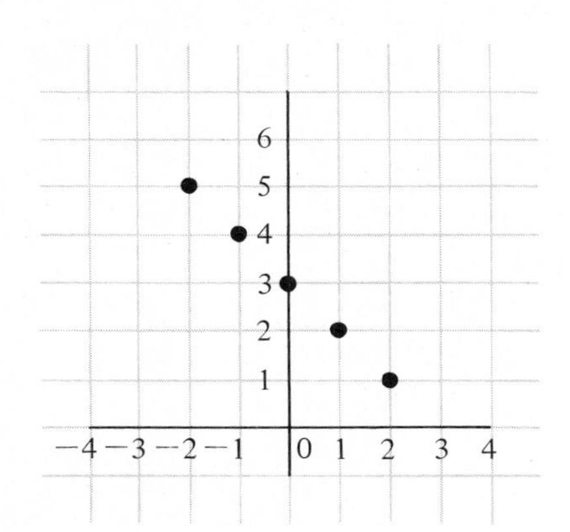

Fig. 8.2

4. Graph the relation $\{(x, y)|x + y = 3\}$.
The equation that defines the relation between independent and dependent variable in this example is the same as that of Example 3. However, the absence of any statement restricting the variables in $\{(x, y)|x + y = 3\}$ implies that both the domain and the range of the variables is the full set of real numbers. Consequently there are an infinite number of ordered pairs in the relation given in this example (partially shown in Fig. 8.3), whereas there are only five ordered pairs in the relation of Example 3 (Fig. 8.2).

8. FUNCTIONS

If no two ordered pairs of a relation have associated with the same first component a different second component, then the relation is called a *function*. Functions are a special kind of relation, and are very important in mathematics. For example,

$$R = \{(0, 1), (1, 2), (2, 3), (3, 4), (4, 5)\}$$

is both a relation and a function because each element in the domain (set of first components) has associated with it one and only one element in the range (set of second components). On the other hand,

$$S = \{(1, 2), (1, 3), (2, 4), (3, 4), (4, 4)\}$$

is a relation but not a function because at least one element in the domain, 1 in this case, has associated with it two or more different elements in the range: 2 and 3.

Notice that

$$T = \{(1, 2), (2, 2), (3, 2), (4, 3), (5, 4)\}$$

is both a relation and a function. Each first component has one and only one element in the range associated with it, even though some elements in the range are associated with more than one element in the domain.

Deciding whether or not a relation is a function can be simplified by studying the graph of the relation. A relation is a function if each abscissa (first component) in the graph has one and only one ordinate (second component). If any abscissa in the graph of a relation has more than one ordinate, the relation is not a function.

The graphs of relations R, S, and T are shown in Figure 8.4. Notice in the cases of R and T, no two pairs of coordinates have the same absicissa and different

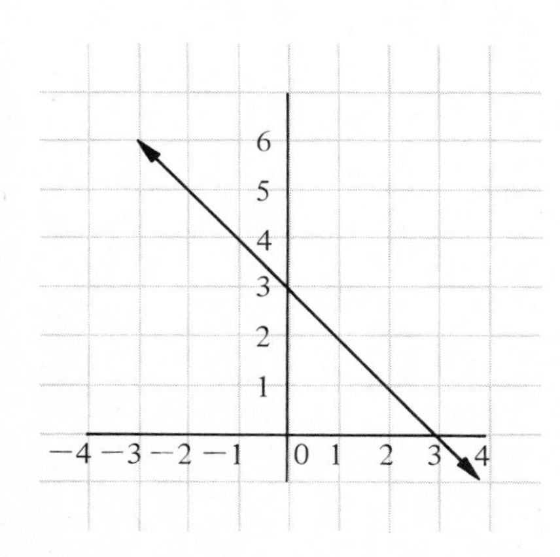

Fig. 8.3

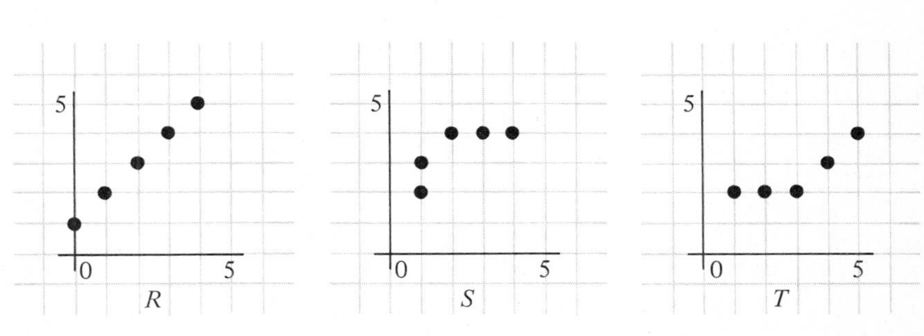

Fig. 8.4

ordinates. Those two relations are functions. In the graph of S, two pairs of coordinates do have the same abscissa and different ordinates: (1, 2) and (1, 3) Therefore, S is a relation but not a function.

In analyzing the graph of a relation to determine whether or not it represents a function, try to visualize a vertical line moving across the graph from left to right. If in its sweep the visualized line never crosses the graph of the relation at more than one point at a time, the relation is a function. On the other hand, if at any place the visualized line crosses the graph of the relation at more than a single point, the relation is not a function.

Consider this test with the relations shown in Figure 8.4. At no place will the visualized vertical line cross more than one point at a time in the graphs of R and T, which are functions. It will, however, cross two points at the same time in the graph of S, at (1, 2) and (1, 3); so S is not a function.

EXAMPLES

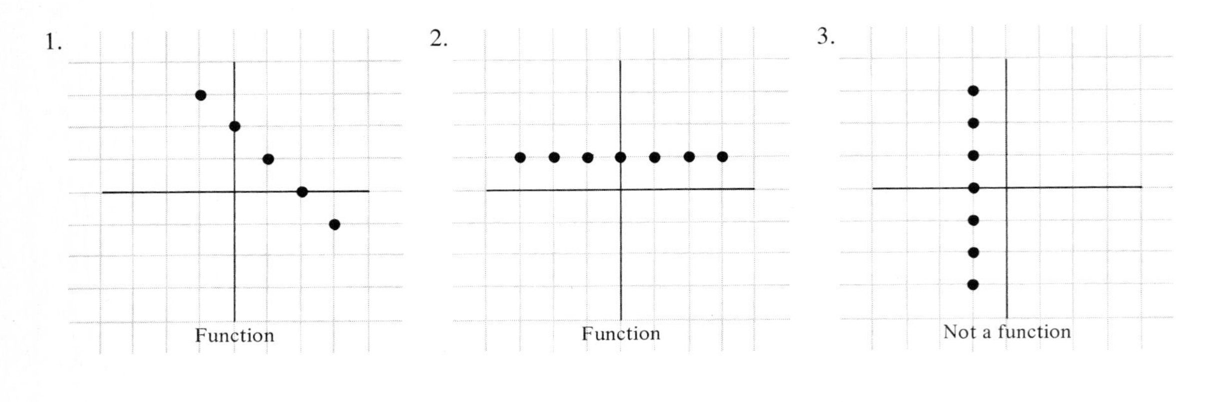

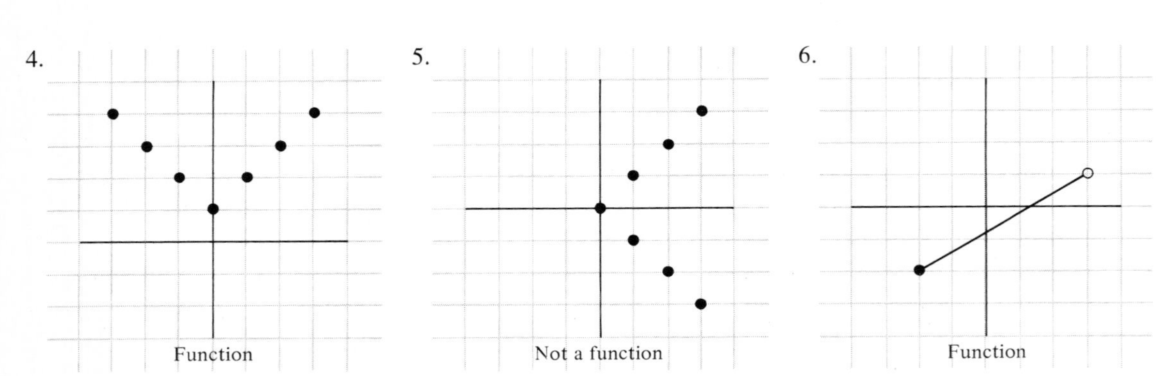

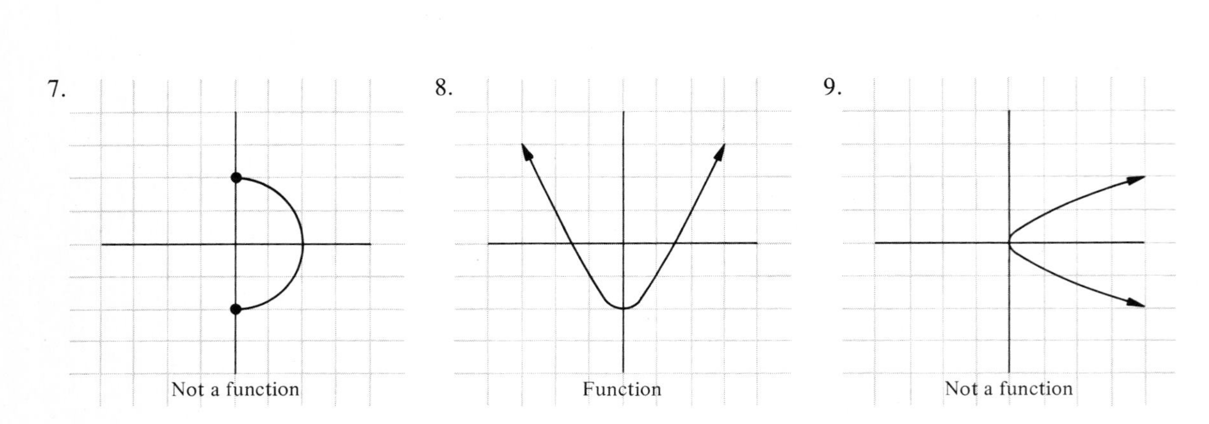

[Note: The open dot, ○, at one end of the graph in Example 6 above means that (3, 1) is not included in the function; the solid dot at the other end, •, means that (−2, −2) is included in the function.]

When we use an equation such as $y = 2x - 1$ to define a function, we say "y is a function of x" (which suggests that "y *depends* upon x"), and may use the notation

$$y = f(x) \quad \text{and} \quad f(x) = 2x - 1$$

The symbol "$f(x)$" is read "f of x," a shortened version of *function of x*.

By extension,

$$\text{if } f(x) = 2x - 1$$

$$\text{then } f(2) = 2(2) - 1 = 3$$

$$f(0) = 2(0) - 1 = -1$$

$$f(-1) = 2(-1) - 1 = -3$$

and so on.

EXAMPLES

1. Given $f(x) = 3x - 4$, find $f(2)$.

$$f(x) = 3x - 4$$

$$f(2) = 3(2) - 4 = 6 - 4 = 2$$

2. Given $f(x) = 2x^2 - x + 5$, find $f(0)$ and $f(3)$.

$$f(x) = 2x^2 - x + 5$$

$$f(0) = 2(0)^2 - (0) + 5 = 0 - 0 + 5 = 5$$

$$f(3) = 2(3)^2 - (3) + 5 = 18 - 3 + 5 = 20$$

3. If $f(x) = x^3 - 2x + 4$, which is greater, $f(0)$ or $f(-1)$?

$$f(x) = x^3 - 2x + 4$$

$$f(0) = (0)^3 - 2(0) + 4 = 4$$

$$f(-1) = (-1)^3 - 2(-1) + 4 = -1 + 2 + 4 = 5$$

Thus, $f(-1) > f(0)$.

(Complete Exercise 8.4)

9. SOLVING RADICAL EQUATIONS

When the variable of an equation appears beneath a radical sign, the equation is called a *radical equation*. When the radical equation contains only second-degree radicals, or square roots, the equation may be solved by appropriately squaring the respective members of the equation, since "the squares of equals are equal." However, when both members are squared there is the risk of introducing *extraneous* (false) *solutions*. Therefore each solution must be verified by substituting it in the given equation.

(8.6) *To solve a radical equation containing square roots:*

Step 1. *Transform the equation so that a single square root radical occurs in one member.*

Step 2. *Square both members.*

Step 3. *Repeat Steps 1 and 2 as many times as necessary to remove all radical signs from the resulting equation.*

Step 4. *Solve the equation of Step 3.*

Step 5. *Verify each solution by substituting it in the given radical equation.*

EXAMPLES

1. Solve: $x - 2 - \sqrt{x + 4} = 0$.

Step 1. $x - 2 - \sqrt{x + 4} = 0$

$$x - 2 = \sqrt{x + 4}$$

Step 2. $(x - 2)^2 = (\sqrt{x + 4})^2$

$$x^2 - 4x + 4 = x + 4$$

Step 4. $x^2 - 4x + 4 - x - 4 = 0$

$$x^2 - 5x = 0$$

$$(x)(x - 5) = 0$$

$$(x) = 0 \qquad (x - 5) = 0$$

$$x = 0 \qquad x - 5 = 0$$

$$x = 5$$

Step 5. $x = 0$: $x - 2 - \sqrt{x + 4} = 0$

$$0 - 2 - \sqrt{0 + 4} \stackrel{?}{=} 0$$

$$-2 - \sqrt{4} \stackrel{?}{=} 0$$

$$-2 - 2 \stackrel{?}{=} 0$$

$$-4 \neq 0$$

$x = 5$: $x - 2 - \sqrt{x + 4} = 0$

$$5 - 2 - \sqrt{5 + 4} \stackrel{?}{=} 0$$

$$5 - 2 - \sqrt{9} \stackrel{?}{=} 0$$

$$5 - 2 - 3 \stackrel{?}{=} 0$$

$$0 = 0$$

There is but *one* solution of the equation: 5.

2. Solve: $\sqrt{2x + 3} + \sqrt{4x - 3} = 6$.

Step 1. $\sqrt{2x + 3} + \sqrt{4x - 3} = 6$

$$\sqrt{2x + 3} = 6 - \sqrt{4x - 3}$$

Step 2. $(\sqrt{2x + 3})^2 = (6 - \sqrt{4x - 3})^2$

$$2x + 3 = 36 - 12\sqrt{4x - 3} + (4x - 3)$$

Step 3. $12\sqrt{4x - 3} = 2x + 30$

$$6\sqrt{4x - 3} = x + 15$$

$$(6\sqrt{4x - 3})^2 = (x + 15)^2$$

$$36(4x - 3) = x^2 + 30x + 225$$

Step 4. $144x - 108 = x^2 + 30x + 225$

$$x^2 - 114x + 333 = 0$$
$$(x - 3)(x - 111) = 0$$
$$(x - 3) = 0 \qquad (x - 111) = 0$$
$$x - 3 = 0 \qquad x - 111 = 0$$
$$x = 3 \qquad x = 111$$

Step 5. $x = 3$: $\sqrt{2x + 3} + \sqrt{4x - 3} = 6$

$$\sqrt{2(3) + 3} + \sqrt{4(3) - 3} \stackrel{?}{=} 6$$
$$\sqrt{9} + \sqrt{9} \stackrel{?}{=} 6$$
$$3 + 3 \stackrel{?}{=} 6$$
$$6 = 6$$

$x = 111$: $\sqrt{2x + 3} + \sqrt{4x - 3} = 6$

$$\sqrt{2(111) + 3} + \sqrt{4(111) - 3} \stackrel{?}{=} 6$$
$$\sqrt{225} + \sqrt{441} \stackrel{?}{=} 6$$
$$15 + 21 \stackrel{?}{=} 6$$
$$36 \neq 6$$

Thus, 3 is a solution, but 111 is not. The solution set: $\{3\}$.

3. Solve: $\sqrt{x - 2} - \sqrt{2x + 3} = 2$.

Step 1. $\sqrt{x - 2} - \sqrt{2x + 3} = 2$

$$\sqrt{x - 2} = 2 + \sqrt{2x + 3}$$

Step 2. $(\sqrt{x - 2})^2 = (2 + \sqrt{2x + 3})^2$

$$x - 2 = 4 + 4\sqrt{2x + 3} + (2x + 3)$$

Step 3. $-4\sqrt{2x + 3} = 4 + 2x + 3 - x + 2$

$$-4\sqrt{2x + 3} = x + 9$$
$$(-4\sqrt{2x + 3})^2 = (x + 9)^2$$
$$16(2x + 3) = x^2 + 18x + 81$$

Step 4. $32x + 48 = x^2 + 18x + 81$

$$0 = x^2 - 14x + 33$$
$$0 = (x - 11)(x - 3)$$
$$x - 11 = 0 \qquad x - 3 = 0$$
$$x = 11 \qquad x = 3$$

Step 5. $x = 11$: $\sqrt{x - 2} - \sqrt{2x + 3} = 2$

$$\sqrt{11 - 2} - \sqrt{2(11) + 3} \stackrel{?}{=} 2$$
$$\sqrt{9} - \sqrt{25} \stackrel{?}{=} 2$$
$$3 - 5 \stackrel{?}{=} 2$$
$$-2 \neq 2$$

$x = 3$: $\sqrt{x - 2} - \sqrt{2x + 3} = 2$

$$\sqrt{3 - 2} - \sqrt{2(3) + 3} \stackrel{?}{=} 2$$
$$\sqrt{1} - \sqrt{9} \stackrel{?}{=} 2$$
$$1 - 3 \stackrel{?}{=} 2$$
$$-2 \neq 2$$

Neither 11 nor 3 is a solution of the equation. The solution set of the given equation is the empty set, $\{\ \}$.

10. SOLVING OTHER EQUATIONS IN QUADRATIC FORM

Certain equations of a degree greater than the second can be solved in the manner of quadratic equations, provided they can be reduced to quadratic form.

EXAMPLES

1. Solve the fourth-degree equation:

$$x^4 - 5x^2 + 4 = 0.$$

If we let $y = x^2$, then $y^2 = x^4$ and the given equation in x can be expressed as an equivalent equation in y:

$$y^2 - 5y + 4 = 0$$

This equation in y is obviously quadratic; it can be solved by factoring (Program 8.2):

$$y^2 - 5y + 4 = 0$$
$$(y - 4)(y - 1) = 0$$
$$y - 4 = 0 \qquad y - 1 = 0$$
$$y = 4 \qquad y = 1$$

But $y = x^2$; therefore:

$$y = x^2 = 4 \qquad y = x^2 = 1$$
$$x = \pm 2 \qquad x = \pm 1$$

The solution set for the given equation, $x^4 - 5x^2 + 4 = 0$, is $\{+2, -2, +1, -1\}$. These solutions can be verified by substituting each into the given fourth-degree equation.

2. Solve the sixth-degree equation:

$$x^6 + 9x^3 + 8 = 0.$$

If we let $y = x^3$, then $y^2 = x^6$, and:

$$x^6 + 9x^3 + 8 = y^2 + 9y + 8 = 0$$

Solving for y:

$$(y + 8)(y + 1) = 0$$
$$y + 8 = 0 \qquad y + 1 = 0$$
$$y = -8 \qquad y = -1$$

Replacing y by its equivalent x^3:

$$y = x^3 = -8 \qquad y = x^3 = -1$$
$$x = -2 \qquad x = -1$$

Because the equation of Example 2 is of the sixth degree, there may exist six solutions. In this case two are real numbers and the other four are imaginary numbers. The solutions determined above, -2, -1, are the two real solutions. All six solutions can be determined by substituting x^3 for y at the step,

$$(y + 8)(y + 1) = (x^3 + 8)(x^3 + 1) = 0$$

and noting that

$$(x^3 + 8) = (x + 2)(x^2 - 2x + 4)$$

and

$$(x^3 + 1) = (x + 1)(x^2 - x + 1)$$

This is to say that

$$x^6 + 9x^3 + 8 = (x + 2)(x + 1)(x^2 - 2x + 4)(x^2 - x + 1) = 0$$

Upon setting each of the factors equal to zero, the two binomial factors will yield the two real solutions and the two trinomial factors will each yield two imaginary solutions apiece, or six solutions in all. The complete solution set is: $\{-2, -1, 1 + i\sqrt{3}, 1 - i\sqrt{3}, \frac{1}{2} + \frac{1}{2}i\sqrt{3}, \frac{1}{2} - \frac{1}{2}i\sqrt{3}\}$.

Equations of an order lower than the second degree are sometimes solvable by quadratic techniques.

EXAMPLE

Solve: $2x = 5\sqrt{x} + 3$.

Set the equation equal to zero:

$$2x - 5\sqrt{x} - 3 = 0$$

Let $y = \sqrt{x}$; then $y^2 = x$. Express the given equation as an equivalent one in y:

$$2x - 5\sqrt{x} - 3 = 2y^2 - 5y - 3 = 0$$

Solve for y:

$$2y^2 - 5y - 3 = 0$$

$$(2y + 1)(y - 3) = 0$$

$$2y + 1 = 0 \qquad y - 3 = 0$$

$$y = -\tfrac{1}{2} \qquad y = 3$$

Replace y by $\sqrt{x}$:

$$y = \sqrt{x} = -\tfrac{1}{2} \qquad y = \sqrt{x} = 3$$

$$x = \tfrac{1}{4} \qquad x = 9$$

Verify the solutions in the given equation:

$$x = \tfrac{1}{4}: \quad 2x = 5\sqrt{x} + 3$$

$$2(\tfrac{1}{4}) \stackrel{?}{=} 5\sqrt{\tfrac{1}{4}} + 3$$

$$\tfrac{1}{2} \stackrel{?}{=} \tfrac{5}{2} + 3$$

$$\tfrac{1}{2} \neq 5\tfrac{1}{2}$$

$$x = 9: \quad 2x = 5\sqrt{x} + 3$$

$$2(9) \stackrel{?}{=} 5\sqrt{9} + 3$$

$$18 \stackrel{?}{=} 15 + 3$$

$$18 = 18$$

Hence 9 is a solution of the equation $2x = 5\sqrt{x} + 3$, but $\frac{1}{4}$ is not.

(Complete Exercise 8.5)

Name.. Date.................... **EXERCISE 8.1**

Solve for x, and verify.

1. $9x^2 = 81$

2. $4x^2 = 100$

3. $3x^2 = 15$

4. $x^2 - 4 = 12$

5. $2x^2 - 9 = x^2$

6. $3x^2 + 8 = 32 - x^2$

7. $6x^2 - 5 = 2x^2 + 20$

8. $3x^2 - 5 = 4x^2 - 5$

9. $ax^2 - b = 0$

10. $\frac{x}{3} + \frac{3}{x} = 4x$

11. $(x - 2)^2 - 16 = -4x$

12. $ax^2 + bx^2 = z$

Show the following to be true.

13. $\sqrt{-9} = 3i$

14. $\sqrt{-50} = 5i\sqrt{2}$ (or $5\sqrt{2}i$)

15. $\sqrt[4]{-32} = 2i\sqrt[4]{2}$

16. $3\sqrt{-12} - 2\sqrt{-3} = 4i\sqrt{3}$

17. $6(2 - \sqrt{-4}) = 12 - 12i$

18. $3i + 2i^2 - 4i^3 = 7i - 2$

Simplify; express in terms of i.

19. $\sqrt{-12}$

20. $(3 + \sqrt{-4})^2$

21. $(1 - \sqrt{-3})^2$

22. $4i + 2i^2 + 3i^3 + i^8$

23. $6 - 2i + 3i^2 + 5i^3 - 7i^6 + 3i^{10}$

Solve, and verify.

24. $x^2 + 3 = 2$

25. $3x^2 + 4 = x^2 - 2$

26. $5x^2 + 16 = x^2$

27. $4x^2 + 49 = 0$

28. $5x + \frac{25}{x} = 0$

29. $\frac{x}{3} - \frac{4}{x} = x$

Name... Date.............................. **EXERCISE 8.2**

Solve by factoring, and verify.

1. $x^2 - 6x + 8 = 0$

2. $x^2 + 2x - 15 = 0$

3. $x^2 + 10x + 24 = 0$

4. $6x^2 - 5x + 1 = 0$

5. $3x^2 + 22x + 35 = 0$

6. $9x^2 + 1 = 6x$

7. $x^2 + 10x + 25 = 0$

8. $x^2 - 2ax - 15a^2 = 0$

9. $25x^2 + 4 = 20x$

10. $3x - 9 + 20x^2 = 0$

11. $4x^2 - 9x = 0$

12. $2x^2 - 8ax = 0$

13. $\dfrac{5}{x+4} = \dfrac{3}{x-2} + 4$

14. $\dfrac{x}{3-x} - \dfrac{x}{x-3} = 0$

Solve.

15. The product of a positive number subtracted from 12 and that number added to 12 is 128. Find the number.

....

16. A tennis court covers 312 sq yd and is 14 yd longer than it is wide. What are the length and width dimensions of the tennis court?

..........., ...

17. The sum of the reciprocals of two consecutive even integers is $\frac{5}{12}$. Find the integers.

....

18. Find the dimensions of a square if increasing its edge by 2 yields an area numerically equal to its new perimeter.

....

19. A contractor estimates that to cover a certain floor with square asphalt tiles it will take 528 standard tiles or 297 larger tiles (which are 3 in. longer and wider than the standard tile). What are the dimensions of the larger tile?

....

20. Within a rectangular garden plot 32 ft by 46 ft there is to be a uniform border around its outer edge for flowers and a rectangular center grass plot of 1,040 sq ft. Find the width of the flower border.

....

Name.. Date.............................. **EXERCISE 8.3**

Solve by means of the quadratic formula; express solutions in simplest radical form.

1. $x^2 - 5x + 6 = 0$

2. $x^2 + 8x + 15 = 0$

3. $x^2 - 6x + 9 = 0$

4. $6x^2 - x - 15 = 0$

5. $6x^2 - 17x + 12 = 0$

6. $3x^2 - 19x = 14$

7. $4x^2 + 6x + 1 = 0$

8. $5x^2 - 12x - 12 = 0$

9. $x^2 - 2x + 5 = 0$

10. $x^2 + 13 = 4x$

11. $3x^2 + 2x = \frac{2}{3}$

12. $2x^2 - 3x + 5 = 0$

13. $9x^2 = 3x + 1$

14. $x^2 - 16 = 0$

15. $5 + 8x^2 = 16x$

16. $\frac{y}{4} + \frac{1}{2y} = 1$

17. $x^2 - 6ax + 8a^2 = 0$

18. $x = 2a + \frac{c}{x}$

Solve.

19. The sum S of the first n counting numbers 1, 2, 3, . . . , n is given by the formula $S = \frac{1}{2}n(n + 1)$. Find n for $S = 105$.

....

20. An open box is formed by cutting a square from each corner of a 5 in. by 7 in. rectangular piece of metal and bending up the edges. If the box is to have a base area of 19 sq in., how long should the corner cuts be?

....

21. What must be the dimensions of a rectangular field of 40 sq rd such that it can be enclosed with 24 lineal rods of fencing?

....

22. The edges of two cubes differ by 2 in. and their volumes differ by 296 cu in. What are the dimensions of each?

....

Name.. Date............................ **EXERCISE 8.4**

Give the domain (D) and range (R) of the following relations.

1. $\{(3, -1), (4, -2), (5, -3), (6, -4)\}$

$D = \{$

$R = \{$

2. $\{(1, 0), (2, 0), (3, 0), (4, 1)\}$

$D = \{$

$R = \{$

3. $\{(-1, 5), (-1, 10), (-1, 15), (-1, 20)\}$

$D = \{$

$R = \{$

4. $\{$(Bill, 2), (Sam, 4), (Ed, 2), (Al, 6)$\}$

$D = \{$

$R = \{$

5. $\{(6, -1), (3, 4), (2, 3), (6, -2)\}$

$D = \{$

$R = \{$

6. Which of the relations (Ex. 1–5) are also functions?

....

Name the set of ordered pairs (x, y) defined by the following.

7. $y = x + 2$, x an integer and $0 < x < 5$.

8. $y = x - 7$, y an integer and $-2 < x < 2$.

9. $\{(x, y)|x + y = 6, x \text{ and } y \text{ positive integers}\}$.

10. $\{(x, y)|2x + y = 4, x \text{ an integer and } -5 \leq x < -1\}$

11. $\{x, y)|y = x^2 - 4, x \text{ an integer and } -2 \leq x \leq 1\}$

12. Write an equation in x and y, x the independent variable, that defines the relation:

$\{(1, 0), (2, 1), (3, 2), (4, 3)\}$

Graph the relations.

13. $\{(x, y) | x - y = 2, x \text{ an integer and } -2 < x < 3\}$

x	y
-1	
0	
1	
2	

14. $\{(x, y) | y + 3 = 2x\}$

x	y

Determine whether or not the graphs represent functions.

15.

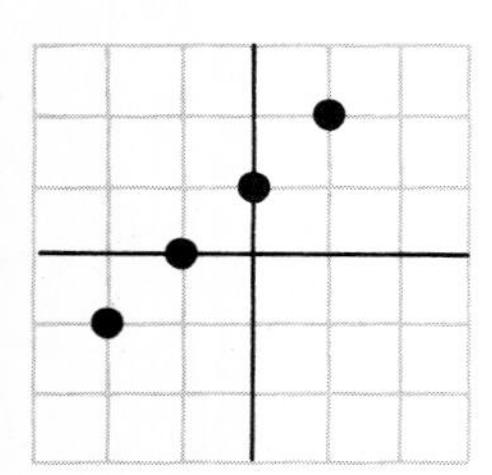

Function? Yes No

16.

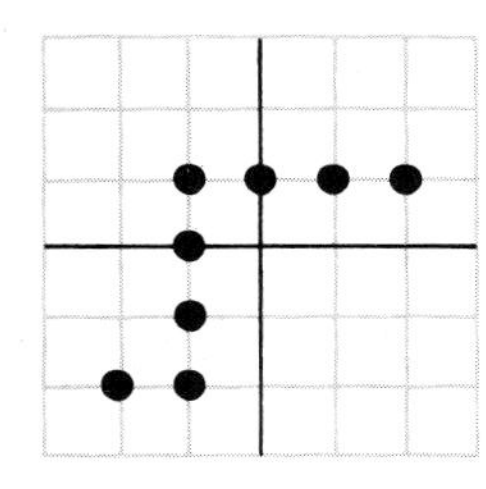

Function? Yes No

17.

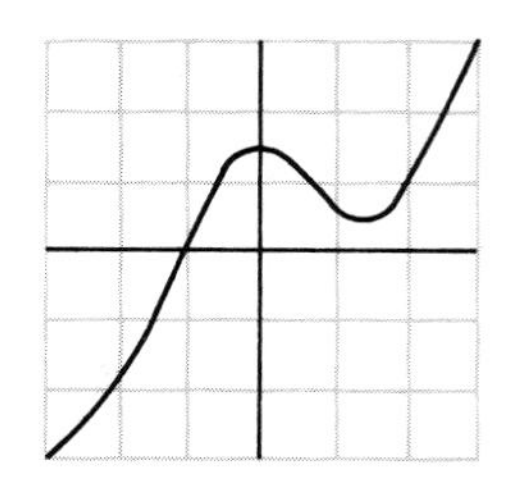

Function? Yes No

18.

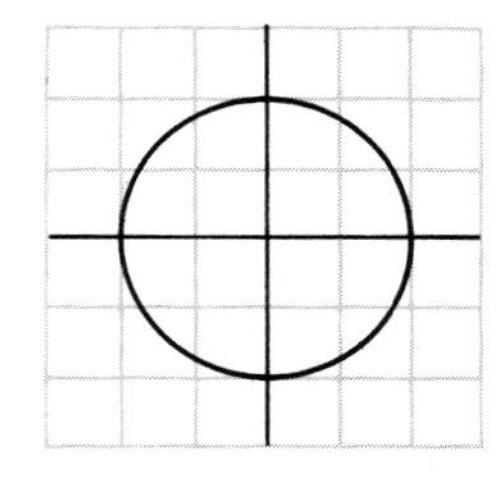

Function? Yes No

Solve.

19. If $f(x) = 2x - 7$, then $f(3) = ?$

20. If $f(x) = 3x^2 - 2x + 1$, compute $f(-4)$.

21. If $f(x) = x^2 + 2x + 3$, which is larger, $f(0)$ or $f(-3)$?

22. If $f(x) = 2x^2 - 3x - 4$, determine $f(x + 1)$.

Name.. Date.............................. **EXERCISE 8.5**

Solve, and verify all solutions.

1. $\sqrt{x+3} = 6$

2. $\sqrt{2x-4} + 2 = 0$

3. $2 + \sqrt{x-2} = 0$

4. $\sqrt{x^2+2} + x = 2$

5. $1 - 2x + \sqrt{4x+1} = 0$

6. $3\sqrt{x-1} = 2\sqrt{2x-1}$

7. $\sqrt{x-2} - 2 = \sqrt{2x+3}$

8. $\sqrt{2x + \sqrt{2x-4}} = 2$

Solve for all real values of x.

9. $x^4 - 10x^2 + 9 = 0$

10. $x^6 + 7x^3 - 8 = 0$

11. $6x + \sqrt{x} - 1 = 0$

12. $2x - 9\sqrt{x} + 10 = 0$

Solve.

13. The diagonal of a rectangle is equal to the square root of the sum of the squares of the two sides. What are the dimensions of the rectangle if its diagonal is 12 and one side exceeds the other by 2?

....

14. The distance in feet, s, that an object will fall in t sec is given by the formula $s = 16t^2$. If a rock is dropped down a mine shaft and its impact at the bottom is heard 5 sec after it is dropped, how deep is the mine shaft? (Assume sound travels 1,120 ft/sec.)

....

Simplify; express in terms of i.

1. $\sqrt{-50} + 2\sqrt{-8} - 3\sqrt{-2}$
2. $(2 - \sqrt{-9})^2$
3. $3i - 2i^2 + 3i^3 - i^4$
4. $i^5 - 2i^3 + 3i^2 + 4i$
5. $(5i^2 - 4i^3)^2$
6. $(3i - 8i^3)^2$
7. $(3i^2 - 2i + i^{13})^2$
8. $(2i^5 + 3i^{10} - 9i^{101})^2$

Solve for x, and verify.

9. $x^2 - 5 = 11$
10. $13 = x^2 - 12$
11. $x^2 - 4 = 1 - 3x^2$
12. $5x^2 - 11 = 2x^2 - 7$
13. $4x^2 + 3 = 3x^2 - 6$
14. $5x^2 - 9 = 8x^2 + 6$
15. $2x^2 + 9a^2 = 5a^2 + x^2$
16. $5x^2 + p^2 = 2x^2$

Solve by factoring, and verify.

17. $x^2 - 4x - 21 = 0$
18. $x^2 - 11x + 30 = 0$
19. $2x^2 + 5x - 3 = 0$
20. $6x^2 + x - 2 = 0$
21. $20x^2 + 9x - 20 = 0$
22. $6x^2 - 9x - 6 = 0$
23. $\frac{x^2}{2} + 7x = 0$
24. $x^2 = \frac{3}{2}x$

Use the quadratic formula to solve for x; express solutions in simplest radical form.

25. $x^2 + 6x + 8 = 0$
26. $x^2 + 3x + 2 = 0$
27. $x^2 + 4x + 1 = 0$
28. $4x^2 + 12x - 3 = 0$
29. $x^2 - 4x + 13 = 0$
30. $x^2 + 2x + 5 = 0$
31. $7x^2 + 5x + 1 = 0$
32. $5x^2 + 3x + 2 = 0$
33. $3x^2 - 12x + 4 = 0$
34. $5x^2 - 10x - 2 = 0$

Solve.

35. A department store buyer paid \$96 for a group of similar articles. If she could have gotten the unit price down by \$4, she could have purchased four more articles for the same money. How many articles did she purchase?

36. The base of a triangle is 4 in. longer than its altitude. If the area of the triangle is 160 sq in., what must be the length of its base?

37. Increasing the square of a certain number by 9 yields the same result as taking 7 from twice the square of that number. Find the number.

38. By cutting a 4-in. square from each corner of a square sheet of metal an open box of 324 cu in. can be formed. How large should the square piece of metal be before the cuts are made?

List the elements in the domain and the range of these relations.

39. $\{(3, 6), (4, 7), (5, 8), (6, 8), (7, 8)\}$
40. $\{(-1, 0), (-2, 1), (-3, 2), (-4, 3)\}$
41. $\{(A, 62 \text{ lb}), (B, 75 \text{ lb}), (C, 32 \text{ lb})\}$
42. {(red, green), (blue, green), (red, yellow)}
43. Is the relation given in Problem 42 a function?
44. Is the relation given in Problem 39 a function?

List the set of ordered pairs defined by each statement.

45. $y = 2 - x$, x an integer between 6 and 10.
46. $y = \frac{1}{2} + x$, x an integer between -8 and -3.
47. $\{(x, y) | 2x + y = 9,\ x \text{ and } y \text{ positive integers}\}$
48. $\{(x, y) | 3y - x = 0,\ y \text{ an integer and } 0 < x < 12\}$
49. $\{(a, b) | b = 2a^2 - 3,\ b \text{ an integer and } -2 \le a \le 1\}$
50. $\{(p, q) | p^2 + q = 0,\ p \text{ an integer and } -5 \le q \le 7\}$

Graph the relations.

51. $\{(x, y) | x + y = 3,\ x \text{ an integer and } -3 \le x \le 1\}$
52. $\{(x, y) | x - y = 2,\ x \text{ an integer and } -2 \le x \le 3\}$
53. $\{(x, y) | 2x - y = 1\}$
54. $\{(x, y) | 2x - 3y = 0\}$
55. If $f(x) = 3x - 5$, then $f(5) = ?$
56. If $f(x) = -2x + 7$, then $f(0) = ?$
57. If $f(x) = 3x^2 - 2x + 5$, then $f(2) = ?$
58. If $f(x) = -x^3 - 2x^2 - x$, then $f(-2) = ?$

Solve, and verify solutions.

59. $5 - \sqrt{x + 2} = 0$
60. $\sqrt{2x + 3} = 3$

61. $\sqrt{x-3}+3=0$

62. $\sqrt{x^2+16}=x+4$

63. $1+\sqrt{2x}=\sqrt{2x+5}$

64. $2+\sqrt{2x+4}=\sqrt{x}$

65. $x^{-4}-5x^{-2}+4=0$

66. $x^6-9x^3+8=0$

67. $\sqrt{4x+3}-\sqrt{6x+2}=0$

68. $3\sqrt{x^2-4}=\sqrt{3x^2+4x+6}$

REVIEW

Simplify; rationalize denominators when appropriate.

1. $(a^{-2}b^2)^3$

2. $(p^2q^{-4})^{-1}$

3. $\left(\frac{a^4b^2}{cd^3}\right)^3$

4. $\left(\frac{x^2y^3}{z^5}\right)^4$

5. $(x^{-1}-y^{-1})^{-1}$

6. $(2x^n)^{n-2}$

7. $\frac{\sqrt[3]{24}}{\sqrt[3]{9}}$

8. $\frac{\sqrt[3]{16}}{\sqrt[3]{25}}$

9. $\frac{1}{\sqrt{3}+2}$

10. $\frac{2}{3-\sqrt{2}}$

11. $\sqrt{4a}+\sqrt{9a}+5\sqrt{a}+\sqrt{16a}$

12. $\sqrt{2}+\sqrt[4]{2^2}+\sqrt[6]{2^3}+\sqrt[8]{2^4}$

13. $10\sqrt{20}-2\sqrt{180}$

14. $5\sqrt{45}-\sqrt{125}$

15. $2\sqrt{3xy}+3\sqrt[6]{x^3y^3}-\sqrt[4]{144x^2y^2}$

16. $3\sqrt{2b}+3\sqrt[6]{8b^9}-\sqrt[4]{64b^6}$

Divide.

17. $\sqrt{2}-\sqrt{5}$ by $\sqrt{2}+\sqrt{5}$

18. $3\sqrt{2}-\sqrt{5}$ by $\sqrt{2}+\sqrt{5}$

19. $4\sqrt{2}-3\sqrt{3}$ by $4\sqrt{2}+2\sqrt{3}$

20. $7\sqrt{5}-5\sqrt{7}$ by $5\sqrt{7}+7\sqrt{5}$

21. $4-\sqrt{5}$ by $\sqrt{2}+\sqrt{6}$

22. $\sqrt{2}+\sqrt{3}$ by $3\sqrt{5}+2\sqrt{3}$

Solve for x.

23. $\frac{1}{3x-2}=\frac{x+4}{3x^2+10x-8}$

24. $\frac{1}{3x-2}=\frac{x+1}{3x^2+10x-8}$

25. $\frac{x+3}{x+1}=\frac{x-1}{x-3}$

26. $\frac{x-1}{x+4}=\frac{x+2}{x-2}$

Graph the solution set.

27. $2x-(x-2)<5$

28. $-(3-x)\geq -1$

29. $x(x+3)\geq x^2+6$

30. $x^2-x(x-3)>-18$

Simplify.

31. $\dfrac{\frac{a}{a+b}-\frac{b-a}{a}}{\frac{a}{a+b}+\frac{b-a}{a}}$

32. $1+\dfrac{2}{a+\dfrac{3}{a+\frac{4}{a}}}$

Name.. Date.............................. Score.....................

ACHIEVEMENT TEST NO. 4 (Chapters 4–8)

Place answers in the blanks provided. (4 points each, total 100)

Simplify:

1. $(5 - [2 + 6(3 - 1) + 2])(4 + |-2|)$ ....

2. $\dfrac{2x^2 - 5x - 3}{3x^2 - 16x + 5} \cdot \dfrac{6x^2 + 7x - 3}{2x^2 - 9 - 3x}$ ....

3. $\dfrac{5m}{m^2 - 4} + \dfrac{2}{m + 2} + \dfrac{m}{2 - m}$ ....

4. $\dfrac{\dfrac{1}{a - b} - \dfrac{1}{a + b}}{\dfrac{2}{a^2 - b^2}}$ ....

5. $[(k - k^{-1})^{-1}]^{-1}$ ....

6. $\sqrt{x^4} + \sqrt[6]{x^8} + 2\sqrt[3]{27x^4}$ ....

7. $(2\sqrt{3} + 3\sqrt{2}) \div (\sqrt{3} - \sqrt{2})$ ....

8. $2i + 3i^2 - 5i^3 + 6i^6$ ....

9. $(1 - \sqrt{-2})^2$ ....

10. $(2 - 3i^3)(4 - i)$ ....

Solve for all real-number solutions:

11. $4x - 4 > 2x - 7$ ....

12. $3 - 4x \leq x - 7$ ....

13. $\dfrac{2a - 1}{a - 1} = \dfrac{2a}{a + 1}$ ....

14. $\begin{cases} 4x + 3y = 3 \\ 5x + 2y + 5 = 0 \end{cases}$ ....

15. $24x^2 - 23x = 12$ ....

16. $2x^2 + 3x + 2 = 0$ ....

17. $x^2 + 4dx + d^2 = 0$ ....

18. $x^6 - 9x^3 + 8 = 0$ ....

19. $\sqrt{x+2} + \sqrt{2x+5} = 5$ ....

20. $\sqrt{x} + \dfrac{4}{\sqrt{x}} + 5 = 0$ ....

21. Evaluate:

$$\begin{vmatrix} -3 & 4 \\ 2 & 0 \end{vmatrix}$$

....

22. What is the point of intersection for the graphs of $x - 2y = 4$ and $y - 3x + 7 = 0$?

....

23. A mother can clean the house in 4 hrs and her daughter can do it in 6 hrs. How long should it take them, working together, to clean the house?

....

24. The difference of two numbers is 11 and the sum of their squares is 61. Find the two numbers.

....

25. Is the graph below that of a function? (Yes or No)

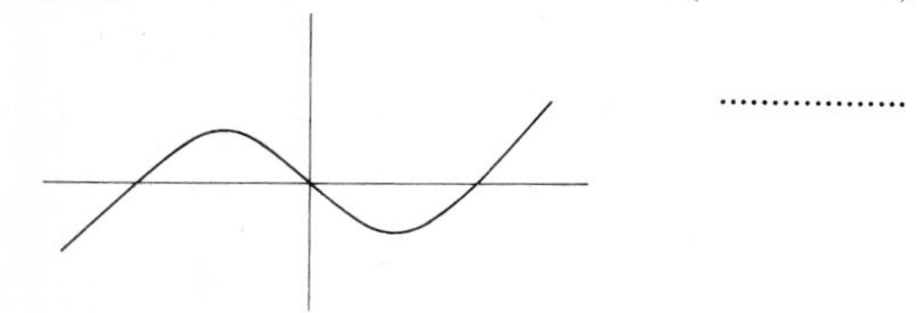

....

PART III

ESSENTIALS OF TRIGONOMETRY

9

RIGHT TRIANGLES

1. DEFINITIONS

Trigonometry is one of the oldest branches of mathematics. Literally it means "three-angle measurement," which indicates the deep historical connection the subject has with triangles. Today broader interpretations have been given to earlier concepts to make trigonometry an indispensable aid to modern scientists and engineers.

A triangle consists of three connected line segments called *sides.* Each pair of sides forms an angle. The point at which a pair of sides meet to form an angle is called a *vertex.* Triangles are usually identified by a capital letter at each vertex, and the sides by the pair of capital letters that mark the vertices at either end. Angles of triangles are normally identified by the letter at the vertex, though at times three letters are used (Fig. 9.1).

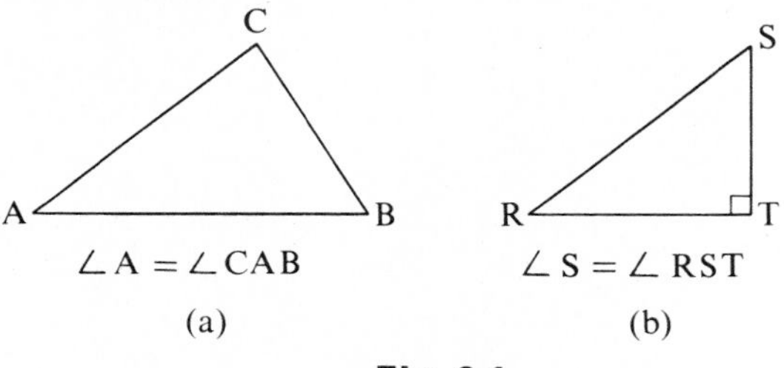

Fig. 9.1

The length or measure of a side of a triangle is also denoted by a letter, generally the lower case version of the capital letter of the opposite vertex (Fig. 9.2).

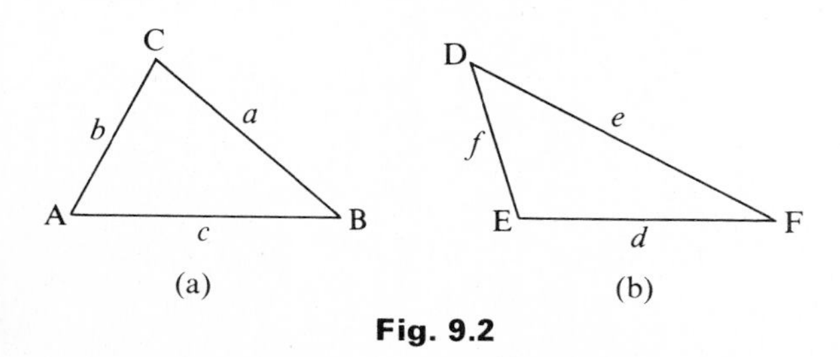

Fig. 9.2

The sum of the measures of the three angles of a triangle lying in a plane is defined to be 180°. An angle whose measure is half that, 90°, is known as a *right angle.* An angle whose measure is less than 90° is called an *acute* angle, and an angle whose measure is greater than 90° is called an *obtuse* angle.

A triangle in which one of the angles is a right angle is called a *right triangle.* Consequently, the sum of the measures of the other two angles must be 90°, since the sum of the measures of all three angles of any triangle is 180°. Figure 9.1b is an example of a right triangle with its right angle at T. The side opposite the right angle of a right triangle is called the *hypotenuse.* The other two sides, those opposite the acute angles, are sometimes referred to as the *legs* of the right triangle.

A triangle in which all three angles are acute is called an *acute triangle* (Fig. 9.1a and 9.2a are examples). A triangle in which one angle is obtuse is called an *obtuse triangle* (Fig. 9.2b, which has an obtuse angle at E, is an example).

Two triangles are *similar* if the measures of their angles are equal and in the same order.

2. PYTHAGOREAN RELATION

One of the most important relationships in mathematics is the *Pythagorean relation:*

If a, b, c denote the measures of the three sides of a right triangle, where c denotes the measure of the hypotenuse, then $c^2 = a^2 + b^2$.

EXAMPLES

1. Express the Pythagorean relation for triangle ABC, a right triangle in which $a = 4$, $b = 3$, and $c = 5$.

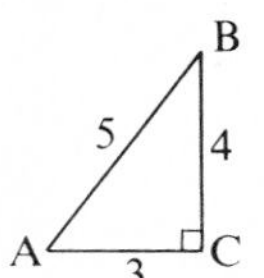

$$c^2 = a^2 + b^2$$
$$(5)^2 = (4)^2 + (3)^2$$
$$25 = 16 + 9$$

2. Find the measure of the hypotenuse of a right triangle if the measures of the legs are 5 and 12.

$$c^2 = (5)^2 + (12)^2$$
$$= 25 + 144$$
$$= 169$$
$$c = \sqrt{169} = 13$$

3. Find the length of the remaining leg of a right triangle in which the length of the hypotenuse is 7 and that of the other leg is 2.

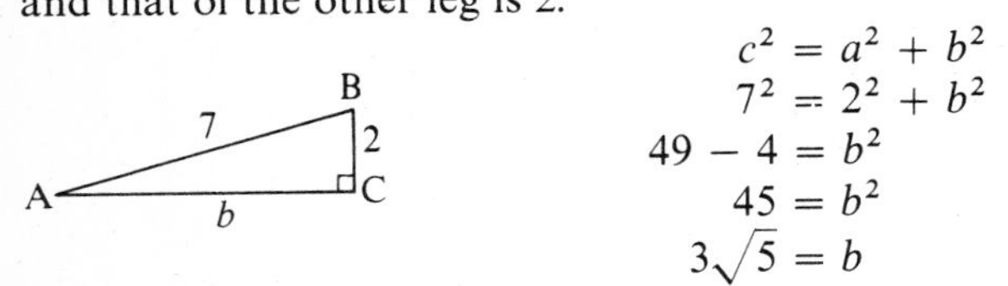

$$c^2 = a^2 + b^2$$
$$7^2 = 2^2 + b^2$$
$$49 - 4 = b^2$$
$$45 = b^2$$
$$3\sqrt{5} = b$$

4. Is a triangle with side measures of 3, 9, and 11 a right triangle?
Any triangle in a plane for which the Pythagorean relation does not hold is not a right triangle. The hypotenuse of a right triangle is always greater in length than either leg. Therefore, if the triangle in question is a right triangle, then $c = 11$ and $c^2 = a^2 + b^2$.
To test the data of the given triangle:
$11^2 \stackrel{?}{=} 3^2 + 9^2$
$121 \stackrel{?}{=} 9 + 81$
$121 \neq 90$
The triangle with side measures of 3, 9, and 11 is not a right triangle.

3. SPECIAL TRIANGLES

By means of the Pythagorean relation we can compute the measures of the sides of other triangles. The two triangles we are about to discuss are especially useful.

A triangle that has two sides of equal length (often referred to as having "equal sides") is called *isosceles*. The angles opposite the equal sides of an isosceles triangle are also equal in measure. Three variations of isosceles triangles are shown in Fig. 9.3.

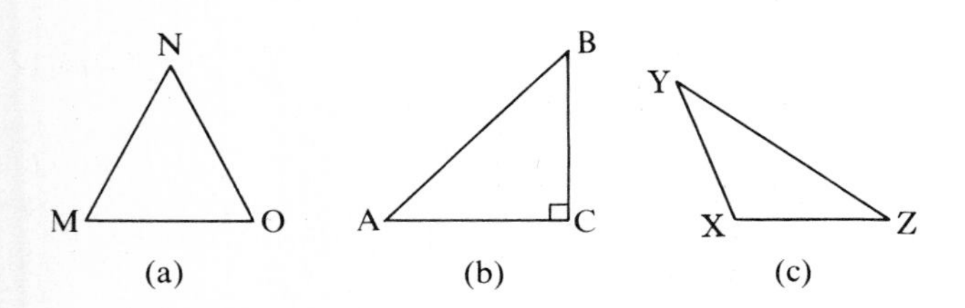

Fig. 9.3

The triangle shown in Fig. 9.3b is a right isosceles triangle—a right triangle whose legs are equal. Because the triangle is isosceles, the angles at A and B are also equal; moreover, these angles must have measures of 45° each, one-half the sum of the measures of the acute angles of every right triangle (90°). If we assume the measure of each leg to be 1 (i.e., $a = b = 1$), then the measure of the hypotenuse c is:

$$c^2 = 1^2 + 1^2 = 2$$
$$c = \sqrt{2}$$

Had the measures of the equal legs been assumed to be 3 or 8—such triangles would be similar to each other—the measures of the three sides are easily computed. They are 3, 3, $3\sqrt{2}$, or 8, 8, $8\sqrt{2}$, each set of three measures in proportion to the 1, 1, $\sqrt{2}$ set.

An *equilateral triangle* is one in which all three sides are of the same length. An equilateral triangle is also an isosceles triangle for any pair of sides, so the measures of each angle must be the same: $\frac{1}{3}$ of 180°, or 60°. (See Fig. 9.4a.)

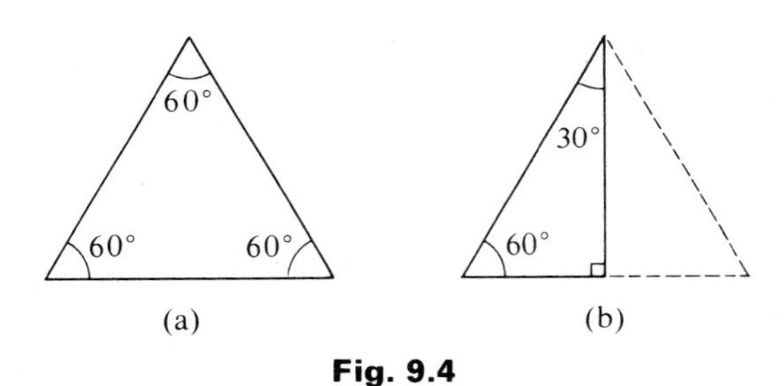

Fig. 9.4

If we cut an equilateral triangle in half by dropping a perpendicular, as in Fig. 9.4b, we produce two 30°–60°–90° triangles. If we assume the measure of the side of the equilateral triangle (now a hypotenuse) to be 2, as in Fig. 9.5, then side AC will have a measure of 1, and the length of side BC can be computed by the Pythagorean relation:

$$a^2 + b^2 = c^2$$
$$a^2 + 1^2 = 2^2$$
$$a^2 = 4 - 1$$
$$a = \sqrt{3}$$

Fig. 9.5

Any other similar 30°–60°–90° triangle will have side measures in proportion to the 1, $\sqrt{3}$, 2 set just derived.

EXAMPLES

1. What is the length of the other two sides of an isosceles right triangle whose leg measures 5?
   Since the triangle is isosceles, $a = b = 5$. By the Pythagorean relation:

$$c = \sqrt{a^2 + b^2} = \sqrt{25 + 25} = \sqrt{50} = 5\sqrt{2}$$

2. What is the measure of the altitude of an equilateral triangle whose side measures 12?
   To find the length of BD:

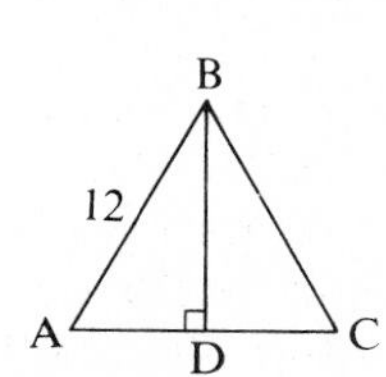

$$\begin{aligned} AD &= \tfrac{1}{2} \times 12 = 6 \\ (AB)^2 &= (AD)^2 + (BD)^2 \\ (BD)^2 &= (AB)^2 - (AD)^2 \\ BD &= \sqrt{(AB)^2 - (AD)^2} \\ &= \sqrt{144 - 36} \\ &= \sqrt{108} = 6\sqrt{3} \end{aligned}$$

(Complete Exercise 9.1)

4. TRIGONOMETRIC FUNCTIONS

With respect to a particular acute angle of a right triangle, the leg that joins the hypotenuse to form the angle is also called the *adjacent side* of the triangle. The leg opposite the angle is called the *opposite side.*

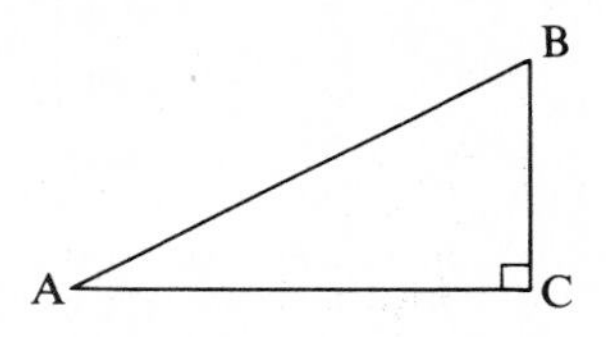

Thus,

with respect to ∠A:	with respect to ∠B:
AB = hypotenuse	AB = hypotenuse
AC = adjacent side	AC = opposite side
BC = opposite side	BC = adjacent side

The measures of the sides of every right triangle, we have noted, are related by the Pythagorean relation, $c^2 = a^2 + b^2$. Relationships also exist between the measures of the sides and the measures of the acute angles. They are given in terms of the *six basic trigonometric functions,** defined as follows:

The *sine* of an acute angle of a right triangle is the ratio of the length of the opposite side to that of the hypotenuse.

* Although the ordered-pair notation is not normally used in this context, each *relation* that associates angle measures and a particular type of ratio—e.g., (A, a/c)—is in fact also a *function.*

The *cosine* of an acute angle of a right triangle is the ratio of the length of the adjacent side to that of the hypotenuse.

The *tangent* of an acute angle of a right triangle is the ratio of the length of the opposite side to that of the adjacent side.

The *cotangent* of an acute angle of a right triangle is the reciprocal of the tangent.

The *secant* of an acute angle of a right triangle is the reciprocal of the cosine.

The *cosecant* of an acute angle of a right triangle is the reciprocal of the sine.

In summary:

sine:	$\dfrac{\text{opposite}}{\text{hypotenuse}}$	cosecant:	$\dfrac{\text{hypotenuse}}{\text{opposite}}$
cosine:	$\dfrac{\text{adjacent}}{\text{hypotenuse}}$	secant:	$\dfrac{\text{hypotenuse}}{\text{adjacent}}$
tangent:	$\dfrac{\text{opposite}}{\text{adjacent}}$	cotangent:	$\dfrac{\text{adjacent}}{\text{opposite}}$

The names of the six functions are usually abbreviated: sin A (sine of angle A), cos A (cosine of angle A), tan A (tangent of angle A), cot A (cotangent of angle A), sec A (secant of angle A), csc A (cosecant of angle A).

With respect to right triangle ABC shown at the right:

$\sin A = \dfrac{a}{c}$ $\quad \sin B = \dfrac{b}{c}$

$\cos A = \dfrac{b}{c}$ $\quad \cos B = \dfrac{a}{c}$

$\tan A = \dfrac{a}{b}$ $\quad \tan B = \dfrac{b}{a}$

$\cot A = \dfrac{b}{a}$ $\quad \cot B = \dfrac{a}{b}$

$\sec A = \dfrac{c}{b}$ $\quad \sec B = \dfrac{c}{a}$

$\csc A = \dfrac{c}{a}$ $\quad \csc B = \dfrac{c}{b}$

EXAMPLES

1. Give the ratios for the six trigonometric functions for A and B in right triangle ABC.

$\sin A = \frac{4}{5}$ $\quad \sin B = \frac{3}{5}$

$\cos A = \frac{3}{5}$ $\quad \cos B = \frac{4}{5}$

$\tan A = \frac{4}{3}$ $\quad \tan B = \frac{3}{4}$

$\cot A = \frac{3}{4}$ $\quad \cot B = \frac{4}{3}$

$\sec A = \frac{5}{3}$ $\quad \sec B = \frac{5}{4}$

$\csc A = \frac{5}{4}$ $\quad \csc B = \frac{5}{3}$

2. State the ratios for the six trigonometric functions for 45° using the isosceles right triangle shown at the right.

$$\sin 45^\circ = \frac{1}{\sqrt{2}} = \frac{1}{2}\sqrt{2} = \frac{\sqrt{2}}{2}$$

$$\cos 45^\circ = \frac{1}{\sqrt{2}} = \frac{1}{2}\sqrt{2} = \frac{\sqrt{2}}{2}$$

$$\tan 45^\circ = \frac{1}{1} = 1$$

$$\cot 45^\circ = \frac{1}{1} = 1$$

$$\sec 45^\circ = \frac{\sqrt{2}}{1} = \sqrt{2}$$

$$\csc 45^\circ = \frac{\sqrt{2}}{1} = \sqrt{2}$$

3. Use the given 30°–60°–90° triangle to record the ratios of the six trigonometric functions for 30° and 60°.

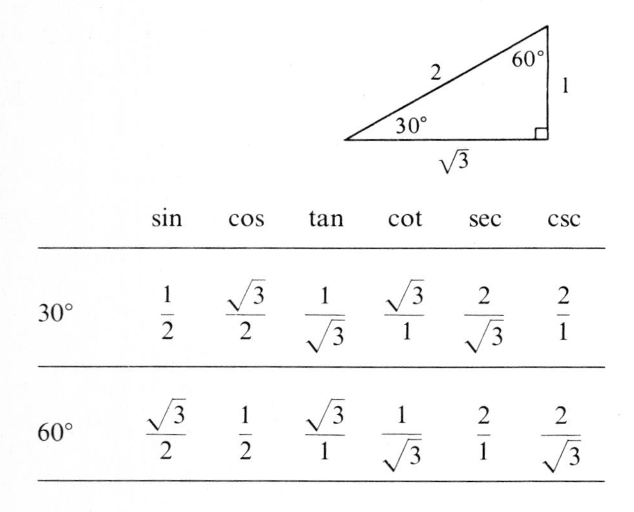

	sin	cos	tan	cot	sec	csc
30°	$\frac{1}{2}$	$\frac{\sqrt{3}}{2}$	$\frac{1}{\sqrt{3}}$	$\frac{\sqrt{3}}{1}$	$\frac{2}{\sqrt{3}}$	$\frac{2}{1}$
60°	$\frac{\sqrt{3}}{2}$	$\frac{1}{2}$	$\frac{\sqrt{3}}{1}$	$\frac{1}{\sqrt{3}}$	$\frac{2}{1}$	$\frac{2}{\sqrt{3}}$

Two angles are said to be *complementary* if the sum of their measures is that of a right angle, 90°. For example, angles of 30° and 60° are complementary; so are angles of 53° and 37°; and 20° and 70°. This relationship may also be stated as "30° is the complement of 60°"; "53° is the complement of 37°"; etc.

Note in Example 1 above that A and B are complementary angles, as are the pairs of acute angles of every right triangle. Note also that sine A = *co*sine B, tangent A = *co*tangent B, secant A = *co*secant B, and vice versa: *co*sine A = sine B, *co*tangent A = tangent B, *co*secant A = secant B. Thus, cosine, cotangent, and cosecant are the *complementary functions* or *co*functions of sine, tangent, and secant, respectively. Similarly, sine, tangent, and secant are cofunctions of cosine, cotangent, and cosecant, respectively. In general:

A trigonometric function of an acute angle is the equivalent of the cofunction of its complementary angle.

This generalization is affirmed in Examples 2 and 3 above: sin 30° = cos(90°–30°) = cos 60°; sec 60° = csc(90°–60°) = csc 30°; tan 45° = cot 45°; etc.

5. TABLE OF VALUES OF TRIGONOMETRIC FUNCTIONS

Associated with each trigonometric function of each acute angle is a number that is equivalent to the defined ratios. These numbers can be identified for some angles by means of special triangles, as was done in Section 4, above. At best, however, this is an incomplete approach. Instead, we use tables, such as the table of Values of Trigonometric Functions (Table B, inside the back cover and on page 252), which have been developed by advanced mathematical techniques.

Because of the relationship between functions and their respective cofunctions, it is possible to reduce the length of such tables by half of what it might otherwise have been. Notice in Table B that angle measures from 0° to 45° are listed in the left column, and that angle measures from 45° through 90° are listed, *from bottom to top*, in the right column.

Table B provides values of the six trigonometric functions at intervals of 1°. Comparable tables are readily available that provide values of the functions at intervals of 10′, or even 1′. For our limited purposes, Table B is sufficiently precise. (It is possible, however, to refine the tabulated data somewhat by "interpolation," as discussed in Section 6, following.)

(9.1) *To use a table of values of trigonometric functions (Table B) to find the value of a trigonometric function of an acute angle:*

Step 1. Locate the row containing the angle measure in either the left or right column.

Step 2. Read the desired function value at the intersection of the row of Step 1 and the appropriate column.

[*Important note:* When the angle measure is read from the *left* column, the function labels appear at the tops of columns; when the angle measure is read from the right column, the function labels appear at the *bottoms* of the columns.]

EXAMPLES

1. Find the value of tan 74° from Table B.
   Step 1. 74° > 45°, so the angle measure will be found in the right column.

   Step 2. Because the angle measure is recorded in the right column, appropriate function labels appear at the bottoms of the columns. At the intersection of the 74°-row and the tangent-column is 3.49. So tan 74° = 3.49.
2. Find the values of sin 38°; cos 52°.
   From Table B:

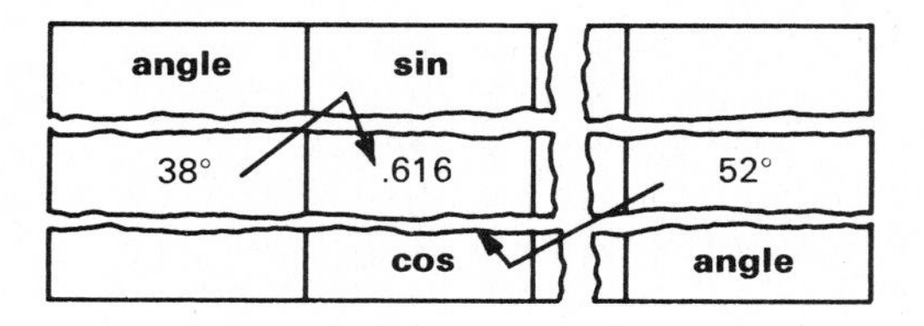

sin 38° = .616
cos 52° = .616

3. Give the values for cos 40°; cos 49°; tan 52°; cot 21°; sec 81°; csc 64°.

cos 40° = .766	cot 21° = 2.61
cos 49° = .656	sec 81° = 6.39
tan 52° = 1.28	csc 64° = 1.11

Notice in Table B that the angle measures tabulated in the left and right columns of each row are complementary, and that the functions named at the top and the bottom of each column are cofunctions.

It is also useful to note that as the angle measures vary increasingly from 0° to 90°, the values of the sine, tangent, and secant functions increase, while those of the cosine, cotangent, and cosecant decrease.

6. INTERPOLATION

To *interpolate* means to compute an intermediate value between two stated or tabulated values. The technique most often used assumes that differences between successive entries in a table are proportional. In the case of the trigonometric functions, this is not actually so; but assuming it to be so over short intervals usually results in an acceptable approximation. If more precise values are needed, more refined tables should be used (which in turn can be further refined by interpolation).

Although angles are more precisely measured in degrees (°), minutes (′) and seconds (″), here we shall use increments of one-tenth degree, the equivalent of 6 minutes. This represents a more appropriate extension of a three-place table, as is Table B, and is sufficient to explain the concept of interpolation.

EXAMPLES

1. Approximate the value of sin 38.4°, using Table B. Related data from Table B are displayed below. The desired interpolated value of sin 38.4° is n, which will be obtained by applying the correction, c, to the value of sin 38°.

		angle	sin		
1°	.4°	38°	.616	c	.013
		38.4°	n		
		39°	.629		(.629 − .616)

Rationale for the procedure: In each column the *differences* between the tabulated and interpolated values (.4° and c) are assumed to bear the same ratio to the respective *differences* between successive entries in the table (1° and .013).
Equating the two ratios produces the proportion:

$$.4°:1° = c:.013$$

Solving the proportion (i.e., "the product of the means equals the product of the extremes"):

$$(1)(c) = (.4)(.013)$$
$$c = .0052$$
$$c = .005 \text{ (rounded to 3 places)}$$

Using c to compute n:

$$n = .616 + c$$
$$= .616 + .005$$
$$= .621$$

Thus, sin 38.4° = .621.

2. Compute the value of cos 56.2°.
   The procedure is the same as that used in Example 1, except the correction is subtracted. When the function is an increasing one, i.e., sine, tangent, and secant, the correction is added. When the function is a decreasing one, i.e., cosine, cotangent, cosecant, the correction is subtracted.

		angle	cos		
1°	.2°	56°	.559	c	.014
		56.2°	n		
		57°	.545		

The proportion:

$$\begin{aligned} .2^\circ : 1^\circ &= c : .014 \\ (1)(c) &= (.2)(.014) \\ c &= .0028 \\ c &= .003 \text{ (rounded to 3 places)} \end{aligned}$$

Applying the correction:

$$\begin{aligned} n &= .559 - .003 \\ &= .556 \end{aligned}$$

Thus, $\cos 56.2^\circ = .556$.

3. If $\tan A = .690$, find A to the nearest tenth of a degree.
In this case the unknown is an angle measure, and the correction will be in tenths of a degree. The data:

		angle	tan		
1°	c	34°	.675	.015	.025
		n	.690		
		35°	.700		

The proportion:

$$\begin{aligned} c : 1^\circ &= .015 : .025 \\ (1)(.015) &= (c)(.025) \\ \frac{.015}{.025} &= c \\ .6 &= c \end{aligned}$$

Applying the correction to find n:

$$\begin{aligned} n &= 34^\circ + c \\ &= 34^\circ + .6^\circ \\ &= 34.6^\circ \end{aligned}$$

Thus, $\tan 34.6^\circ = .690$.

4. If $\cot B = 3.87$, find B, to the nearest tenth degree.

		angle	cot		
1°	c	14°	4.01	.14	.28
		n	3.87		
		15°	3.73		

The proportion:

$$\begin{aligned} c : 1^\circ &= .14 : .28 \\ (c)(.28) &= (1)(.14) \\ .28c &= .14 \\ c &= \frac{.14}{.28} = .5 \end{aligned}$$

Thus, $\cot 14.5^\circ = 3.87$.

(Complete Exercise 9.2)

7. SOLVING RIGHT TRIANGLES

To *solve a triangle* means to find the measures of its three sides and three angles. In order to do so, measures of three of the six parts must be known—one of which must be the measure of a side. (Three angles whose measures add to 180° do not determine a unique triangle, but a set of similar triangles.)

When the triangle to be solved is a right triangle, one angle is known to measure 90°; so to know also the measures of one of the acute angles and a side, or the measures of two sides, is sufficient to complete the solution.

(9.2) *To solve a right triangle:*

Step 1. *Involve one of the unknown measures with two of the known measures in an equation based upon either the Pythagorean relation or a trigonometric function.*

Step 2. *Solve the equation of Step 1 for the unknown measure.*

Step 3. *Repeat Steps 1 and 2 for each unknown measure, preferably using given data rather than derived data.*

(The suggestion in Step 3 to use given data rather than derived or computed data is to avoid compounding errors. An error in a derived measure will lead to other errors in measures computed from it.)

EXAMPLES

1. Solve right triangle ABC in which $A = 50^\circ$ and $c = 20$.

To find side a:

$$\sin 50^\circ = \frac{a}{20}$$

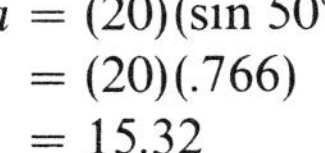

$$\begin{aligned} a &= (20)(\sin 50^\circ) \\ &= (20)(.766) \\ &= 15.32 \end{aligned}$$

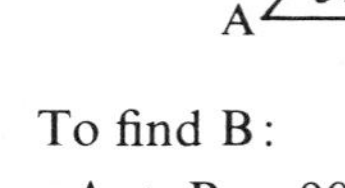

To find side b:

$$\cos 50^\circ = \frac{b}{20}$$

$$\begin{aligned} b &= (20)(\cos 50^\circ) \\ &= (20)(.643) \\ &= 12.86 \end{aligned}$$

To find B:

$$\begin{aligned} A + B &= 90^\circ \\ 50^\circ + B &= 90^\circ \\ B &= 40^\circ \end{aligned}$$

2. Solve right triangle ABC in which $a = 4$, $b = 8$.

To find c:

$c^2 = a^2 + b^2$

$= 16 + 64$

$c = \sqrt{80}$

$= 8.9$ (Table A)

To find A:

$\tan A = \frac{4}{8} = .5$

$A = 27°$ (nearest degree)

To find B:

$\tan B = \frac{8}{4} = 2$

$B = 63°$ (nearest degree)

Using interpolation:

$\tan A = 26.5°$ $\qquad$ $\tan B = 63.4°$

(The sum of the measures of A and B should be 90°. Here, it is $26.5° + 63.4° = 89.9°$; the error is due to rounding.)

3. Solve right triangle ABC in which $B = 22°$ and $a = 12$.

$A = 90° - 22° = 68°$

$\sec 22° = \frac{c}{12}$

$c = (12)(\sec 22°)$

$= (12)(1.08)$

$= 12.96$

$\tan 22° = \frac{b}{12}$

$b = (12)(.404)$

$= 4.848$

When solving for the measure of a side of a right triangle by means of a trigonometric function, two choices are always possible. In one, the unknown measure will occur in the numerator of the ratio fraction; in the other, the unknown measure will occur in the denominator of the ratio fraction. The former is usually preferable, computationally, because it leads to a multiplication instead of a division. For instance, in Example 3 above, c could have been determined either by using sec 22° or cos 22°:

$\sec 22° = \frac{c}{12}$ $\qquad$ $\cos 22° = \frac{12}{c}$

$c = (12)(\sec 22°)$ $\qquad$ $c = \frac{12}{\cos 22°}$

$c = 12 \times 1.08$ $\qquad$ $c = \frac{12}{.927}$

Unless one is using a calculator, it is easier to compute the product, 12×1.08, than to compute the quotient, $12 \div .927$.

(Complete Exercise 9.3)

8. TRIGONOMETRIC IDENTITIES

Among the advantages which trigonometric functions bring to mathematics is their great flexibility. By means of *identities* it is possible to restate trigonometric expressions equivalently in many different ways. For example, the generalization about a function and its respective cofunction given at the end of Section 4 leads to these six identities in which θ (theta) denotes an angle measure:

$\sin \theta = \cos (90° - \theta)$ $\qquad$ $\cot \theta = \tan (90° - \theta)$

$\cos \theta = \sin (90° - \theta)$ $\qquad$ $\sec \theta = \csc (90° - \theta)$

$\tan \theta = \cot (90° - \theta)$ $\qquad$ $\csc \theta = \sec (90° - \theta)$

Another important group of identities is the reciprocal identities. By our original definitions of the six trigonometric functions (Section 4), sine and cosecant are reciprocals, cosine and secant are reciprocals, and tangent and cotangent are reciprocals. Thus:

$\sin \theta = \frac{1}{\csc \theta}$ $\qquad$ $\cot \theta = \frac{1}{\tan \theta}$

$\cos \theta = \frac{1}{\sec \theta}$ $\qquad$ $\sec \theta = \frac{1}{\cos \theta}$

$\tan \theta = \frac{1}{\cot \theta}$ $\qquad$ $\csc \theta = \frac{1}{\sin \theta}$

Several more identities may be derived by considering a right triangle such as the one following, with acute angle θ and sides whose measures are denoted by x, y, and r:

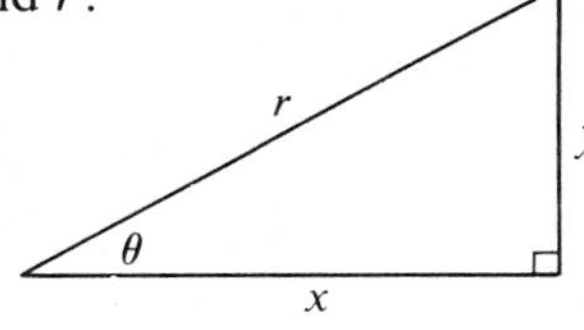

Note that $\sin \theta = y/r$, $\cos \theta = x/r$, and $\tan \theta = y/x$. The ratio $\sin \theta / \cos \theta$ can be shown to be equivalent to $\tan \theta$ in the following way:

$$\frac{\sin \theta}{\cos \theta} = \frac{y/r}{x/r} = \frac{y}{x} = \tan \theta$$

Similarly,

$$\frac{\cos \theta}{\sin \theta} = \frac{x/r}{y/r} = \frac{x}{y} = \cot \theta$$

Thus, we have two important ratio identities:

$$\frac{\sin \theta}{\cos \theta} = \tan \theta \qquad \frac{\cos \theta}{\sin \theta} = \cot \theta$$

For the triangle above, the Pythagorean relation holds that:

$$x^2 + y^2 = r^2$$

If we divide each term by r^2, we have:

$$\frac{x^2}{r^2} + \frac{y^2}{r^2} = \frac{r^2}{r^2}$$

$$\left(\frac{x}{r}\right)^2 + \left(\frac{y}{r}\right)^2 = 1$$

Since $\sin \theta = y/r$ and $\cos \theta = x/r$, the following identity results:

$$(\sin \theta)^2 + (\cos \theta)^2 = 1$$

which is usually written:

$$\sin^2 \theta + \cos^2 \theta = 1$$

This important identity is one of the three Pythagorean identities; the other two are:

$$\tan^2 \theta + 1 = \sec^2 \theta$$

$$1 + \cot^2 \theta = \csc^2 \theta$$

which may be derived in essentially the same way, except by dividing the Pythagorean relation $x^2 + y^2 = r^2$ by x^2 and y^2, respectively, instead of by r^2. An alternative method is to use previously derived identities, as shown in Example 1, following.

EXAMPLES

1. Use $\sin^2 \theta + \cos^2 \theta = 1$ to show:
   (a) $\tan^2 \theta + 1 = \sec^2 \theta$;
   (b) $1 + \cot^2 \theta = \csc^2 \theta$.
   (a) Divide each term of $\sin^2 \theta + \cos^2 \theta = 1$ by $\cos^2 \theta$:

$$\frac{\sin^2 \theta}{\cos^2 \theta} + \frac{\cos^2 \theta}{\cos^2 \theta} = \frac{1}{\cos^2 \theta}$$

$$\left(\frac{\sin \theta}{\cos \theta}\right)^2 + 1 = \left(\frac{1}{\cos \theta}\right)^2$$

$$\tan^2 \theta + 1 = \sec^2 \theta$$

   (b) Divide each term of $\sin^2 \theta + \cos^2 \theta = 1$ by $\sin^2 \theta$:

$$\frac{\sin^2 \theta}{\sin^2 \theta} + \frac{\cos^2 \theta}{\sin^2 \theta} = \frac{1}{\sin^2 \theta}$$

$$1 + \cot^2 \theta = \csc^2 \theta$$

2. Express $\cos \theta \tan \theta \sec \theta$ (i.e., $\cos \theta \cdot \tan \theta \cdot \sec \theta$) in terms of a single trigonometric function.

One approach (refer to the right triangle on page 213):

$$\cos \theta = \frac{x}{r}, \quad \tan \theta = \frac{y}{x}, \quad \sec \theta = \frac{r}{x}$$

$$\cos \theta \cdot \tan \theta \cdot \sec \theta = \frac{\cancel{x}}{\cancel{r}} \cdot \frac{y}{\cancel{x}} \cdot \frac{\cancel{r}}{x} = \frac{y}{x} = \tan \theta$$

Another approach—express all functions in terms of sines and cosines:

$$\cos \theta \cdot \tan \theta \cdot \sec \theta = \cancel{\cos \theta} \cdot \frac{\sin \theta}{\cancel{\cos \theta}} \cdot \frac{1}{\cos \theta}$$

$$= \frac{\sin \theta}{\cos \theta} = \tan \theta$$

3. Show $\cos x + (\sin x \tan x) = \sec x$.
   Start with the left member:

$$\cos x + (\sin x \tan x) = \cos x + \left(\sin x \cdot \frac{\sin x}{\cos x}\right)$$

$$= \cos x + \left(\frac{\sin^2 x}{\cos x}\right)$$

$$= \frac{\cos^2 x + \sin^2 x}{\cos x}$$

$$= \frac{1}{\cos x} = \sec x$$

4. Verify: $\dfrac{\sec A}{\tan A + \cot A} = \sin A.$

$$\frac{\sec A}{\tan A + \cot A} = \frac{\frac{1}{\cos A}}{\frac{\sin A}{\cos A} + \frac{\cos A}{\sin A}}$$

$$= \frac{\frac{1}{\cos A}}{\frac{\sin^2 A + \cos^2 A}{\sin A \cos A}}$$

$$= \frac{\frac{1}{\cos A}}{\frac{1}{\sin A \cos A}}$$

$$= \frac{1}{\cancel{\cos A}} \cdot \frac{\sin A \cancel{\cos A}}{1}$$

$$= \sin A$$

There is no readily prescribed program by which trigonometric identities, such as those in Examples 3 and 4 above, can be verified. In general, it is usually advantageous to attack the more complicated member and to try to transform it into the expression of the other member by various substitutions. Transforming all terms to sines and cosines (as in Examples 2, 3, and 4) often proves fruitful. While it is always possible to express the trigonometric terms as ratios of triangular measures (as was done in Example 2) and to verify the identity that way, for later applications in mathematics it is better to learn to work entirely with trigonometric identities.

(Complete Exercise 9.4)

Name.. Date.............................. **EXERCISE 9.1**

Given right triangle ABC, right angle at C. Use the Pythagorean relation to find the measure of the third side. Use Table A (page 251 and inside front cover) to express square roots decimally.

1. $a = 6, b = 8, c = ?$

2. $a = 5, b = ?, c = 13$

3. $a = ?, b = 24, c = 25$

4. $a = 9, b = 13, c = ?$

5. $a = ?, b = 9, c = 17$

6. $a = b = ?, c = 14$

7. $a = c = 5, b = ?$

8. $a = ?, b = 2.1, c = 7.5$

9. $a = \frac{1}{3}, b = \frac{1}{4}, c = ?$

Solve, using radical notation.

10. What is the length of the hypotenuse of an isosceles right triangle whose leg measures 7?

....

11. What is the measure of the altitude of an equilateral triangle whose side is 3?

....

12. The diagonal distance between vertices of a square is 36 ft. What is the edge dimension of the square?

....

13. In a 30°–60°–90° triangle, the length of the shortest side is 5. What are the lengths of the other two sides?

....

14. The sides of a triangle measure d, $3d$, and $2\frac{1}{2}d$. Is the triangle a right triangle?

....

15. The area of a triangle is equal to $\frac{1}{2}$ the product of the base and altitude measures. What is the area of an isosceles right triangle whose hypotenuse is 14?

....

16. Write a formula for the area of an equilateral triangle in terms of the measure of its side, s.

....

17. The length of a rectangle is three times its width. If its diagonal is 20 inches long, what are the length and width measurements?

....

18. Compute the length of CD if the length of AB is 4. (Use radical notation.)

....

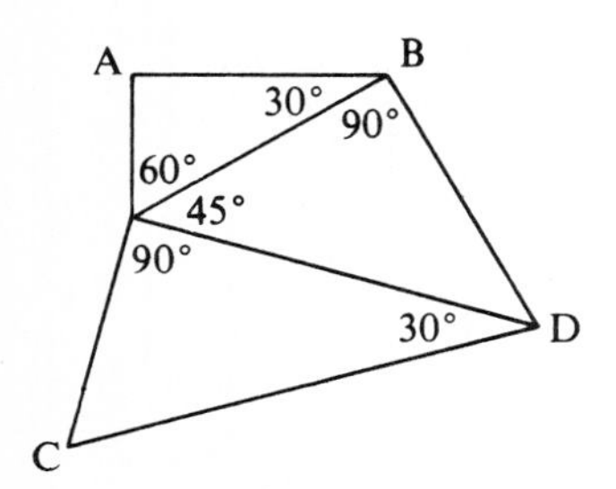

Name.. Date.............................. **EXERCISE 9.2**

1. Complete the following with respect to right triangle MNO whose sides measure m, n, o.

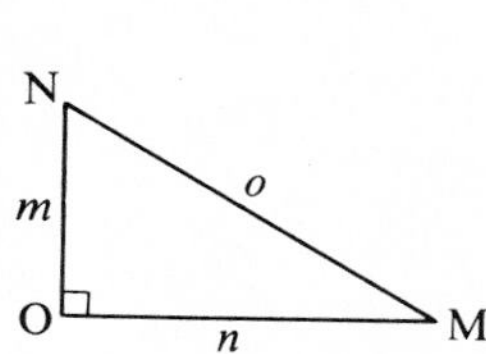

(a)	(b)	(c)	(d)
sin M =	sec N =	............M $= \frac{m}{n}$	............M $= \frac{o}{n}$
cos M =	csc M =	............M $= \frac{n}{m}$	............N $= \frac{n}{o}$
tan N =	cot N =	............N $= \frac{o}{n}$	............N $= \frac{m}{o}$

Express the six trigonometric ratios for the angle given in the right triangle.

2.

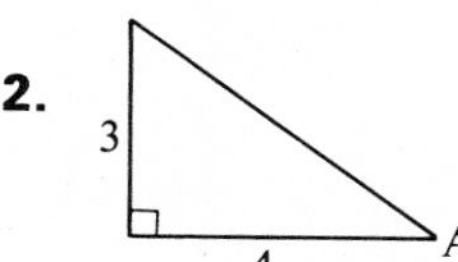
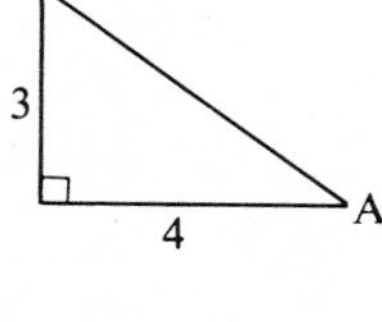

(a)	(b)	(c)
sin A =	tan A =	sec A =
cos A =	cot A =	csc A =

3.

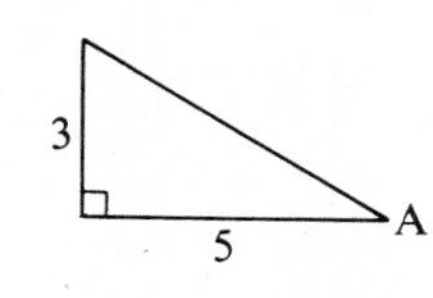

(a)	(b)	(c)
sin A =	tan A =	sec A =
cos A =	cot A =	csc A =

4.

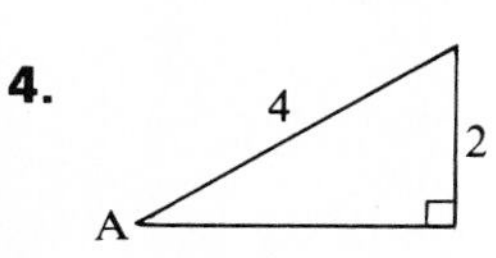

(a)	(b)	(c)
sin A =	tan A =	sec A =
cos A =	cot A =	csc A =

5.

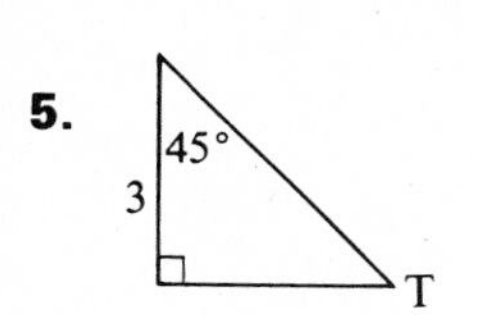

(a)	(b)	(c)
sin T =	tan T =	sec T =
cos T =	cot T =	csc T =

6.

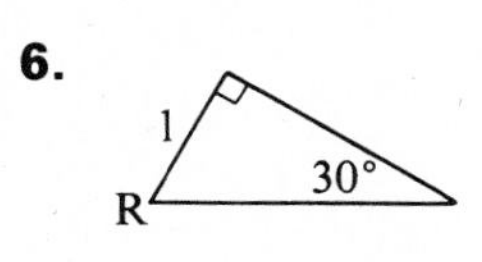

(a)	(b)	(c)
sin R =	tan R =	sec R =
cos R =	cot R =	csc R =

7. Express the trigonometric ratios decimally for sin A, cos A, and tan A in right triangles ABC, ADE, and AFG.

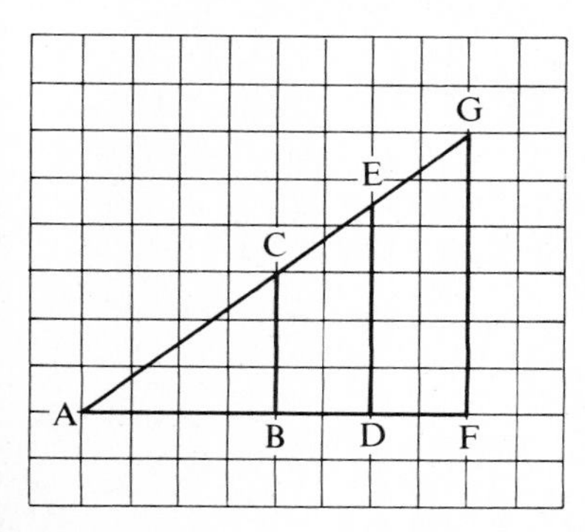

	△ABC	△ADE	△AFG
sin A			
cos A			
tan A			

8. Give the complementary angle for:

(a)		(b)		(c)		(d)		(e)	
60°	....	84°	....	35°	....	2°	....	49°	....
45°	....	17°	....	55°	....	28°	....	78°	....
58°	....	90°	....	0°	....	$18\frac{1}{2}°$	....	62°30′	....

9. Complete:

(a)	(b)	(c)
sin 24° = cos	tan .... = cot 89°	cot 57° = ... 33°
sec 15° = csc	tan 40° = ... 50°	csc 45° = ... 45°

10. Use Table B to complete the following.

(a)	(b)	(c)
sin 15° =	sec 80° =	cos ....° = .139
cos 28° =	csc 64° =	sec ....° = 2.46
tan 37° =	sec 26° =	cot ....° = 1.28
cot 42° =	tan 53° =	sin ....° = .985
sec 12° =	sin ....° = .122	tan ....° = 1.54
csc 28° =	cos ....° = .978	csc ....° = 4.13
sin 50° =	tan ....° = .869	sin ....° = 1.00

Use Table B and interpolation to compute these.

11. sin 29.4° =

12. tan 73.7° =

13. cos 46.3° =

14. sin A = .807

A = ...

15. cot B = 2.17

B = ...

16. sec M = 3.89

M = ...

Name.. Date.............................. **EXERCISE 9.3**

Measures given are for right triangle ABC, right angle at C. Find the measure of the indicated part. Express side measures decimally and angle measures to the nearest degree.

1. $b = 10$, $B = 53°$, $a = ?$ **2.** $A = 19°$, $c = 8$, $b = ?$ **3.** $a = 14$, $A = 42°$, $c = ?$

4. $a = 16$, $b = 25$, $A = ?$ **5.** $b = 13$, $c = 18$, $B = ?$ **6.** $c = 28$, $a = 7$, $B = ?$

Solve the following right triangles ABC, right angle at C. Express angle measures to nearest tenth of a degree.

7. $c = 52$, $B = 34.0°$ **8.** $a = 28$, $b = 45$

9. $a = 74$, $A = 65.6°$

10. $a = 12$, $c = 37$

Solve.

11. A driveway rises at an angle of 10° from where it joins the street. A point up the driveway 35′ from the street is how high above the level of the street?

....

12. A guy wire supporting a radio tower is to make an angle with the ground of 55° and connect to the tower 185′ above the ground. How long must the wire be?

....

13. A diagonal of a rectangle, 22 × 65, makes two angles with the sides. What are the measures of the angles, to the nearest degree?

....

Name.. Date.............................. **EXERCISE 9.4**

Use the identities of Section 8 to verify the following identities.

1. $\sin\theta \sec\theta = \tan\theta$

2. $\sin A \sec A \cot A = 1$

3. $\sec B \cot B = \csc B$

4. $\cos A \tan A = \dfrac{1}{\csc A}$

5. $\dfrac{\sin^2 X + \cos^2 X}{\cos X} = \sec X$

6. $\dfrac{1 + \sin A}{\sin A} = 1 + \csc A$

7. $\sin^2\theta + \sin^2\theta \cot^2\theta = 1$

8. $\cos^2 A - \sin^2 A = 2\cos^2 A - 1$

9. $\dfrac{1 - \sin C}{\cos C} = \sec C - \tan C$

10. $(\csc^2 T)(1 - \cos^2 T) = 1$

11. $\dfrac{\cos x \csc x}{\cot x} = 1$

12. $\tan \theta + \cot \theta = \csc \theta \sec \theta$

13. $(\cot A + 1)^2 = \csc^2 A + 2 \cot A$

14. $\cot^2 \theta \cos^2 \theta = \cot^2 \theta - \cos^2 \theta$

15. $\dfrac{\sin A + \tan A}{1 + \cos A} = \tan A$

16. $\dfrac{1 - \cos \theta}{\sin^2 \theta} = \dfrac{\tan \theta}{\tan \theta + \sin \theta}$

Decide whether or not each of the sets of three numbers could be the measures of the sides of a right triangle.

1. 3, 4, 5

2. 6, 8, 10

3. 6, 13, 14

4. 5, 12, 13

5. 4, 7, 8

6. 12, 25, 37

In the following, a and b are the measures of the legs of a right triangle, and c the hypotenuse. Compute the missing measure.

7. $a = 8, b = 15, c = ?$

8. $a = ?, b = 35, c = 37$

9. $a = 30, b = ?, c = 50$

10. $a = 45, b = 28, c = ?$

11. $a = 6, b = ?, c = 9$

12. $a = ?, b = 3, c = 7$

13. $a = 5, b = 10\sqrt{2}, c = ?$

14. $a = 5, b = ?, c = 5\sqrt{2}$

Conjecture: For any two positive numbers, m and n, the terms $m^2 - n^2$, $m^2 + n^2$, and $2mn$ will be the side measures of a right triangle.

15. Test the conjecture by using three pairs of numbers for m and n.

16. Prove the conjecture algebraically.

Find the measure of the missing part of the isosceles right triangle.

17. legs = 7; hypotenuse = ?

18. legs = 3; hypotenuse = ?

19. legs = ?; hypotenuse = $9\sqrt{2}$

20. legs = ?; hypotenuse = $\sqrt{72}$

Find the measures of the missing parts of the equilateral triangle.

21. side = 8; altitude = ?

22. side = $2\sqrt{3}$; altitude = ?

23. side = ?; altitude = 5

24. side = ?; altitude = $3\sqrt{3}$

25. A rope 73 ft long is attached to the top of a flagpole, and can reach the ground 48 ft from the base of the pole. If the pole is perpendicular to level ground, how high is the pole?

26. A tether ball swings on a rope 8 ft long attached to the top of a pole 12 ft high. How far is the ball from the pole when it is 7 ft above the ground?

Express the six trigonometric ratios for angle A.

27.

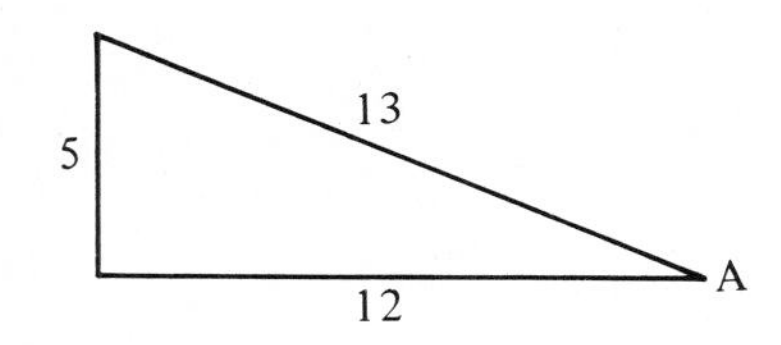

28.

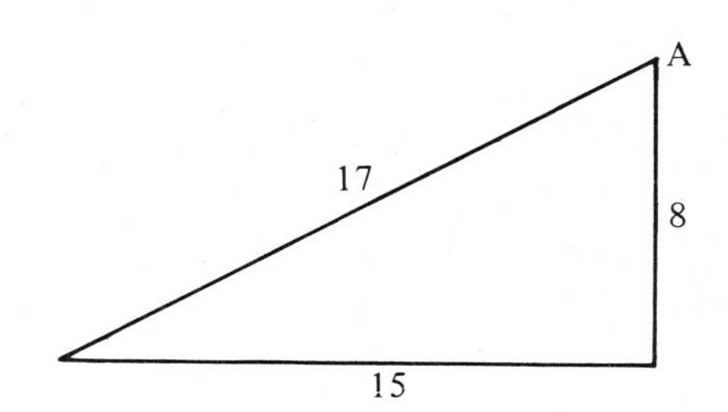

Give the complementary angle for these.

	(a)	(b)	(c)
29.	30°	40°	70°
30.	15°	65°	85°
31.	45°	67°	8°
32.	77°	0°	21°

Use Table B to write values for these.

	(a)	(b)
33.	sin 30°	cos 30°
34.	sin 45°	cos 45°
35.	sin 68°	cos 68°
36.	sin 37°	tan 47°
37.	cos 29°	cot 64°
38.	sec 57°	csc 70°
39.	tan 88°	sec 8°
40.	cot 23°	tan 23°

Use Table B to write the angle measure of A.

	(a)	(b)
41.	sin A = .276	cos A = .819
42.	tan A = .249	cot A = .196
43.	tan A = 1.80	cot A = .424
44.	sin A = .755	cos A = .438
45.	sec A = 1.08	csc A = 1.17
46.	csc A = 2.67	sec A = 1.89

	(a)	(b)
47.	$\tan A = .754$	$\cos A = .017$
48.	$\sin A = .070$	$\cot A = .466$

Use Table B and interpolation to write values for these.

	(a)	(b)
49.	$\sin 26.5°$	$\tan 32.2°$
50.	$\cos 15.8°$	$\cot 19.4°$
51.	$\cos 48.3°$	$\sec 62.1°$
52.	$\sin 68.4°$	$\csc 82.9°$

Give the angle measure of A to the nearest tenth degree.

	(a)	(b)
53.	$\cot A = .920$	$\sin A = .979$
54.	$\sec A = 1.36$	$\tan A = 2.02$
55.	$\csc A = 1.40$	$\cot A = .192$
56.	$\cos A = .397$	$\sin A = .127$

Triangle ABC is a right triangle, right angle at C. Find the missing measure. Express angles to nearest degree.

57. $b = 44$, $B = 17°$, $a = ?$

58. $A = 40°$, $c = 100$, $b = ?$

59. $a = 38$, $A = 41°$, $c = ?$

60. $a = 24$, $b = 16$, $A = ?$

61. $b = 10$, $c = 13$, $B = ?$

62. $c = 12$, $a = 3$, $B = ?$

Solve right triangle ABC, right angle at C, using the given data. Express angle measures to the nearest tenth of a degree.

63. $c = 12$, $B = 16.4°$

64. $a = 11$, $b = 7$

65. $a = 6$, $c = 18$

66. $A = 45.5°$, $a = 9.1$

Use the identities of Section 8 to verify the following.

67. $2 \cos A \sec A - \tan A \cot A = 1$

68. $\sec^2 B = \csc B \sin B + \dfrac{1}{\cot^2 B}$

69. $\sec A - \tan A \sin A = \cos A$

70. $\cot C = \dfrac{\cos C \sec C}{\tan C}$

71. $\sec^2 A - \sin^2 A = \cos^2 A + \tan^2 A$

72. $\dfrac{\sin B + \tan B}{1 + \sec B} = \sin B$

73. $\dfrac{\csc C}{\cos C} = (1 + \tan^2 C)(\cot C)$

74. $\sin^4 A - \cos^4 A = \sin^2 A - \cos^2 A$

75. $(\sin A + \cos A)^2 = 2 \sin A \cos A + 1$

REVIEW

Simplify

1. $\sqrt{-12} + \sqrt{-75} + \sqrt{-3}$

2. $\sqrt{-2} + \sqrt{-8} - \sqrt{-72}$

3. $3i - 2i^3 + i^4 - 3i^2$

4. $6 - 2i^3 + 3i - 5i^4$

5. $(3i^5 + 2i^3)^2$

6. $(2i^7 + 3i^4)^2$

Solve for x.

7. $x^2 + 3x = 10$

8. $x^2 + 3x = 4$

9. $x^2 - 10x + 21 = 0$

10. $x^2 - 6x = 0$

11. $x^2 + 4x = 0$

12. $9x^2 - 3x - 2 = 0$

13. $8x^2 + 26x + 15 = 0$

14. $x^2 + 3x + 1 = 0$

15. $x^2 + 6x - 4 = 0$

16. $2x^2 - 12x + 3 = 0$

17. $5x^2 + 3x = 1$

18. $2x^2 - x + 1 = 0$

19. $x^2 - 4x + 13 = 0$

20. $x^2 + 18 = 0$

Complete.

21. If $f(x) = 3x^2 - 4x + 7$, then $f(2) = ?$

22. If $f(x) = 4x^3 - 3x + 2$, then $f(-3) = ?$

23. If $f(a) = 2a^2 - 3a$, then $f(a + 3) = ?$

24. If $f(c) = 3c^2 + c^{-1}$, then $f(2) = ?$

25. If $f(x) = (x - 3)^{-2}$, then $f(-1) = ?$

10

GENERAL TRIANGLE

1. GENERAL TRIGONOMETRIC FUNCTIONS

The six basic trigonometric functions were defined in Section 4 of Chapter 9 in terms of the acute angles of a right triangle. By means of the Cartesian coordinate system we may extend those definitions to angles of any size. If a triangle is placed in the first quadrant of the Cartesian coordinate system with one leg along the horizontal axis and the vertex of one of the acute angles at the origin (0, 0), then the vertex of the other acute angle has as coordinates the measures of the two legs. The distance between the two vertices will be the measure of the hypotenuse. This is illustrated in Fig. 10.1a with a specific right triangle.

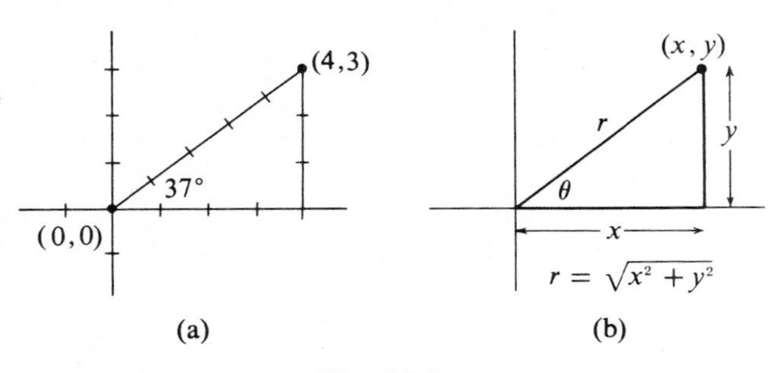

Fig. 10.1

In Fig. 10.1b, the specific dimensions of Fig. 10.1a are replaced by variables: x (abscissa), y (ordinate), r (distance between (0, 0) and (x, y)), and θ (angle). For any acute angle θ, then, the six trigonometric functions may be defined as ratios involving abscissas, ordinates and distances:

$$\sin\theta = \frac{y \text{ (ordinate)}}{r \text{ (distance)}} \qquad \cot\theta = \frac{x \text{ (abscissa)}}{y \text{ (ordinate)}}$$

$$\cos\theta = \frac{x \text{ (abscissa)}}{r \text{ (distance)}} \qquad \sec\theta = \frac{r \text{ (distance)}}{x \text{ (abscissa)}}$$

$$\tan\theta = \frac{y \text{ (ordinate)}}{x \text{ (abscissa)}} \qquad \csc\theta = \frac{r \text{ (distance)}}{y \text{ (ordinate)}}$$

A ray with its terminal point at the origin of the coordinate system and containing some other specified point in the plane forms an angle with the positive side of the horizontal axis. Such angles are said to be in *standard position*. Four angles in standard position—one in each quadrant—are shown in Fig. 10.2.

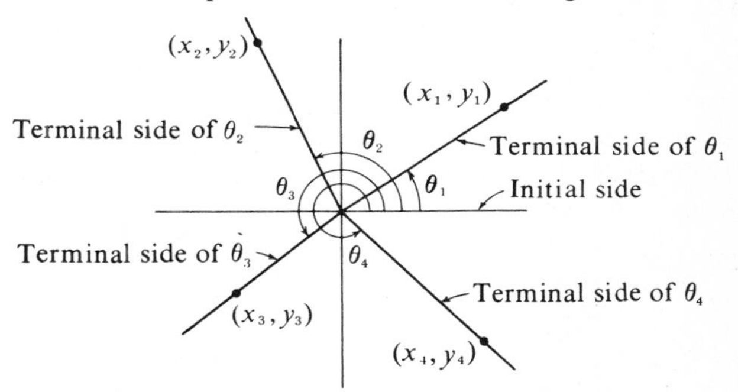

Fig. 10.2

When an angle is in standard position, the positive side of the horizontal axis is referred to as the *inital side*; the other side, which contains the origin and the point in the plane, is called the *terminal side* (see Fig. 10.2). In trigonometry it is useful to think of the measure of an angle in standard position as being related to the amount of rotation, about the origin, necessary to send the initial side into the terminal side. For instance, moving a clock hand backwards from 3 to 12 generates an angle of 90°; from 3 to 1 generates an angle of 60° (i.e., $\frac{2}{12} \times 360°$); from 3 back to 11 generates an angle of 120° (i.e., $\frac{4}{12} \times 360°$); and so on.

By defining angles in this way, and by using the abscissa-ordinate-distance definitions of the trigonometric functions, these functions can be extended in meaning and application to angles whose measures are greater than 90°.

To select one function as an example, the values of the tangent of the four angles displayed in Fig. 10.2 are given by the ratios:

$$\tan\theta_1 = \frac{y_1}{x_1} \qquad \tan\theta_3 = \frac{y_3}{x_3}$$

$$\tan\theta_2 = \frac{y_2}{x_2} \qquad \tan\theta_4 = \frac{y_4}{x_4}$$

EXAMPLES

1. Write the values of the six trigonometric functions for β (beta), an angle in standard position, if $(-3, 2)$ is on the terminal side of β.

$(-3, 2)$

$$r = \sqrt{(-3)^2 + (2)^2} = \sqrt{9 + 4} = \sqrt{13}$$

$$\sin\beta = \frac{\text{ord}}{\text{dis}} = \frac{2}{\sqrt{13}}$$

$$\cos\beta = \frac{\text{abs}}{\text{dis}} = \frac{-3}{\sqrt{13}} = -\frac{3}{\sqrt{13}}$$

$$\tan\beta = \frac{\text{ord}}{\text{abs}} = \frac{2}{-3} = -\frac{2}{3}$$

$$\cot\beta = \frac{\text{abs}}{\text{ord}} = \frac{-3}{2} = -\frac{3}{2}$$

$$\sec\beta = \frac{\text{dis}}{\text{abs}} = \frac{\sqrt{13}}{-3} = -\frac{\sqrt{13}}{3}$$

$$\csc\beta = \frac{\text{dis}}{\text{ord}} = \frac{\sqrt{13}}{2}$$

Note that $90^\circ < \beta < 180^\circ$.

2. Write the values of the six trigonometric functions for ϕ (phi), in standard position, if $(-1, -1)$ is on the terminal side.

$(-1, -1)$

$$r = \sqrt{(-1)^2 + (-1)^2} = \sqrt{1 + 1} = \sqrt{2}$$

$$\sin\phi = \frac{y}{r} = -\frac{1}{\sqrt{2}} \qquad \cot\phi = \frac{x}{y} = 1$$

$$\cos\phi = \frac{x}{r} = -\frac{1}{\sqrt{2}} \qquad \sec\phi = \frac{r}{x} = -\sqrt{2}$$

$$\tan\phi = \frac{y}{x} = 1 \qquad \csc\phi = \frac{r}{y} = -\sqrt{2}$$

Note that $180^\circ < \phi < 270^\circ$.

3. Write the values of the six trigonometric functions for γ (gamma), in standard position, if $(4, -3)$ is on the terminal side.

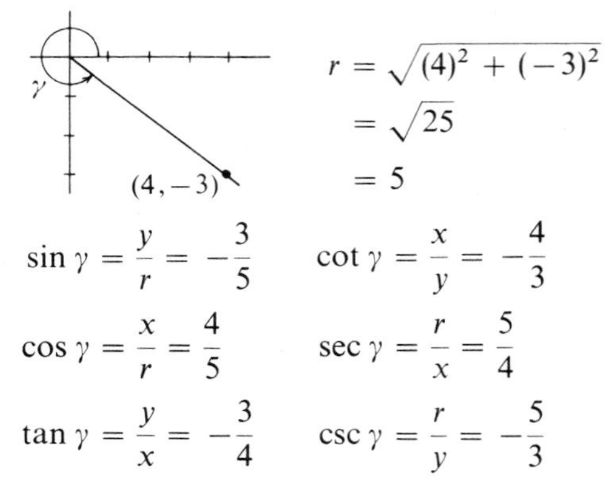

$$r = \sqrt{(4)^2 + (-3)^2} = \sqrt{25} = 5$$

$$\sin\gamma = \frac{y}{r} = -\frac{3}{5} \qquad \cot\gamma = \frac{x}{y} = -\frac{4}{3}$$

$$\cos\gamma = \frac{x}{r} = \frac{4}{5} \qquad \sec\gamma = \frac{r}{x} = \frac{5}{4}$$

$$\tan\gamma = \frac{y}{x} = -\frac{3}{4} \qquad \csc\gamma = \frac{r}{y} = -\frac{5}{3}$$

Note that $270^\circ < \gamma < 360^\circ$.

2. REFERENCE ANGLE

Numerical equivalents of the trigonometric ratios for angles greater than 90° have been compiled in tables. While it is possible to extend Table B, Values of Trigonometric Functions (on page 252 and inside back cover), to include angles from 0° through 360°, it is also possible to operate with Table B alone. To do so requires a knowledge of the signs (+ or −) of the values of the functions in the various quadrants and how to compute the measure of a reference angle.

With respect to the signs, the values of all six trigonometric functions are positive for first-quadrant angles—that is, for angles in standard position with terminal sides in the first quadrant, or $0^\circ < \theta < 90^\circ$. This is so because all abscissas, ordinates, and distances are positive in the first quadrant. (The distance is positive in every quadrant, since r is equal to the principal root of $\sqrt{(\text{abscissa})^2 + (\text{ordinate})^2}$.)

In the second quadrant (i.e., $90^\circ < \theta < 180^\circ$), the abscissas are negative, while ordinates and distances are positive. Therefore the cosine (x/r) and tangent (y/x) functions, as well as their reciprocals, secant and cotangent, have negative ratio values in the second quadrant; the other two functions—sine and cosecant—have positive values. (See Example 1, Section 1.)

In the third quadrant (i.e., $180^\circ < \theta < 270^\circ$), both abscissas and ordinates are negative while distances are positive. Therefore all values of functions of angles in the third quadrant are negative except for tangent and cotangent. (See Example 2, Section 1.)

In the fourth quadrant (i.e., $270^\circ < \theta < 360^\circ$), ordinates are negative while abscissas and distances are positive. So functions of angles that have the ordinate as part of their ratio (sine, tangent, cotangent, cosecant) have negative values in the fourth quadrant. (See Example 3, Section 1.) These generalizations are summarized in Fig. 10.3.

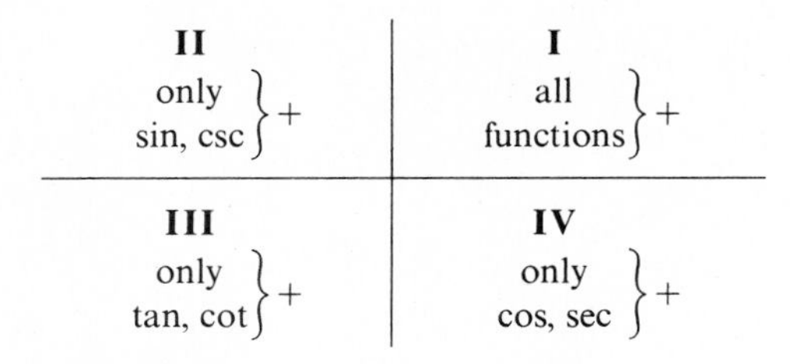

Fig. 10.3

The *reference angle* for an angle in standard position is the *acute* angle formed by the terminal side of the angle in standard position and the horizontal axis.

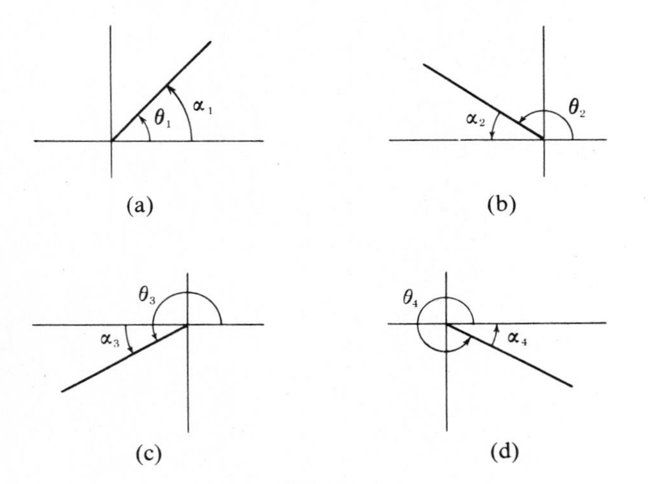

Fig. 10.4

Note in Fig. 10.4a that the measure of the reference angle, α (alpha), for an angle in the first quadrant is that of the angle itself.

For an angle in the second quadrant (Fig. 10.4b), the reference angle (α_2) has the measure of the *supplement* of the given angle (θ_2), i.e., $\alpha_2 = 180° - \theta_2$.

For an angle in the third quadrant (Fig. 10.4c), the measure of the reference angle (α_3) is the difference between that of the given angle (θ_3) and 180°, i.e., $\alpha_3 = \theta_3 - 180°$.

For an angle in the fourth quadrant (Fig. 10.4d), the measure of the reference angle (α_4) is the difference between 360° and the measure of the given angle, θ_4, i.e., $\alpha_4 = 360° - \theta_4$.

EXAMPLES

1. What are the measures of the reference angles for (a) 62°; (b) 140°; (c) 200°; (d) 300°?

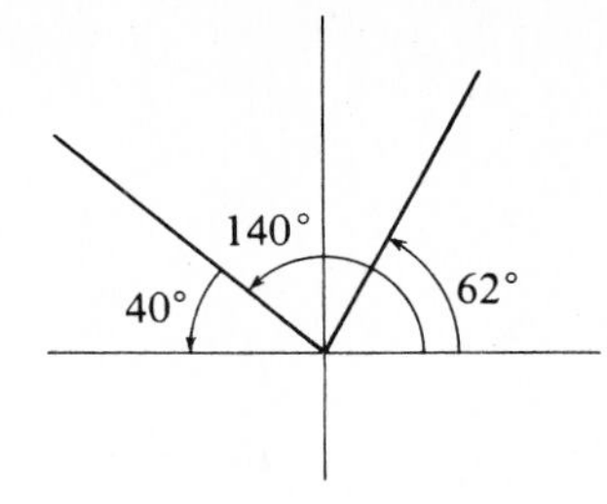

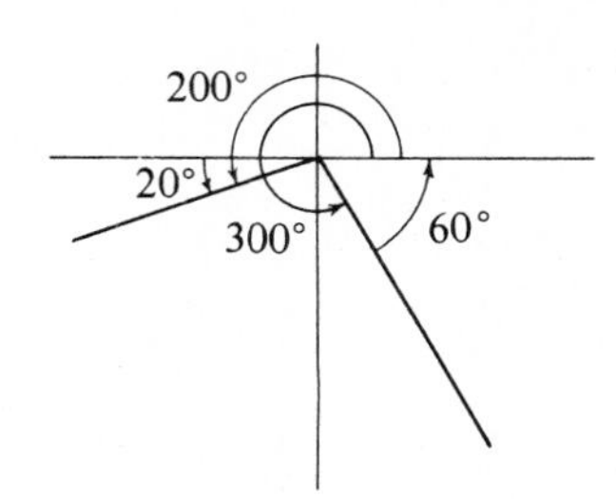

(a) 62° is a first-quadrant angle; its reference angle has the same measure: 62°.
(b) 140° is a second-quadrant angle; its reference angle has the measure of its supplement:
$180° - 140° = 40°$.
(c) 200° is a third-quadrant angle; the measure of its reference angle is equal to the difference between that of the given angle and 180°:
$200° - 180° = 20°$.
(d) 300° is a fourth-quadrant angle; the measure of its reference angle is equal to the difference between 360° and that of the given angle:
$360° - 300° = 60°$.

2. What are the measures of the reference angles for angles of (a) 122°; (b) 317°; (c) 97°; (d) 265°?
(a) 122° is in quadrant II; so
$180° - 122° = 58°$ (ref. angle).
(b) 317° is in quadrant IV; so
$360° - 317° = 43°$ (ref. angle).
(c) 97° is in quadrant II; so
$180° - 97° = 83°$ (ref. angle).
(d) 265° is in quadrant III; so
$265° - 180° = 85°$ (ref. angle).

(Complete Exercise 10.1)

3. TRIGONOMETRIC VALUES FROM TABLES

The generalizations advanced in Section 2 concerning the reference angle and the sign of the trigonometric ratios of angles of the various quadrants allow us to use Table B (on page 252 and inside back cover) to express the values of the trigonometric functions of any angle.

(10.1) *To use a table of values of trigonometric functions (Table B) to find the value of a trigonometric function of an angle:*
Step 1. Determine the quadrant of the given angle and the sign (+ or −) of the given trigonometric function in that quadrant.
Step 2. Compute the reference angle for the given angle.

Step 3. *Find the entry in Table B for the reference angle of Step 2.*
Step 4. *Affix the sign determined in Step 1 to the entry found in Step 3 for the value of the given trigonometric function of the given angle.*

EXAMPLES

1. Write the value of cos 140°.
   Step 1. 140° is in quadrant II; the cosine function is negative in II
   Step 2. Reference angle for 140°: $180° - 140° = 40°$
   Step 3. $\cos 40° = .766$ (Table B)
   Step 4. $\cos 140° = -(\cos 40°) = -.766$
2. What is the value of sin 330°?
   Step 1. 330° is in quadrant IV: sine is negative in IV
   Step 2. $360° - 330° = 30°$ (reference angle)
   Step 3. $\sin 30° = .500$ (from Table B)
   Step 4. $\sin 330° = -(\sin 30°) = -.500$
3. $\tan 215° = ?$
   215° is in III; tan is positive in III
   $215° - 180° = 35°$
   $\tan 35° = .700$
   $\tan 215° = \tan 35° = .700$
4. $\sec 109.2° = ?$
   109.2° is in II: sec is negative in II
   $180° - 109.2° = 70.8°$
   $\sec 109.2° = -(\sec 70.8°) = -3.04$

When the circumstances are reversed, that is, when we have the value of a trigonometric function and wish to find the corresponding angle measure, we reverse the procedure of Program 10.1. In most cases, however, two angles less than 360° will have the same trigonometric value. For example:

$$\sin 30° = \sin 150° = +.500$$

$$\sin 210° = \sin 330° = -.500$$

(10.2) *To use a table of values of trigonometric functions (Table B) to find the angle measures less than 360° that have a given trigonometric value:*

Step 1. *Determine from the sign of the given trigonometric value the two quadrants in which the given trigonometric function would be valid.*
Step 2. *Find the angle measure in Table B that has the same trigonometric value as the absolute value of the given trigonometric value.*
Step 3. *Compute the measures of angles in the quadrants noted in Step 1 that have the angle measure of Step 2 as a reference angle.*

EXAMPLES

1. For what values of A is $\cos A = .940$?
   Step 1. Cosine is positive in I and IV
   Step 2. $|.940| = .940 = \cos 20°$ (from Table B)
   Step 3. 20° is reference angle for 20° in I, and for $360° - 20° = 340°$ in IV
   Thus, $\cos 20° = .940$
   $\cos 340° = .940$
2. For what values of θ is $\tan \theta = -.577$?
   Step 1. Tangent is negative in II and IV
   Step 2. $|-.577| = .577 = \tan 30°$
   Step 3. II: $180° - 30° = 150°$
   IV: $360° - 30° = 330°$
   Thus, $\tan 150° = \tan 330° = -.577$
3. For what value of ϕ is $\sec \phi = -2.37$?
   Secant is negative in II and III
   $|-2.37| = 2.37 = \sec 65°$
   II: $180° - 65° = 115°$
   III: $180° + 65° = 245°$
   $\sec 115° = \sec 245° = -2.37$
4. For what value of B is $\tan B = 1.19$ if sin B is negative?
   Tangent is positive in I and III; sine is negative in III and IV. Therefore B must be in III.
   $|1.19| = 1.19 = \tan 50°$ (reference angle)
   III: $180° + 50° = 230°$
   $\tan 230° = 1.19$, sin 230° is negative;
   $\tan 50° = 1.19$ also, but sin 50° is positive;
   so only $B = 230°$ satisfies all the conditions of the problem.

Angles of 0°, 90°, 180°, and 270° are called *quadrantal angles.* When in standard position they are not considered a part of any quadrant because their terminal sides fall on the borders between quadrants. The trigonometric values for these angles can easily be determined by the ratio method used in Section 1. Note in Fig. 10.5 that the terminal side of each of the four quadrantal angles has a measure of 1. For an angle of 0°, the terminal side coincides with the initial side.

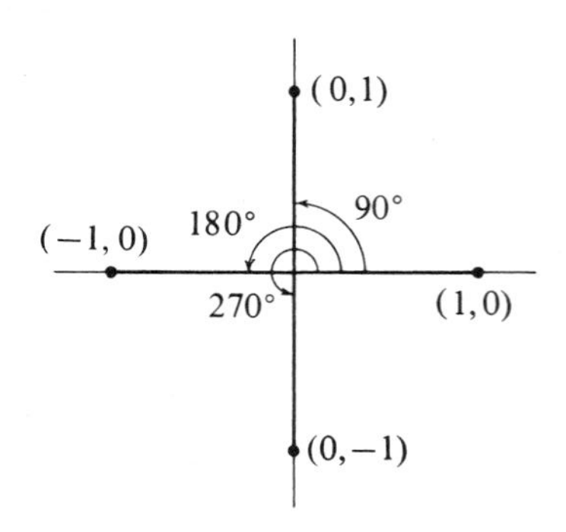

Fig. 10.5

The following table of trigonometric values can be verified by the data of Fig. 10.5. Ratio fractions with denominators of 0 are meaningless, and are indicated in the table by the abbreviation "und" for "undefined."

	0°	90°	180°	270°
sin	$\frac{0}{1} = 0$	$\frac{1}{1} = 1$	$\frac{0}{1} = 0$	$\frac{-1}{1} = -1$
cos	$\frac{1}{1} = 1$	$\frac{0}{1} = 0$	$\frac{-1}{1} = -1$	$\frac{0}{1} = 0$
tan	$\frac{0}{1} = 0$	$\frac{1}{0}$ und	$\frac{0}{-1} = 0$	$\frac{-1}{0}$ und
cot	$\frac{1}{0}$ und	$\frac{0}{1} = 0$	$\frac{-1}{0}$ und	$\frac{0}{-1} = 0$
sec	$\frac{1}{1} = 1$	$\frac{1}{0}$ und	$\frac{1}{-1} = -1$	$\frac{1}{0}$ und
csc	$\frac{1}{0}$ und	$\frac{1}{1} = 1$	$\frac{1}{0}$ und	$\frac{1}{-1} = -1$

(Complete Exercise 10.2)

4. LAW OF SINES

A very important set of ratios in trigonometry is known as the *Law of Sines*:

$$\frac{a}{\sin A} = \frac{b}{\sin B} = \frac{c}{\sin C}$$

Expressed in words: *The measures of the sides of a triangle are proportional to the sines of the angles opposite those sides.*

The Law of Sines may be developed in several ways, one of which is the following.

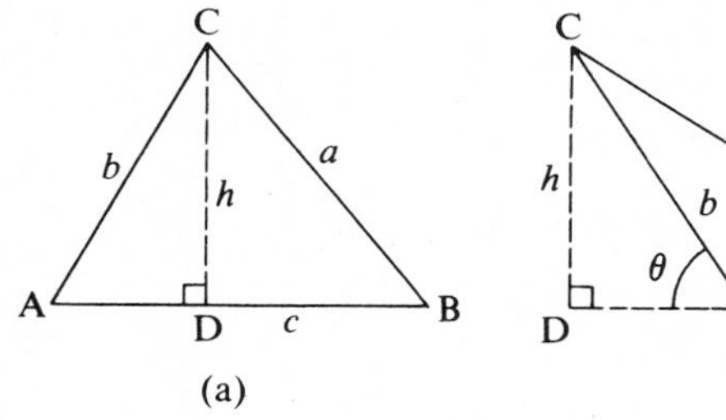

Fig. 10.6

Consider the two variations of the general triangle in Fig. 10.6—the acute triangle in Fig. 10.6a and the obtuse triangle in Fig. 10.6b. In both, a right triangle appears, ADC, having h as the length of perpendicular CD. In right triangle ADC of Fig. 10.6a

$$\sin A = \frac{h}{b}, \quad \text{or} \quad h = b \sin A$$

In right triangle ADC of Fig. 10.6b

$$\sin \theta = \frac{h}{b}, \quad \text{or} \quad h = b \sin \theta$$

Since $A = 180° - \theta$, $\sin \theta = \sin A$, and for the obtuse triangle of Fig. 10.6b

$$h = b \sin \theta = b \sin A$$

Similarly, in both variations of the general triangle in Fig. 10.6 there is right triangle BDC. In each

$$\sin B = \frac{h}{a}, \quad \text{or} \quad h = a \sin B$$

Equating the two respective expressions for h we obtain

$$b \sin A = a \sin B$$

or

$$\frac{a}{\sin A} = \frac{b}{\sin B}$$

which holds for both types of triangle.

When a perpendicular is extended from vertex A to the opposite side, a similar line of reasoning leads to:

$$\frac{b}{\sin B} = \frac{c}{\sin C}$$

In general, then:

$$\frac{a}{\sin A} = \frac{b}{\sin B} = \frac{c}{\sin C}$$

The Law of Sines is particularly useful for solving general triangles when the known measures are those of:

- two angles and one side;
- two sides and the angle opposite one of them.

In the second case—two sides and an opposite angle—two solutions *may* be possible. (See Examples 2 and 3.)

EXAMPLES

(Note: Angle measures are to nearest degree.)

1. Solve triangle ABC in which $A = 70°$, $B = 30°$, and $b = 100$.

To find C:

$$C = 180° - (A + B) = 180° - (70° + 30°) = 80°$$

To find a:

$$\frac{a}{\sin A} = \frac{b}{\sin B}$$

$$a = \frac{b \sin A}{\sin B} = \frac{(100)(\sin 70°)}{\sin 30°} = \frac{(100)(.940)}{.500} = \frac{94.0}{.5} = 188$$

To find c:

$$\frac{b}{\sin B} = \frac{c}{\sin C}$$

$$c = \frac{b \sin C}{\sin B} = \frac{(100)(\sin 80°)}{\sin 30°} = \frac{(100)(.985)}{.500} = \frac{98.5}{.500} = 197$$

2. Solve triangle ABC in which $A = 40°, a = 7, b = 10$.
To find B:

$$\frac{a}{\sin A} = \frac{b}{\sin B}$$

$$\sin B = \frac{b \sin A}{a} = \frac{(10)(\sin 40°)}{7} = \frac{(10)(.643)}{7} = .919$$

If $\sin B = .919$, then $B = 67°$, also $113°$.

If $B = 67°$, then:
$C = 180° - (A + B)$
$= 180° - (40° + 67°)$
$= 73°$

If $B = 113°$, then:
$C = 180° - (A + B)$
$= 180° - (40° + 113°)$
$= 27°$

Consequently, two different triangles fit the given data.

When $C = 73°$:

$$\frac{c}{\sin C} = \frac{a}{\sin A}$$

$$c = \frac{a \sin C}{\sin A} = \frac{(7)(\sin 73°)}{\sin 40°} = \frac{(7)(.956)}{.643} = 10.4$$

When $C = 27°$:

$$\frac{c}{\sin C} = \frac{a}{\sin A}$$

$$c = \frac{a \sin C}{\sin A} = \frac{(7)(\sin 27°)}{\sin 40°} = \frac{(7)(.454)}{.643} = 4.9$$

Triangle I: $A = 40°, B = 67°, C = 73°, a = 7, b = 10, c = 10.4$

Triangle II: $A = 40°, B = 113°, C = 27°, a = 7, b = 10, c = 4.9$

3. Solve triangle ABC in which $C = 70°, b = 20, c = 25$.
To find B:

$$\sin B = \frac{b \sin C}{c} = \frac{(20)(\sin 70°)}{25} = \frac{(20)(.940)}{25} = .752$$

$B = 49°, 131°$

If $B = 49°$, then:
$A = 180° - (B + C)$
$= 180° - (49° + 70°)$
$= 61°$

If $B = 131°$, then:
$A = 180° - (B + C)$
$= 180° - (131° + 70°)$
$= 180° - 201°$
$= -21°$

Since a measure of $-21°$ for A has no meaning in this situation, only one triangle fits the given data; its angle measures are $A = 61°, B = 49°, C = 70°$.
To find a:

$$a = \frac{c \sin A}{\sin C} = \frac{(25)(\sin 61°)}{\sin 70°} = \frac{(25)(.875)}{.940} = 23$$

(Complete Exercise 10.3)

5. LAW OF COSINES

The Law of Sines is inadequate to solve a triangle when the known measures are those of:

- two sides and the included angle;
- three sides.

Instead we may use the *Law of Cosines*, which may be stated:

The square of the measure of the side of a triangle is equal to the sum of the squares of the measures of the two other sides minus twice the product of the measures of these two sides multiplied by the cosine of their included angle.

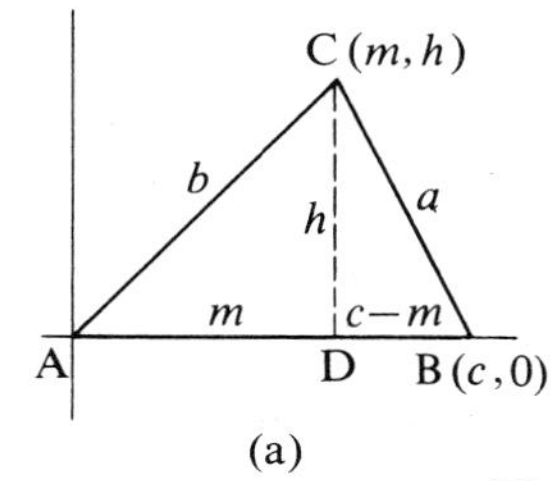

(a)

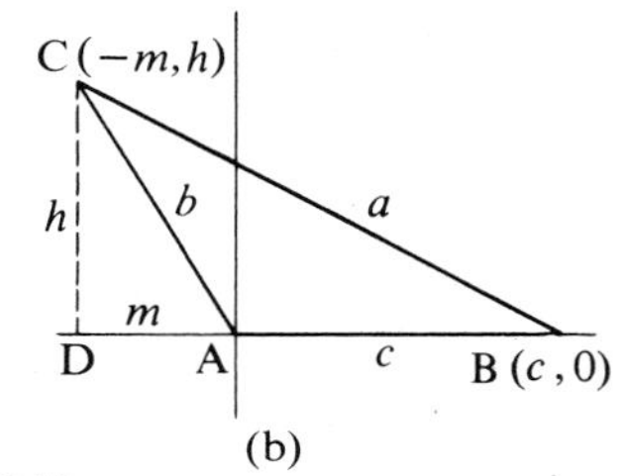

(b)

Fig. 10.7

For either version of the general triangle shown in Fig. 10.7, the Law of Cosines states:

$$a^2 = b^2 + c^2 - 2bc \cos A$$

$$b^2 = a^2 + c^2 - 2ac \cos B$$

$$c^2 = a^2 + b^2 - 2ab \cos C$$

This set of expressions of the Law of Cosines may be derived in various ways. One is to place the triangle in the Cartesian plane so that the vertex of one of its angles is in standard position at the origin, as has been done with the acute and obtuse triangles of Fig. 10.7. In both cases a right triangle, ADC, may be observed in which the measure of AD is m. In Fig. 10.7a the coordinates of vertex C are clearly (m, h), and

$$\cos A = \frac{m}{b}$$

or

$$m = b \cos A$$

Continuing with the acute triangle of Fig. 10.7:

$$h^2 = b^2 - m^2 \qquad \text{(in } \triangle \text{ADC)}$$

and

$$h^2 = a^2 - (c - m)^2 \qquad \text{(in } \triangle \text{DBC)}$$
$$= a^2 - c^2 + 2cm - m^2$$

Equating the expressions for h^2, simplifying, and substituting ($b \cos A$) for m:

$$a^2 - c^2 + 2cm - m^2 = b^2 - m^2$$

$$a^2 = b^2 + c^2 - 2cm$$

$$a^2 = b^2 + c^2 - 2c(b \cos A)$$

$$a^2 = b^2 + c^2 - 2bc \cos A$$

In the obtuse triangle of Fig. 10.7, the coordinates of C are $(-m, h)$, and so

$$\cos A = \frac{-m}{b}$$

or

$$m = -b \cos A$$

Again:

$$h^2 = b^2 - m^2 \qquad \text{(in } \triangle \text{ADC);}$$

but in this obtuse version of the general triangle:

$$h^2 = a^2 - (m + c)^2 \qquad \text{(in } \triangle \text{DBC)}$$
$$= a^2 - m^2 - 2mc - c^2$$

Equating the expressions for h^2, simplifying, and substituting ($-b \cos A$) for m, we arrive eventually at the same expression for the Law of Cosines as above:

$$a^2 - m^2 - 2mc - c^2 = b^2 - m^2$$

$$a^2 = b^2 + c^2 + 2mc$$

$$a^2 = b^2 + c^2 + 2c(-b \cos A)$$

$$a^2 = b^2 + c^2 - 2bc \cos A$$

A similar line of reasoning can be used to derive

$$b^2 = a^2 + c^2 - 2ac \cos B$$

and

$$c^2 = a^2 + b^2 - 2ab \cos C$$

by locating B and C, respectively, at the origin.

EXAMPLES

(Note: Angle measures are to nearest degree.)

1. Solve triangle ABC in which $a = 5$, $b = 6$, and $C = 60°$.

   To find c:

   $$c^2 = a^2 + b^2 - 2ab \cos C$$
   $$= (5)^2 + (6)^2 - 2(5)(6)(\cos 60°)$$
   $$= 25 + 36 - 2(5)(6)(.500)$$
   $$= 25 + 36 - 30$$
   $$= 31$$

   $$c = \sqrt{31} = 5.6 \qquad \text{(from Table A)}$$

   To find A:

   $$\frac{a}{\sin A} = \frac{c}{\sin C}$$

   $$\sin A = \frac{a \sin C}{c}$$
   $$= \frac{(5)(\sin 60°)}{5.6}$$
   $$= .773$$
   $$A = 51°$$

   To find B:

   $$B = 180° - (A + C)$$
   $$= 180° - (51° + 60°)$$
   $$= 180° - 111°$$
   $$= 69°$$

2. To solve the triangle ABC in which $a = 5$, $b = 7$, $c = 9$.

   To find A:

   $$a^2 = b^2 + c^2 - 2bc \cos A$$

   $$\cos A = \frac{b^2 + c^2 - a^2}{2bc}$$
   $$= \frac{7^2 + 9^2 - 5^2}{2(7)(9)}$$
   $$= \frac{49 + 81 - 25}{126}$$

   $$\cos A = \frac{105}{126} = .833$$

   $$A = 34°$$

   To find B:

   $$b^2 = a^2 + c^2 - 2ac \cos B$$

   $$\cos B = \frac{a^2 + c^2 - b^2}{2ac}$$
   $$= \frac{5^2 + 9^2 - 7^2}{2(5)(9)}$$
   $$= \frac{25 + 81 - 49}{90}$$

   $$\cos B = \frac{57}{90} = .633$$

   $$B = 51°$$

   To find C:

   $$\cos C = \frac{a^2 + b^2 - c^2}{2ab} = \frac{25 + 49 - 81}{2(5)(7)} = \frac{-7}{70}$$
   $$= -.100$$

   $$C = 180° - 84° \text{ (reference angle)} = 96°$$

   [Note: The sum of the computed angle measures, 34°, 51°, and 96°, is 181°. Discrepancies of this sort are due to rounding errors in computing the measures of the angles.]

(Complete Exercise 10.4)

Name.. Date.............................. **EXERCISE 10.1**

The capital letter designates an angle in standard position, the terminal side of which contains the point whose abscissa and ordinate are given. Make a rough sketch of the angle, compute the distance ($r = \sqrt{(\text{abscissa})^2 + (\text{ordinate})^2}$) and write the six trigonometric ratios for the angle.

1. A; (−4, 3)

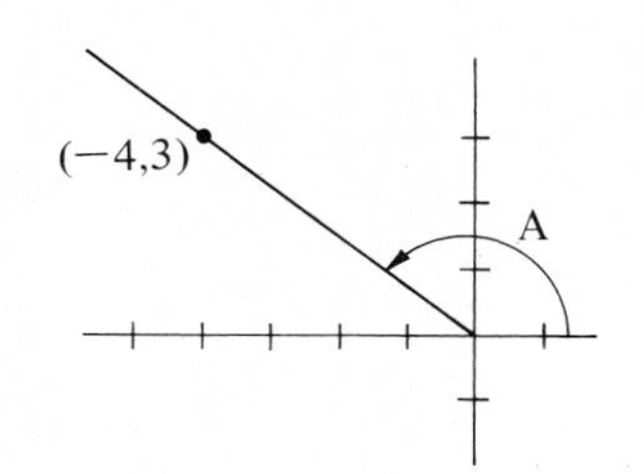

sin A =

cos A =

tan A =

cot A =

sec A =

csc A =

2. B; (5, 12)

3. C; (8, −15)

4. D; (−6, −8)

5. E; (−2, 1)

6. F; (−3, −5)

Make a rough sketch of each of the angles to show the position of the given angle and its reference angle. Determine the measure of the reference angle.

7. 120°.................

8. 195°.................

9. 80°.................

10. 208° ....

11. 312° ....

12. 13°.....................

13. 345° ....

14. 181° ....

15. $162\frac{1}{2}$° ....

Make a sketch of the given angle, determine the measure of its reference angle, and predict the sign (+ or −) of the value for each of the six trigonometric functions of the given angle.

16. 223° Ref. ∠: ....

+ or −

sin 223° ......−......

cos 223° ....

tan 223° ......+......

cot 223° ....

sec 223° ....

csc 223° ....

17. 154° Ref. ∠: ....

+ or −

sin 154° ....

cos 154° ....

tan 154° ....

cot 154° ....

sec 154° ....

csc 154° ....

18. 75° Ref. ∠: ....

+ or −

sin 75° ....

cos 75° ....

tan 75° ....

cot 75° ....

sec 75° ....

csc 75° ....

19. 330° Ref ∠: ....

+ or −

sin 330° ....

cos 330° ....

tan 330° ....

cot 330° ....

sec 330° ....

csc 330° ....

Name.. Date.............................. **EXERCISE 10.2**

Use Table B to find the trigonometric value.

1. $\sin 140° =$

2. $\tan 350° =$

3. $\sec 300° =$

4. $\cos 315° =$

5. $\cot 230° =$

6. $\csc 100° =$

7. $\sin 243° =$

8. $\csc 307° =$

9. $\sin 180° =$

10. $\sec 270° =$

11. $\tan 180° =$

12. $\cos 90° =$

Give the two angle measures less than 360° for θ to make the equation true.

13. $\sin \theta = .423$

$\theta =$ ...., ...

14. $\tan \theta = -.839$

$\theta =$ ...., ...

15. $\csc \theta = -1.84$

$\theta =$ ...., ...

16. $\sec \theta = 1.72$

$\theta =$ ...., ...

17. $\cos \theta = 0$

$\theta =$ ...., ...

18. $\cot \theta =$ undefined

$\theta =$ ...., ...

Solve.

19. If A < 360°, sec A = 1.29 and tan A < 0 (i.e., negative), then A = ?

....

20. If B < 360°, tan B = −2.05 and sin B > 0 (i.e., positive), then B = ?

....

21. If C < 360°, sin C = −.743 and cot C > 0, then C = ?

....

22. If D < 360°, cos D = −1, then D = ?

....

23. Complete the tables for each of the equations, then graph each with a smooth continuous curve. Round off y-values to the nearest hundredth.

x	0°	15°	30°	45°	60°	75°	90°	105°	120°	135°	150°	165°
$y = \sin x$												
$y = \cos x$												

x	180°	195°	210°	225°	240°	255°	270°	285°	300°	315°	330°	345°	360° = 0°
$y = \sin x$													
$y = \cos x$													

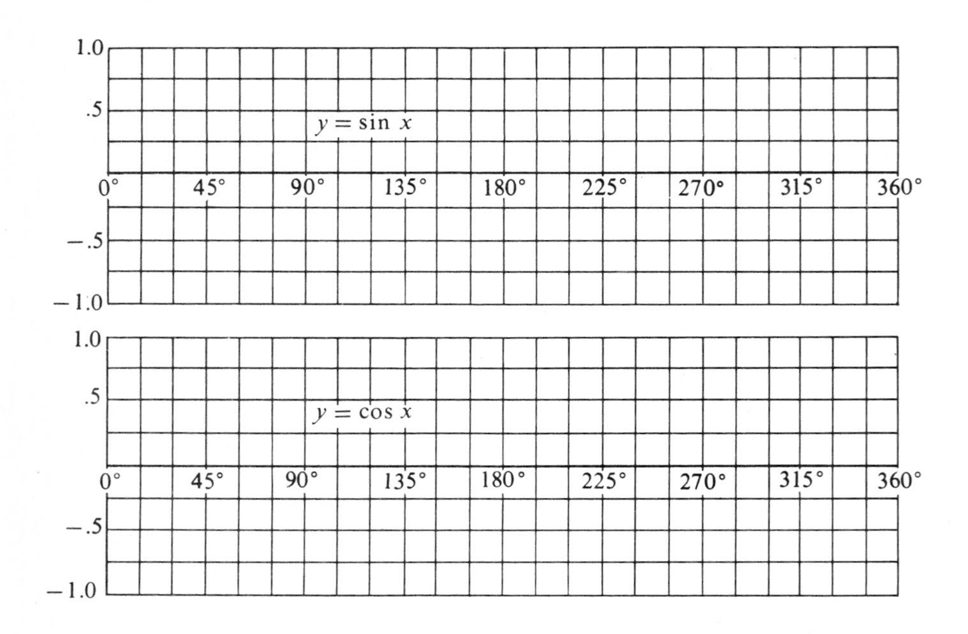

Name.. Date.............................. **EXERCISE 10.3**

Solve triangle ABC for which the given data are known. Compute angle measures to nearest degree.

1. $c = 25$, $A = 68°$, $B = 35°$

2. $B = 23°$, $b = 40$, $c = 50$

3. $a = 19$, $A = 26°$, $C = 71°$

4. $A = 24°$, $a = 20$, $b = 13$

5. The following two area formulas apply to triangles whose vertices are A, B, C: K (area) $= \frac{1}{2}ab \sin C$, and $K = \frac{1}{2}a^2 \frac{\sin B \sin C}{\sin A}$. Write the two remaining versions of these formulas by noting a pattern in each.

$K = \frac{1}{2}ac \sin$ ....

$K =$ .... $\sin A$

$K = \frac{1}{2}b^2 \frac{\qquad}{\sin B}$

$K = \frac{1}{2}c^2 \frac{\sin \qquad}{\qquad}$

Solve. Compute angle measures to nearest degree.

6. Find the area of triangle ABC in which $a = 24$; $b = 15$; $C = 77°$.

....

7. Compute the area of triangle PQR in which $p = 20$, $r = 31$, $P = 24°$, $R = 141°$.

....

8. Ship A is 80 miles west of ship B. If A is sailing N 35° E and B is sailing N 15° W, how far must A sail along its course to cross the path of B?

....

9. A 20-foot ladder just reaches the sill of a window when it makes an angle of 65° with the ground. What angle must a 30-foot ladder make with the ground in order to just reach the sill of the same window?

....

Name.. Date.............................. **EXERCISE 10.4**

Solve triangle ABC for which the given data are known. Compute angle measures to nearest degree.

1. $a = 7, b = 9, C = 60°$

2. $a = 7, b = 8, c = 10$

3. $A = 136°, b = 24, c = 4$

4. $a = 9, b = 12, c = 5$

Solve. Compute angle measures to nearest degree.

5. Use the Law of Cosines to find c in the *right* triangle ABC in which $C = 90°$, $a = 5$, $b = 12$.

....

6. Use the Law of Cosines to tell why it is impossible to have a triangle with side-measures of 3, 9, and 15.

....

7. $K = \sqrt{s(s-a)(s-b)(s-c)}$, where $s = \frac{1}{2}(a+b+c)$, is called Hero's formula for the area of a triangle, and requires knowing the measures of the three sides. Use it to find the area of the triangle of Exercise 4.

....

8. How much does the area of a triangle increase when the lengths of its sides are doubled?

....

9. The lengths of the diagonals of a parallelogram are 80 and 50. They cross at an angle of 67°. What are the approximate lengths of the sides of the parallelogram?

.................., ...

Write the six trigonometric functions for each angle.

1. (−3,4) A

2.

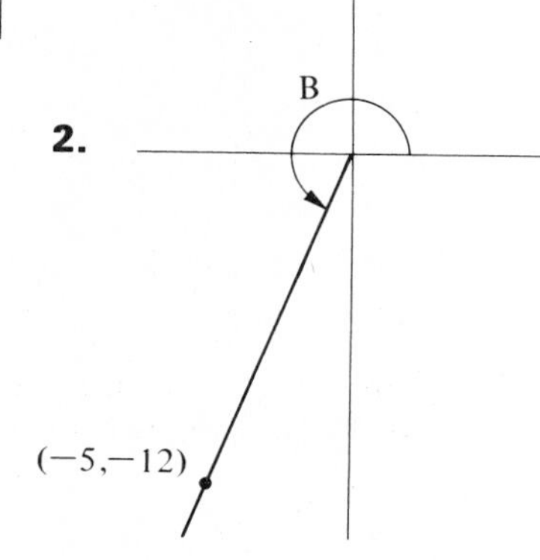

3.

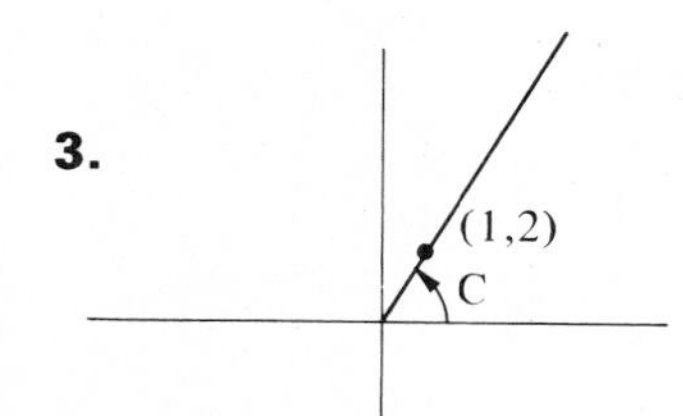

4.

5.

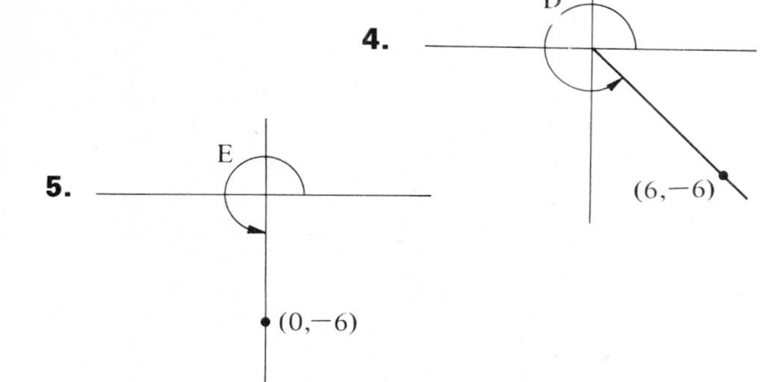

6.

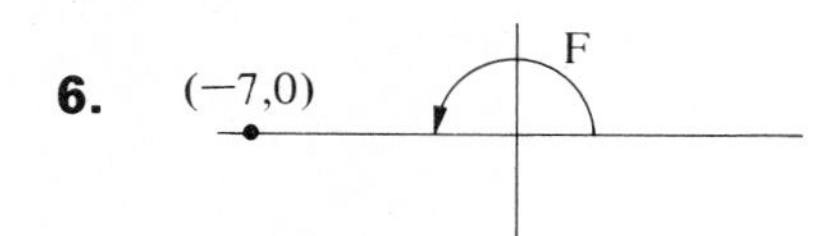

Give the reference angle.

7. 130° **8.** 205°
9. 73° **10.** 118°
11. 216° **12.** 27°
13. 314° **14.** 269°
15. 182° **16.** 272°

Complete the table (+ or −).

	17. 132°	**18.** 17°	**19.** 316°	**20.** 155°	**21.** 212°	**22.** 340°
sin	+					
cos	−					
tan						
cot						
sec						
csc						

Use Table B to find the trigonometric value.

23. sin 150° **24.** tan 340°
25. cos 233° **26.** sec 295°
27. csc 312° **28.** cot 226°
29. sin 268° **30.** csc 286°
31. sin 270° **32.** sec 90°
33. tan 0° **34.** sin 90°
35. sin 117.5° **36.** 228.6°
37. cot 316.8° **38.** tan 112.3°

Give a pair of angle measures for θ, such that $0° \leq \theta < 360°$

39. $\cos\theta = -.927$ **40.** $\sin\theta = .574$
41. $\tan\theta = 1.38$ **42.** $\cot\theta = -.810$
43. $\sec\theta = -3.24$ **44.** $\csc\theta = -1.41$
45. $\sin\theta = -.834$ **46.** $\tan\theta = 3.82$
47. $\tan\theta =$ undef. **48.** $\csc\theta =$ undef.

Solve.

49. $0° \leq A < 360°$
$\tan A = -.754$
$\sec A > 0$
$A = ?$

50. $0° \leq B < 360°$
$\sin B = .669$
$\cos B < 0$
$B = ?$

51. $0° \leq C \leq 180°$
$\csc C = 1.10$
$\tan C > 0$
$C = ?$

52. $90° \leq D \leq 270°$
$\cot D = -.700$
$\cos D > 0$
$D = ?$

53. $0° \leq E \leq 180°$
$\sin E = -.342$
$\tan E < 0$
$E = ?$

54. $180° \leq F \leq 360°$
$\csc F = -1$
$\tan F =$ undef.
$F = ?$

Given the following data with respect to triangle ABC; determine the measure called for.

55. $a = 18$, $A = 45°$, $B = 105°$, $c = ?$
56. $A = 120°$, $a = 8$, $b = 3$, $B = ?$

57. $a = 21$, $A = 51°$, $c = 15$, $C = ?$
58. $b = 73$, $B = 19°$, $C = 27°$, $c = ?$
59. $b = 28$, $B = 48°$, $c = 35$, $C = ?$
60. $b = 10$, $c = 8$, $C = 43°$, $B = ?$
61. $b = 16$, $c = 21$, $C = 122°$, $a = ?$
62. $a = 32$, $b = 47$, $B = 39°$, $c = ?$

Compute the area of triangle ABC, if

63. $a = 6$, $c = 10$, $B = 45°$
64. $a = 27$, $b = 40$, $C = 37°$
65. $b = 56$, $c = 41$, $A = 74°$
66. $b = 21$, $c = 12$, $A = 32°$

Determine the measure called for with respect to triangle ABC.

67. $b = 10$, $c = 20$, $A = 50°$, $a = ?$
68. $a = 10$, $b = 15$, $C = 52°$, $c = ?$
69. $a = 4$, $b = 9$, $c = 11$, $C = ?$
70. $a = 17$, $b = 19$, $c = 5$, $A = ?$
71. $b = 70$, $c = 60$, $A = 32°$, $a = ?$
72. $a = 12$, $c = 4.5$, $B = 122°$, $b = ?$
73. $a = 24$, $b = 25$, $c = 7$, $B = ?$
74. $a = 15$, $b = 20$, $c = 24$, $C = ?$

Use Hero's formula, $K = \sqrt{s(s-a)(s-b)(s-c)}$, to compute the area of triangle ABC in which

75. $a = 5$, $b = 6$, $c = 7$
76. $a = 13$, $b = 17$, $c = 10$
77. $a = 10$, $b = 3$, $c = 5$
78. $a = 16$, $b = 9$, $c = 7$
79. Two ships pass each other. One is sailing west at 18 mph, and the other northeast at 15 mph. To the nearest mile, how far apart will the ships be two hours after they pass?
80. A ship sails 16 miles due east from its mooring, then along a course S 55° E. After 18 miles along this course, how far is the ship from its mooring?

REVIEW

Use the identities of Section 8, Chapter 9, to verify the following.

1. $\cos^2 A = (1 + \sin A)(1 - \sin A)$
2. $(\sec A + \tan A)(\sec A - \tan A) = 1$
3. $\sin B \csc^2 B = \csc B$
4. $\tan^2 A \cot A = \tan A$
5. $\sec^2 X \csc^2 X = \sec^2 X + \csc^2 X$
6. $\dfrac{\sin Y}{1 - \cos Y} = \dfrac{1 + \cos Y}{\sin Y}$
7. $\dfrac{\sin A \sec A}{\tan A} = 1$
8. $(\cos B)(\tan B)(\csc B) = 1$
9. $\dfrac{1}{1 + \sin B} + \dfrac{1}{1 - \sin B} = 2 \sec^2 B$
10. $\dfrac{1 + \tan^2 Z}{\csc^2 Z} = \tan^2 Z$

Express each of the following equivalently in terms of $\cos\theta$ only; simplify.

11. $\tan\theta \sin\theta + \cos\theta$
12. $1 - \dfrac{\sin^2\theta}{1 + \cos\theta}$
13. $\dfrac{\cos\theta \sec\theta}{1 + \tan^2\theta}$
14. $\dfrac{\cos\theta \tan\theta + \sin\theta}{\tan\theta}$
15. $\dfrac{2 - \tan^2\theta}{\sec^2\theta} - 1 + 2\sin^2\theta$
16. $\dfrac{2\cos^2\theta - \sin^2\theta + 1}{\cos\theta}$

Compute the length of CD (use radical notation).

17.

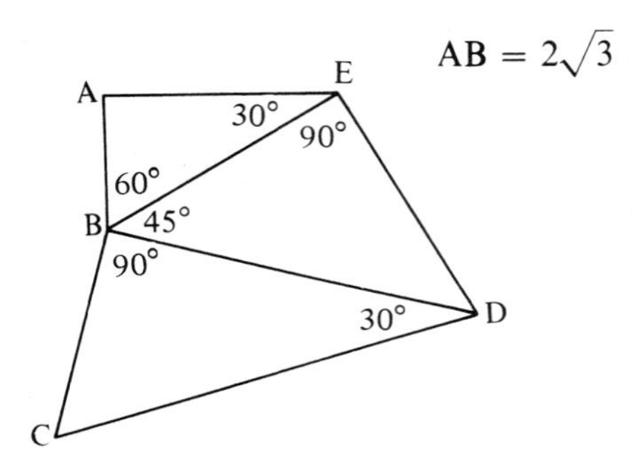

18.

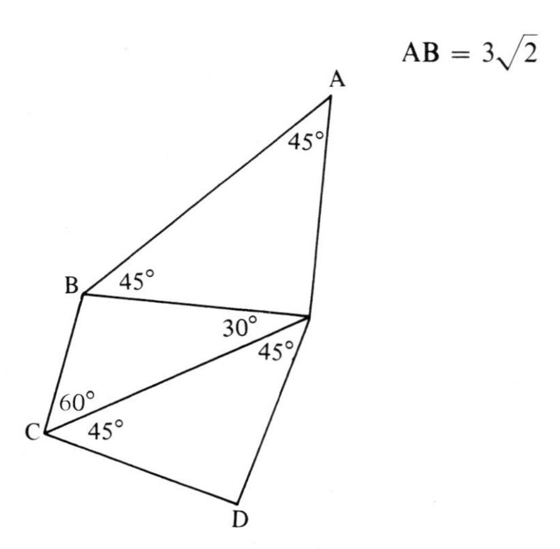

Name.. Date.............................. Score......................

ACHIEVEMENT TEST NO. 5 (Chapters 9 and 10)

True–False. (2 points each, total 10)

1. It is possible to have an oblique triangle containing two obtuse angles. ....

2. At least two angles are equal in an isosceles triangle. ....

3. Sine and cosecant are cofunctions. ....

4. Sine values never go beyond the interval bounded by $+1$ and -1. ....

5. The cotangent function has different signs in quadrants II and IV. ....

Complete the sentence. (3 points each, total 30)

6. The complementary angle for 56° is __?__. ....

7. The supplementary angle for 85° is __?__. ....

8. $\sin 48° = \cos$ __?__. ....

9. $\sec 39° =$ __?__ 51°. ....

10. The tangent function has negative values in quadrants __?__ and __?__. ....

11. The measure of the reference angle for 187° is __?__. ....

12. Two quadrantal angles are __?__ and __?__. ....

13. The ratio $\frac{\text{distance}}{\text{ordinate}}$ is associated with the __?__ function. ....

For $\triangle PQR$, having side measures p, q, and r:

14. The Law of Sines states that $q =$ __?__. ....

15. The Law of Cosines states that $p^2 =$ __?__. ....

Solve. (4 points each, total 12)

16. Express $\frac{\cos A \sec A}{\tan^2 A}$ in terms of cot A. ....

17. Express $\frac{\sin \theta}{1 + \cos \theta} + \frac{1 + \cos \theta}{\sin \theta}$ in terms of csc θ.

....

18. Is $\frac{\csc \theta}{\cos \theta}$ identical to $\cot \theta (1 + \tan^2 \theta)$?

....

Use Table B to complete. (2 points each correct answer, total 20)

19. $\sin 152° = \underline{\ ?\ }$. ....

20. $\tan 357° = \underline{\ ?\ }$. ....

21. $\sec 59.5° = \underline{\ ?\ }$. ....

22. $\cot \underline{\ ?\ } = -.650$ ...., ...

23. $\cos \underline{\ ?\ } = -.485$ ...., ...

24. $\tan \underline{\ ?\ } = 1.000$ ...., ...

25. If $\theta < 360°$, $\sec \theta = -.307$, and $\tan \theta > 0$, then $\theta = \underline{\ ?\ }$.

....

Solve. Compute angle measures to nearest degree. (7 points each problem, total 28)

26. In $\triangle ABC$, $A = 42°$, $C = 58°$, and $a = 10$. Find b.

....

27. In $\triangle ABC$, $C = 35°$, $b = 40$, $c = 25$. Find two measures for B.

.........., ...

28. In $\triangle ABC$, $a = 4$, $b = 7$, $c = 5$. Find A.

....

29. A pole leans 10° from the vertical toward the sun and casts a shadow 30 ft long when the angle of elevation of the sun is 40° above the horizon. How long is the pole?

....

Name.. Date.............................. Score......................

100 PROBLEM FINAL ACHIEVEMENT TEST

1. $.0632 + .804 + 2.786 + 42.3 + 186 =$ ....

2. $43.0 - .0762 =$ ....

3. $.036 \times .008754 =$ ....

4. $.019199 \div .0526 =$ ....

5. Arrange in order from least to greatest:
$\frac{6}{5}$, 1.19, $\frac{5}{6}$, $1\frac{1}{2}\%$
........., ..., ..., ...

6. Compute: $18\frac{3}{4} - 6\frac{2}{3} + 14\frac{3}{8} - 4\frac{1}{12} =$ ....

7. $144\frac{2}{3} \times 3{,}246\frac{3}{16} =$ ....

8. $8\frac{4}{7} \div 1\frac{1}{14} =$ ....

9. Simplify: $\dfrac{1}{1 + \dfrac{\frac{3}{4}}{1 - \frac{1}{2}}}$ ....

10. 28 is $3\frac{1}{2}$ times the size of what number? ....

11. Replace □ with < or >:
.625 □ $\frac{43}{72}$ ....

12. What number, when divided by $\frac{3}{8}$, yields a quotient of $\frac{4}{9}$ and a remainder of $\frac{1}{4}$? ....

13. If vegetables, when dried, lose $\frac{9}{14}$ of their fresh weight, how much must $4\frac{2}{7}$ lb of dried vegetables have weighed when fresh? ....

14. Express .004 as a fraction. ....

15. Express $\frac{7}{16}$ as a per cent. ....

16. Express 702.4% as a numeral in mixed form. ....

17. Express $\frac{1}{2}\%$ as a decimal. ....

18. Express $\frac{3}{400}$ as a decimal. ....

19. 70% of $3\frac{1}{3}$ is what number? ....

20. 16 is what per cent greater than 12? ....

21. .096 is 12% of what number? ....

Name.. Date..............................

22. 18% less than 6.50 is what number? ....

23. A sample of salt weighed 252 grams before drying and 231 grams after drying. What per cent moisture did the sample contain? ....

24. What was the price of an article before taxes if it cost $141.68, including a 10% tax? ....

25. A coat, reduced 20% from its original price, is now on sale for $64.80. What was its original price? ....

26. A library contains 16,350 volumes, 16% of which are novels. If 25% of the novels were published in 1970 or later, how many novels are there in the library with publication dates prior to 1970? ....

Simplify:

27. $4[(5 + 2)^2 - (3 + |-2|)] - |9 - 12|$ ....

28. $(3a^2 + 2b - c) - (4a^2 + 3b - c) + (c - 3b - 2a^2) =$ ....

29. $(4x^2 - 3x + 2x^3)(x^2 - 2x) =$ ....

30. $2a - \{6b - 3[2b - 4 - (2a - 1)] - 2\} =$ ....

31. $\dfrac{(3)(-8)(2)}{(-6)(4)(-1)} =$ ....

32. $\dfrac{-12a^3x^3 + 7a^6x^0 - 42ax^4}{-14a^2x^3} =$ ....

33. $(4x^3 + 5y^3 - 13xy^2 + 4x^2y) \div (5y + 2x) =$ ....

Solve:

34. $4x - 3(3 + 4x) = 3$ ....

35. $\dfrac{3x}{4} + \dfrac{1}{6} = \dfrac{x}{3} - 1\dfrac{1}{2}$ ....

36. $2x - 5 < 4x + 3$ ....

Name.. Date..............................

37. $6x + 4 \geq 2(x + 5)$ ....

38. $\begin{cases} x + 2y = 6 \\ 4x - 4 = 3y \end{cases}$ ....

39. Evaluate $\begin{vmatrix} -3 & 7 \\ 2 & -4 \end{vmatrix}$ ....

40. Solve by Cramer's Rule

$\begin{cases} x + 4y = 0 \\ 3x + 14y = -1 \end{cases}$ ....

41. At what point will the graphs of $2x - 3y = 9$ and $3x + y = 8$ intersect? ....

42. At what point will the graph of $2x - 3y = 8$ cross the x-axis? ....

43. Find two numbers, one $\frac{3}{5}$ of the other, whose sum is 60. ...., ...

44. The sum of the digits of a two-digit numeral is 10. If the digits are reversed, the new numeral represents a number 36 larger than the original. What must have been the original numeral? ....

45. Part of \$2,500 is invested at 4% and the rest of it at 6%. If the total yearly income from these investments is \$129, how much is invested at each rate? ....

46. If $S = \frac{rl - a}{r - 1}$, find r when $S = 84$, $a = 3$, $l = 54$. ....

Expand and simplify:

47. $st[s^2 - t(3s - 5t)^2] =$ ....

48. $(a + b)^2 + b(2a + b) + (b - a)^2 =$ ....

49. $\left(\frac{a}{4} - \frac{b}{3}\right)\left(\frac{a}{2} + \frac{b}{6}\right) =$ ....

Factor completely:

50. $243x^4y^4 - 3$ ....

51. $100x^2 + 30x - 4$ ....

52. $2a^2x + 3a^2y - 2b^2x - 3b^2y$ ....

53. $a^4 - 13a^2 + 36$ ....

Name.. Date..............................

Simplify:

54. $\dfrac{2x^2 - 7x + 6}{2x^2 - 6x - 20} \cdot \dfrac{x^2 + x - 2}{3 - 5x + 2x^2} =$ ....

55. $\dfrac{3x^2 - 10x + 8}{x + 7} \div \dfrac{28 - 25x + 3x^2}{x^2 - 49} =$ ....

56. $\dfrac{3}{a^2 - 9} - \dfrac{4}{a^2 - 16} + \dfrac{2}{a^2 - a - 12} =$ ....

57. $\dfrac{\dfrac{1}{a} - \dfrac{1}{b}}{\dfrac{1}{a^2} - \dfrac{1}{b^2}} =$ ....

Solve:

58. $\dfrac{1}{2x - 1} - \dfrac{1}{3 - x} = \dfrac{2}{3x - 9}$ ....

59. $\dfrac{2a + 4}{a^2 + 2a - 8} - \dfrac{a + 6}{a^2 + 3a - 4} = \dfrac{a - 4}{a^2 - 3a + 2}$ ....

60. $\begin{cases} \dfrac{1}{x} + \dfrac{1}{y} = 5 \\ \dfrac{1}{x} - \dfrac{1}{y} = -3 \end{cases}$ ....

61. The sum of two numbers is $3\frac{1}{3}$ and the sum of their reciprocals is $1\frac{1}{4}$. Find the two numbers. ....

Simplify:

62. $(a^2b^0c^{-3})^{-2} =$ ....

63. $\dfrac{x^{-1} + y^{-1}}{x + y} =$ ....

64. $(x^{-2} + y^{-1})^2 =$ ....

65. $\left(\dfrac{a^{2/3}y^{-3/4}}{a^{-1/2}y^{1/2}}\right)^6 =$ ....

66. $\sqrt{\sqrt[3]{\sqrt{64}}} =$ ....

67. $\dfrac{36}{\sqrt[3]{6}} =$ ....

Name.. Date...............................

68. $\dfrac{\sqrt{4}}{1 + \sqrt{7}} =$ ....

69. $(\sqrt{3})(\sqrt{12x} - \sqrt{3x} + \sqrt{27x}) =$ ....

70. $3i^3 - 2i^2 + 5i^5 + 4i^4 =$
$(i^2 = -1)$ ....

71. $(3i - \sqrt{-3})^2 =$ ....

72. If $f(x) = 2x^2 - 3x + 1$, then $f(\sqrt{3} - 2) = ?$

....

73. Name the ordered pairs, (x, y), defined by $\{(x, y)|y = 3x - 2, x, y \text{ integers}, -2 < x < 3\}$

....

Solve:

74. $6x^2 = 15 + x$ ....

75. $x^2 - 4x - 1 = 0$ ....

76. $3x^2 + 2x + 1 = 0$ ....

77. $2x^4 + 7x^2 - 15 = 0$ ....

78. $\sqrt{2x - 5} = 2 + \sqrt{x - 2}$ ....

79. Solve for x: $x^2 + 6mx + m^2 = 0$

....

80. A certain rectangle is twice as long as it is wide. If each dimension were increased by 3 in., then its new area would be 104 sq in. What are its dimensions?

81.

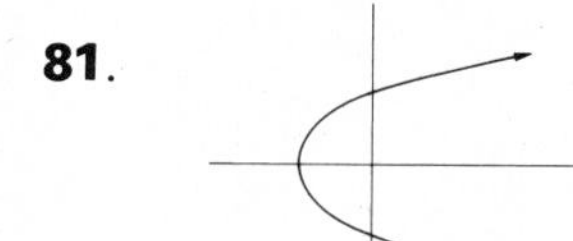

Is the graphed relation also a function? (Yes, No)

....

In right triangle ABC, with side measures a, b, c:

82. $\cos A = ?$ ....

83. $a = 8, c = 17, b = ?$ ....

84. $\csc B = . \ ? \ . \ . \ A$ ....

85. $a = 5, b = 12, B = ?$
(nearest 0.1°) ....

86. $A = 49°, b = 20, c = ?$ ....

Name.. Date..............................

87. Express in terms of cosecant:

$$\frac{\sin x}{1 + \cos x} + \cot x$$ ....

88. Is $2\cos^2 x - 1$ identical to

$$\frac{1 - \tan^2 x}{1 + \tan^2 x}?$$ ....

89. What trigonometric function has for its ratio:

$$\frac{\text{distance}}{\text{ordinate}}?$$ ....

90. What is the measure of the reference angle for 217°? ....

91. In what quadrant(s) is the tangent function negative? ....

92. The terminal side of what angle in standard position contains $(1, -1)$? ....

93. $\cot A = .105$ and $\sin A < 0$; $A = ?$ ....

94. $\sin B = 0$ and $\cos B = -1$; $B = ?$ ....

In general triangle ABC, with side measures a, b, c (compute angle measures to nearest degree):

95. $a = 100$, $B = 37°$, $C = 110°$, $c = ?$ ....

96. $b = 30$, $c = 20$, $C = 30°$, $B = ?$ ....

97. $a = 9$, $b = 11$, $c = 15$, $B = ?$ ....

98. $A = 120°$, $b = 6$, $c = 3$, $a = ?$ ....

99. A carrier is 200 miles north of a tanker. If the carrier sails S 45° E (i.e., SE) and the tanker sails N 10° E, how far must the carrier sail to cross the route of the tanker? ....

100. A golfer hits his drive off the tee 220 yd, but to the left of the line between tee and hole. If the distance from tee to hole is 420 yd, and the golfer is 250 yd from the hole after his drive, at what angle was his drive from the direct line? ....

TABLE A: POWERS—ROOTS—RECIPROCALS

n	n^2	n^3	$\sqrt{n}$	$\sqrt[3]{n}$	$1/n$
1	1	1	1.000	1.000	1.0000
2	4	8	1.414	1.260	.5000
3	9	27	1.732	1.442	.3333
4	16	64	2.000	1.587	.2500
5	25	125	2.236	1.710	.2000
6	36	216	2.449	1.817	.1667
7	49	343	2.646	1.913	.1429
8	64	512	2.828	2.000	.1250
9	81	729	3.000	2.080	.1111
10	100	1,000	3.162	2.154	.1000
11	121	1,331	3.317	2.224	.0909
12	144	1,728	3.464	2.289	.0833
13	169	2,197	3.606	2.351	.0769
14	196	2,744	3.742	2.410	.0714
15	225	3,375	3.873	2.466	.0667
16	256	4,096	4.000	2.520	.0625
17	289	4,913	4.123	2.571	.0588
18	324	5,832	4.243	2.621	.0556
19	361	6,859	4.359	2.668	.0526
20	400	8,000	4.472	2.714	.0500
21	441	9,261	4.583	2.759	.0476
22	484	10,648	4.690	2.802	.0455
23	529	12,167	4.796	2.844	.0435
24	576	13,824	4.899	2.884	.0417
25	625	15,625	5.000	2.924	.0400
26	676	17,576	5.099	2.962	.0385
27	729	19,683	5.196	3.000	.0370
28	784	21,952	5.292	3.037	.0357
29	841	24,389	5.385	3.072	.0345
30	900	27,000	5.477	3.107	.0333
31	961	29,791	5.568	3.141	.0323
32	1,024	32,768	5.657	3.175	.0312
33	1,089	35,937	5.745	3.208	.0303
34	1,156	39,304	5.831	3.240	.0294
35	1,225	42,875	5.916	3.271	.0286
36	1,296	46,656	6.000	3.302	.0278
37	1,369	50,653	6.083	3.332	.0270
38	1,444	54,872	6.164	3.362	.0263
39	1,521	59,319	6.245	3.391	.0256
40	1,600	64,000	6.325	3.420	.0250
41	1,681	68,921	6.403	3.448	.0244
42	1,764	74,088	6.481	3.476	.0238
43	1,849	79,507	6.557	3.503	.0233
44	1,936	85,184	6.633	3.530	.0227
45	2,025	91,125	6.708	3.557	.0222
46	2,116	97,336	6.782	3.583	.0217
47	2,209	103,823	6.856	3.609	.0213
48	2,304	110,592	6.928	3.634	.0208
49	2,401	117,649	7.000	3.659	.0204
50	2,500	125,000	7.071	3.684	.0200

n	n^2	n^3	$\sqrt{n}$	$\sqrt[3]{n}$	$1/n$
51	2,601	132,651	7.141	3.708	.0196
52	2,704	140,608	7.211	3.733	.0192
53	2,809	148,877	7.280	3.756	.0189
54	2,916	157,464	7.348	3.780	.0185
55	3,025	166,375	7.416	3.803	.0182
56	3,136	175,616	7.483	3.826	.0179
57	3,249	185,193	7.550	3.849	.0175
58	3,364	195,112	7.616	3.871	.0172
59	3,481	205,379	7.681	3.893	.0169
60	3,600	216,000	7.746	3.915	.0167
61	3,721	226,981	7.810	3.936	.0164
62	3,844	238,328	7.874	3.958	.0161
63	3,969	250,047	7.937	3.979	.0159
64	4,096	262,144	8.000	4.000	.0156
65	4,225	274,625	8.062	4.021	.0154
66	4,356	287,496	8.124	4.041	.0152
67	4,489	300,763	8.185	4.062	.0149
68	4,624	314,432	8.246	4.082	.0147
69	4,761	328,509	8.307	4.102	.0145
70	4,900	343,000	8.367	4.121	.0143
71	5,041	357,911	8.426	4.141	.0141
72	5,184	373,248	8.485	4.160	.0139
73	5,329	389,017	8.544	4.179	.0137
74	5,476	405,224	8.602	4.198	.0135
75	5,625	421,875	8.660	4.217	.0133
76	5,776	438,976	8.718	4.236	.0132
77	5,929	456,533	8.775	4.254	.0130
78	6,084	474,552	8.832	4.273	.0128
79	6,241	493,039	8.888	4.291	.0127
80	6,400	512,000	8.944	4.309	.0125
81	6,561	531,441	9.000	4.327	.0123
82	6,724	551,368	9.055	4.344	.0122
83	6,889	571,787	9.110	4.362	.0120
84	7,056	592,704	9.165	4.380	.0119
85	7,225	614,125	9.220	4.397	.0118
86	7,396	636,056	9.274	4.414	.0116
87	7,569	658,503	9.327	4.431	.0115
88	7,744	681,472	9.381	4.448	.0114
89	7,921	704,969	9.434	4.465	.0112
90	8,100	729,000	9.487	4.481	.0111
91	8,281	753,571	9.539	4.498	.0110
92	8,464	778,688	9.592	4.514	.0109
93	8,649	804,357	9.644	4.531	.0108
94	8,836	830,584	9.695	4.547	.0106
95	9,025	857,375	9.747	4.563	.0105
96	9,216	884,736	9.798	4.579	.0104
97	9,409	912,673	9.849	4.595	.0103
98	9,604	941,192	9.899	4.610	.0102
99	9,801	970,299	9.950	4.626	.0101
100	10,000	1,000,000	10.000	4.642	.0100

TABLE B: VALUES OF TRIGONOMETRIC FUNCTIONS

angle	sin	cos	tan	cot	sec	csc	angle
0°	.000	1.00	.000	. .	1.00	. .	90°
1°	.017	1.00	.017	57.3	1.00	57.3	89°
2°	.035	.999	.035	28.6	1.00	28.7	88°
3°	.052	.999	.052	19.1	1.00	19.1	87°
4°	.070	.998	.070	14.3	1.00	14.3	86°
5°	.087	.996	.087	11.4	1.00	11.5	85°
6°	.105	.995	.105	9.51	1.01	9.57	84°
7°	.122	.993	.123	8.14	1.01	8.21	83°
8°	.139	.990	.141	7.12	1.01	7.19	82°
9°	.156	.988	.158	6.31	1.01	6.39	81°
10°	.174	.985	.176	5.67	1.02	5.76	80°
11°	.191	.982	.194	5.14	1.02	5.24	79°
12°	.208	.978	.213	4.70	1.02	4.81	78°
13°	.225	.974	.231	4.33	1.03	4.45	77°
14°	.242	.970	.249	4.01	1.03	4.13	76°
15°	.259	.966	.268	3.73	1.04	3.86	75°
16°	.276	.961	.287	3.49	1.04	3.63	74°
17°	.292	.956	.306	3.27	1.05	3.42	73°
18°	.309	.951	.325	3.08	1.05	3.24	72°
19°	.326	.946	.344	2.90	1.06	3.07	71°
20°	.342	.940	.364	2.75	1.06	2.92	70°
21°	.358	.934	.384	2.61	1.07	2.79	69°
22°	.375	.927	.404	2.48	1.08	2.67	68°
23°	.391	.921	.424	2.36	1.09	2.56	67°
24°	.407	.914	.445	2.25	1.09	2.46	66°
25°	.423	.906	.466	2.14	1.10	2.37	65°
26°	.438	.899	.488	2.05	1.11	2.28	64°
27°	.454	.891	.510	1.96	1.12	2.20	63°
28°	.469	.883	.532	1.88	1.13	2.13	62°
29°	.485	.875	.554	1.80	1.14	2.06	61°
30°	.500	.866	.577	1.73	1.15	2.00	60°
31°	.515	.857	.601	1.66	1.17	1.94	59°
32°	.530	.848	.625	1.60	1.18	1.89	58°
33°	.545	.839	.650	1.54	1.19	1.84	57°
34°	.559	.829	.675	1.48	1.21	1.79	56°
35°	.574	.819	.700	1.43	1.22	1.74	55°
36°	.588	.809	.727	1.38	1.24	1.70	54°
37°	.602	.799	.754	1.33	1.25	1.66	53°
38°	.616	.788	.781	1.28	1.27	1.62	52°
39°	.629	.777	.810	1.23	1.29	1.59	51°
40°	.643	.766	.839	1.19	1.31	1.56	50°
41°	.656	.755	.869	1.15	1.33	1.52	49°
42°	.669	.743	.900	1.11	1.35	1.49	48°
43°	.682	.731	.933	1.07	1.37	1.47	47°
44°	.695	.719	.966	1.04	1.39	1.44	46°
45°	.707	.707	1.00	1.00	1.41	1.41	45°
angle	**cos**	**sin**	**cot**	**tan**	**csc**	**sec**	**angle**

ANSWERS TO EXERCISES

EXERCISE 1.1

1. 9527 **2.** 16,420 **3.** 76,307 **4.** 600,350 **5.** 22,986,045 **6.** Six million, eight hundred forty-two thousand, three hundred seventy-five **7.** Forty-two million, three hundred twenty thousand, seven hundred fifty **8.** Sixty million, four thousand, three hundred twenty-six **9.** Seven hundred million **10.** Six billion, three hundred million, seven **11.** 8 **12.** 2 **13.** 0 **14.** 0 **15.** 1 **16.** 2 **17.** 2 **18.** 2 **19.** 3 **20.** 904 **21.** 2868 **22.** 4830 **23.** 18,989 **24.** 155,609 **25.** 1685 **26.** 19,496 **27.** 103,980 **28.** 173,362 **29.** 21,128 **30.** 159,157 **31.** 1,455,773 **32.** 1287 **33.** 2609 **34.** 13,425 **35.** 28,611 **36.** 21,742 **37.** 110,462 **38.** 105,901 **39.** 29,599,511 **40.** (a) Don, 556 (b) 754 (c) 2,153 (d) 2,153 **41.** 6974 **42.** (a) 409 mi; (b) 255 mi

EXERCISE 1.2

1. 32 **2.** 23 **3.** 70 **4.** 25 **5.** 13 **6.** 47 **7.** 5 **8.** 29 **9.** 5 **10.** 12 **11.** 69 **12.** 27 **13.** 24 **14.** 59 **15.** 194 **16.** 270 **17.** 613 **18.** 250 **19.** 156 **20.** 159 **21.** 112 **22.** 138 **23.** 573 **24.** 562 **25.** 376 **26.** 277 **27.** 445 **28.** 34 **29.** 579 **30.** 5222 **31.** 1415 **32.** 32,957 **33.** 70,576 **34.** 847,909 **35.** 2475 **36.** 42,382 **37.** 64,078 **38.** 1384 **39.** 38,975 **40.** 406,278 **41.** 491,456 **42.** 33 **43.** 558 **44.** 277 **45.** $6861 **46.** 12,085 **47.** $201,358 **48.** 144,538 **49.** 10,862 sq mi **50.** $28 **51.** (a) 12,636 kwh; (b) 5182 kwh; (c) 6274 kwh

EXERCISE 1.3

1. 84 **2.** 99 **3.** 144 **4.** 348 **5.** 185 **6.** 288 **7.** 162 **8.** 504 **9.** 364 **10.** 720 **11.** 3224 **12.** 4375 **13.** 3123 **14.** 3412 **15.** 12,366 **16.** 32,256 **17.** 118,156 **18.** 483,392 **19.** 20,794 **20.** 226,706 **21.** 302,895 **22.** 2,283,750 **23.** 55,004,001 **24.** 137,199,335 **25.** 13,262,613 **26.** 1,523,475,415 **27.** 1575 mi **28.** 19,920 m **29.** 33,538,320 mi **30.** 323 mi **31.** 4046 mi **32.** $3348 **33.** $329.04 **34.** 10,080 min **35.** 10,500 gal **36.** A, by 176 bu

EXERCISE 1.4

1. 138 (0) **2.** 685 (18) **3.** 1433 (318) **4.** 161 (21) **5.** 921 (0) **6.** 246 (24) **7.** 164 (0) **8.** 45 (306) **9.** 4718 (189) **10.** 6046 (558) **11.** 548 (3,420) **12.** 53 (11,223) **13.** 64; 512; 700; 8,426,816 **14.** 87; 8472; 6,745,212; 621,621,126 **15.** 724; 8384; 81,000; 42,367,300; 14 billion **16.** 545; 38,420; 7770; 84,625 **17.** 4326; 5328; 462; 4,268,754; 62,010,030 **18.** 441; 574; 1904; 3535; 294,224 **19.** 4320; 36,848; 81,648; 93,808; 10,096 **20.** 8424; 79,362; 4005; 627,498; 777,777,777 **21.** 963 **22.** 28,472 **23.** 96 times **24.** 17 days

CHAPTER 1 SUPPLEMENTARY EXERCISES

1. 72,408 **3.** 1,207,615 **5.** 9,023,000,000 **7.** 5 **9.** 5 **11.** 7 **13.** 6 **15.** 77 **17.** 116 **19.** 123 **21.** 257 **23.** 303 **25.** 6020 **27.** 1890 **29.** 119,818 **31.** 16,136 **33.** 25,803 **35.** 82,310 **37.** 67,423 **39.** 46 **41.** 28 **43.** 5 **45.** 331 **47.** 149 **49.** 108 **51.** 307 **53.** 2408 **55.** 2579 **57.** 8290 **59.** 20,321 **61.** 64,775 **63.** 28,854 **65.** 18,542 **67.** 484,978 **69.** 126 **71.** 342 **73.** 1652 **75.** 16,205 **77.** 350 **79.** 1206 **81.** 32,016 **83.** 2048 **85.** 96 **87.** 17,145 **89.** 90,522 **91.** 127,820 **93.** 290,460 **95.** 1,618,800 **97.** 8,361,458 **99.** 2,781,864 **101.** 23 **103.** 123 **105.** 52 **107.** 56 (Rem. 4) **109.** 50 **111.** 578 **113.** 3 **115.** 3 (Rem. 29) **117.** 35 (Rem. 1) **119.** 19 (Rem. 40) **121.** 148 (Rem. 17) **123.** 24 (Rem. 32) **125.** 39 (Rem. 658) **127.** 140 (Rem. 1,789) **129.** 375; 6429 **131.** 6245; 874,130 **133.** 3374; 4263; 346,381 **135.** 3465; 38,880 **137.** 13,449 **139.** Alf, 39 **141.** 11,419 **143.** 200,354 sq mi **145.** 1644 **147.** 23 hr **149.** 234 pages

EXERCISE 2.1

1. = **2.** $<$ **3.** $>$ **4.** $<$ **5.** = **6.** $>$ **7.** $<$ **8.** = **9.** $>$ **10.** $>$ **11.** $<$ **12.** $>$ **13.** $\frac{2}{3}$ **14.** $\frac{3}{4}$ **15.** $\frac{3}{5}$ **16.** $\frac{1}{2}$ **17.** $\frac{7}{2}$ **18.** $\frac{1}{5}$ **19.** $\frac{6}{5}$ **20.** $\frac{7}{9}$ **21.** $\frac{5}{16}$ **22.** $\frac{9}{4}$ **23.** $\frac{31}{63}$ **24.** $\frac{17}{63}$ **25.** $\frac{4}{6}$ **26.** $\frac{60}{100}$ **27.** $\frac{210}{240}$ **28.** $\frac{9}{93}$ **29.** $\frac{20}{16}$ **30.** $\frac{132}{84}$ **31.** $\frac{9}{24}$ **32.** $\frac{4}{26}$ **33.** $\frac{6}{9}$ **34.** $\frac{21}{49}$ **35.** $\frac{14}{35}$ **36.** $\frac{96}{216}$ **37.** $\frac{75}{50}$ **38.** $\frac{54}{213}$ **39.** $\frac{10}{35}$ **40.** $\frac{4}{12}$ **41.** $3\frac{2}{5}$ **42.** $3\frac{1}{9}$ **43.** $4\frac{4}{7}$ **44.** $9\frac{2}{5}$ **45.** $2\frac{1}{150}$ **46.** $3\frac{10}{11}$ **47.** $3\frac{33}{125}$ **48.** $15\frac{7}{75}$ **49.** $\frac{20}{3}$ **50.** $\frac{61}{8}$ **51.** $\frac{30}{7}$ **52.** $\frac{49}{4}$ **53.** $\frac{113}{3}$ **54.** $\frac{731}{9}$ **55.** $\frac{283}{7}$ **56.** $\frac{1514}{5}$ **57.** $\frac{3}{16}, \frac{1}{4}, \frac{5}{16}, \frac{3}{8}$ **58.** $\frac{1}{6}, \frac{1}{2}, \frac{2}{3}, \frac{5}{6}$ **59.** $\frac{2}{3}, \frac{5}{6}, \frac{7}{8}, \frac{11}{12}$ **60.** $\frac{17}{4}, \frac{22}{5}, \frac{31}{7}, \frac{9}{2}$ **61.** $\frac{2}{9}, \frac{2}{7}, \frac{3}{10}, \frac{3}{8}$ **62.** $\frac{25}{6}, \frac{21}{5}, \frac{13}{3}, \frac{9}{2}$ **63.** (a) $\frac{12}{7}$; (b) $\frac{20}{7}$; (c) $\frac{3}{16}$; (d) $\frac{3}{4}$

EXERCISE 2.2

1. 240 **2.** 4050 **3.** 240 **4.** 8736 **5.** $\frac{17}{24}$ **6.** $\frac{7}{8}$ **7.** $\frac{25}{24}$ **8.** $\frac{3}{4}$ **9.** $\frac{11}{8}$ **10.** $\frac{49}{30}$ **11.** $\frac{83}{126}$ **12.** $\frac{7}{20}$ **13.** $\frac{5}{8}$ **14.** $\frac{1}{10}$ **15.** $\frac{11}{54}$ **16.** $\frac{13}{209}$ **17.** $\frac{29}{210}$ **18.** $35\frac{1}{24}$ **19.** $53\frac{1}{4}$ **20.** $119\frac{11}{16}$ **21.** $171\frac{16}{25}$ **22.** $120\frac{91}{135}$ **23.** $102\frac{11}{792}$ **23.** $102\frac{17}{792}$ **24.** $6\frac{1}{6}$ **25.** $5\frac{5}{12}$ **26.** $211\frac{9}{40}$ **27.** $31\frac{5}{8}$ **28.** $22\frac{1}{4}$ **29.** $400\frac{45}{56}$ **30.** $21\frac{3}{16}$ **31.** $4732\frac{96}{113}$ **32.** $6\frac{18}{35}$ **33.** $\frac{5}{8}$ **34.** $1\frac{1}{10}$ **35.** $2\frac{3}{5}$ **36.** $\frac{13}{24}$ **37.** $\frac{7}{24}$ **38.** $1\frac{11}{12}$ **39.** $\frac{1}{15}$ **40.** $2\frac{13}{32}$ in.

EXERCISE 2.3

1. $\frac{1}{6}$ **2.** $\frac{1}{5}$ **3.** $1\frac{3}{7}$ **4.** $\frac{16}{25}$ **5.** $\frac{3}{4}$ **6.** $\frac{9}{50}$ **7.** 4 **8.** $1\frac{53}{112}$ **9.** $7\frac{7}{15}$ **10.** $33\frac{1}{2}$ **11.** 0 **12.** $4\frac{7}{8}$ **13.** $1\frac{7}{8}$ **14.** $\frac{3}{4}$ **15.** $2\frac{1}{4}$ **16.** $1\frac{3}{4}$ **17.** $\frac{1}{9}$ **18.** $10\frac{2}{3}$ **19.** $\frac{3}{8}$ **20.** $5\frac{1}{7}$ **21.** $17\frac{1}{2}$ **22.** $\frac{1}{4}$ **23.** $\frac{81}{112}$ **24.** $\frac{16}{19}$ **25.** 2 **26.** $\frac{3}{5}$ **27.** 15 **28.** $344\frac{6}{7}$ **29.** $513\frac{9}{20}$ **30.** $11{,}295\frac{3}{28}$ **31.** $1544\frac{2}{15}$ **32.** $\frac{3}{28}$ **33.** 4 **34.** $\frac{8}{9}$ **35.** $\frac{1}{2}$ **36.** $3\frac{1}{3}$ **37.** $\frac{1}{2}$ **38.** $\frac{2}{45}$ **39.** 96 **40.** $3\frac{4}{5}$ **41.** $\frac{5}{11}$ **42.** $1\frac{3}{5}$ **43.** 12 **44.** 45 kg **45.** 90

CHAPTER 2 SUPPLEMENTARY EXERCISES

1. $<$ **3.** $>$ **5.** $>$ **7.** $>$ **9.** $>$ **11.** $\frac{2}{5}$ **13.** $\frac{6}{7}$ **15.** $\frac{3}{14}$ **17.** $\frac{23}{29}$ **19.** $\frac{8}{12}$ **21.** $\frac{8}{24}$ **23.** $\frac{3}{8}$ **25.** $\frac{28}{32}$ **27.** $\frac{3}{18}$ **29.** $1\frac{3}{7}$ **31.** $4\frac{1}{2}$ **33.** $1\frac{1}{3}$ **35.** $1\frac{1}{4}$ **37.** $\frac{11}{5}$ **39.** $\frac{17}{7}$ **41.** $\frac{7}{2}$ **43.** $\frac{57}{11}$ **45.** $\frac{247}{3}$ **47.** $\frac{1}{2}; \frac{2}{3}; \frac{5}{6}; \frac{11}{12}$ **49.** $\frac{2}{3}; \frac{4}{5}; \frac{5}{6}; \frac{7}{8}$ **51.** $\frac{13}{16}$ **53.** $2\frac{1}{4}$ **55.** $\frac{37}{48}$ **57.** $1\frac{5}{6}$ **59.** $\frac{7}{20}$ **61.** $\frac{2}{21}$ **63.** $\frac{1}{2}$ **65.** $\frac{7}{40}$ **67.** $\frac{2}{7}$ **69.** $\frac{1}{20}$ **71.** $7\frac{5}{6}$ **73.** $13\frac{31}{35}$ **75.** $13\frac{5}{12}$ **77.** $3\frac{1}{3}$ **79.** $8\frac{15}{28}$ **81.** $6\frac{1}{4}$ **83.** $15\frac{1}{3}$ **85.** $13\frac{25}{42}$ **87.** $14\frac{16}{21}$ **89.** $21\frac{1}{40}$ **91.** $\frac{3}{35}$ **93.** 1 **95.** $\frac{9}{55}$ **97.** $\frac{1}{25}$ **99.** $\frac{9}{16}$ **101.** $\frac{55}{192}$ **103.** $6\frac{5}{12}$ **105.** 0 **107.** $77\frac{5}{8}$ **109.** $3585\frac{13}{35}$ **111.** $5{,}441\frac{1}{16}$ **113.** $1\frac{1}{4}$ **115.** $\frac{4}{5}$ **117.** $3\frac{1}{2}$ **119.** $\frac{3}{4}$ **121.** $1\frac{13}{77}$ **123.** $\frac{75}{256}$ **125.** $10\frac{1}{2}$ **127.** 70 **129.** $2\frac{1}{2}$ **131.** $\frac{2}{9}$ **133.** $1\frac{13}{36}$ **135.** 12 **137.** $\frac{1}{3}$ **139.** $\frac{98}{135}$ **141.** $1\frac{13}{32}$ **143.** $\frac{14}{29}$ **145.** $5\frac{11}{12}$ yd **147.** $\frac{7}{8}$ in. **149.** $2\frac{7}{8}$ in. **151.** $4.24 **153.** $11\frac{7}{10}$ mph **155.** 36 **157.** $8\frac{2}{3}$ ft **159.** $8\frac{1}{4}$ in.

CHAPTER 2 REVIEW

1. 1232 **3.** 44,510 **5.** 1116 **7.** 3702 **9.** 17,884 **11.** 31,416 **13.** 517,678 **15.** 1,946,817 **17.** 51 (Rem. 36) **19.** 201 (Rem. 10) **21.** 648 (Rem. 113) **23.** $>$ **25.** $<$

EXERCISE 3.1

1. $>$ **2.** $<$ **3.** $>$ **4.** $<$ **5.** $>$ **6.** $<$ **7.** $>$ **8.** $<$ **9.** .3, .36, 3, 3.528, 3.62 **10.** .304, .340, .431, 3.04, 3.25 **11.** 6.01, 6.03, 6.19, 6.2, 6.214 **12.** .000102, .0003, .0006, .00099, .002 **13.** 74.495 **14.** 144.400 **15.** 472.232 **16.** 42.7079 **17.** 5029.2077 **18.** 3.05325 **19.** 28.107 **20.** 321.39 **21.** 353.265 **22.** 67.457 **23.** .8257 **24.** .000113 **25.** 96.9381 **26.** 273.0702 **27.** 1.895866 **28.** 3.31356 **29.** 2.3188 **30.** 1.2037028 **31.** .0000465 **32.** .030576 **33.** 3 min 33.3 sec **34.** 129.682 **35.** 43.39 acres **36.** 34.316 ft **37.** 127.26 ft **38.** 100.5 lb

EXERCISE 3.2

1. 5463.25; 5463.2; 5463; 5460; 5500 **2.** 398.00; 398.0; 398; 400; 400 **3.** 37.50; 37.5; 38; 40; 0 **4.** 12.78 **5.** 15.21 **6.** 2.76 **7.** .02 **8.** 1609.36 **9.** .00 **10.** $\frac{23}{50}$ **11.** $\frac{4}{25}$ **12.** $\frac{8}{25}$ **13.** $\frac{7}{8}$ **14.** $\frac{37}{100}$ **15.** $\frac{7}{100}$ **16.** $\frac{7}{5}$ (or $1\frac{2}{5}$) **17.** $\frac{293}{8}$ (or $36\frac{5}{8}$) **18.** $\frac{7}{4}$ (or $1\frac{3}{4}$) **19.** $\frac{7}{4}$ (or $1\frac{3}{4}$) **20.** $\frac{7}{200}$ **21.** $\frac{7}{2000}$ **22.** $\frac{5}{8}$ **23.** $\frac{5}{8}$ **24.** $\frac{5}{8}$ **25.** .291 **26.** .375 **27.** .34 **28.** .12 **29.** .136 **30.** 3.8 **31.** .25 **32.** .231 **33.** 4.286 **34.** .286 **35.** 1.25 **36.** .047 **37.** 43.5 mph **38.** $40.10 **39.** 40.1 **40.** (a) 14 trips (b) 13 trips

EXERCISE 3.3

1. 35%; 87%; 43%; 52% **2.** 90%; 30%; 60%; 10% **3.** 9%; 4%; 3.6%; 187% **4.** .32%; 100%; 400%; 106%; **5.** 10.05%; 627.4%; 130%; .06% **6.** .24%; 100.24%; 365%; 2% **7.** 60%; 62.5%; 75% **8.** 125%; 150%; 112.5% **9.** 28.57%; 68%; 36.36% **10.** 22.5%; 5.5%; 41.67% **11.** .09; .006; .35; .18 **12.** 2.2; 1; .0004; 3 **13.** 1.65; .165; .0165; .00165 **14.** .005; .0075; .001; .103 **15.** $\frac{13}{50}$; $\frac{8}{25}$; $\frac{2}{125}$ **16.** $\frac{5}{4}$; $\frac{1}{8}$; $\frac{1}{80}$ **17.** $\frac{33}{40}$; $\frac{27}{100}$; $\frac{1}{20}$ **18.** $\frac{3}{8}$; $\frac{1}{2000}$; $\frac{17}{400}$ **19.** .4; 40% **20.** .125; 12.5% **21.** $\frac{3}{50}$; 6% **22.** $\frac{7}{8}$; 87.5% **23.** $\frac{12}{25}$; .48 **24.** $\frac{1}{6}$; .1667 **25.** .2667; 26.67% **26.** 1.75; 175% **27.** $\frac{1}{2000}$; .05% **28.** $1\frac{2}{25}$; 108% **29.** $\frac{1}{500}$; .002 **30.** $1\frac{3}{5}$; 1.6 **31.** 2.5; 250% **32.** $\frac{1}{4}$; 25% **33.** $\frac{1}{40}$; .025 **34.** .2222; 22.22% **35.** .3125; 31.25% **36.** $\frac{147}{2000}$; 7.35% **37.** $1\frac{7}{25}$; 128% **38.** $\frac{1}{25000}$; .0000 **39.** $\frac{1}{800}$; .0013 **40.** 2.3125; 231.25% **41.** $\frac{21}{400}$; 5.25% **42.** $\frac{3}{10000}$; .0003 **43.** .0688; 6.88% **44.** $\frac{11}{5000}$; .22% **45.** $\frac{1}{300}$; .0033 **46.** $\frac{9}{700}$; .0129

EXERCISE 3.4

1. 25% **2.** 60% **3.** 112 **4.** 32 **5.** 12.5% **6.** 98 **7.** 6.12 **8.** 60% **9.** 10.58 **10.** 700 **11.** 12.5% **12.** .12 **13.** 20.4 **14.** 66.7% **15.** 2.23 **16.** .02 **17.** 80% **18.** .01 **19.** $2\frac{6}{7}$ **20.** 10% **21.** 132 **22.** .008 **23.** $\frac{7}{20}$ **24.** 78.9% **25.** 6.75 **26.** 12 **27.** $\frac{3}{3200}$ **28.** 20% **29.** 9,400 **30.** .00 **31.** $13.20 **32.** 15% **33.** $5600 **34.** $43,710 **35.** $7\frac{1}{3}$% **36.** $198/wk **37.** $334.55 **38.** 75 lb **39.** $81.25 **40.** 12 hits

EXERCISE 3.5

1. 1.25; 125% **2.** 1.375; 137.5% **3.** $1\frac{3}{5}$; 160% **4.** $1\frac{1}{50}$; 102% **5.** 3; 3.00 **6.** $2\frac{1}{4}$; 2.25 **7.** 15.5; 1,550% **8.** 16; 1600% **9.** $3\frac{1}{3}$; 3.33+ **10.** 1.714; 171.4% **11.** $1\frac{13}{25}$; 152% **12.** 120% **13.** 20 **14.** 9 **15.** 600% **16.** 106 **17.** $\frac{1}{4}$ **18.** 33 **19.** 102% **20.** 2% **21.** $\frac{1}{20}$ **22.** $\frac{4}{7}$ **23.** 25% **24.** $2511 **25.** 130% **26.** $681.00 **27.** 75% **28.** 77° **29.** 91 lb **30.** $301.42 **31.** 111.5 + %

CHAPTER 3 SUPPLEMENTARY EXERCISES

1. $<$ **3.** $>$ **5.** $>$ **7.** 14.19 **9.** 87.534 **11.** 71.025 **13.** 43.4927 **15.** 351.7551 **17.** 37.666 **19.** 83.16 **21.** 19.16 **23.** 249.31 **25.** 23.92 **27.** 13.184 **29.** 2.025 **31.** 3.8454 **33.** 13.41065 **35.** (a) 35 (b) 30 **37.** (a) 46.28 (b) 46.280 **39.** (a) .365 (b) 0 **41.** .26 **43.** 93.28 **45.** 45.94 **47.** 2.81 **49.** 19.80 **51.** 13,409.29 **53.** 625; $62\frac{1}{2}\%$ **55.** $\frac{16}{25}$; 64% **57.** $\frac{97}{100}$; 97% **59.** $\frac{3}{500}$; .006 **61.** .028; $2\frac{4}{5}\%$ **63.** $1\frac{4}{5}$; 1.8 **65.** $24\frac{0}{10}$; 2400% **67.** 5.58 **69.** 300 **71.** 28.8 **73.** 700 **75.** .0175 **77.** 72.8 **79.** .052 **81.** 30 **83.** 114% **85.** 48 **87.** \$20 **89.** 8 **91.** \$34,400 **93.** \$656.88 **95.** 266.7% **97.** \$31.40 **99.** \$64 **101.** 120% **103.** \$5.25 **105.** 30.48%

CHAPTER 3 REVIEW

1. forty-six thousand, two hundred thirty-one **3.** four million, three hundred twenty-six thousand, five hundred eleven **5.** thirty-six million **7.** (a) 7 (b) 1 (c) 4 (d) 6(e) 5 **9.** .5; $\frac{5}{8}$; $\frac{2}{3}$; $\frac{3}{4}$ **11.** $\frac{2}{7}$; $\frac{3}{10}$; $\frac{4}{11}$; $\frac{3}{8}$ **13.** $2\frac{11}{24}$ **15.** $2\frac{57}{70}$ **17.** $\frac{1}{12}$ **19.** $1\frac{3}{14}$ **21.** $\frac{5}{12}$ **23.** $8\frac{3}{4}$ **25.** 8 **27.** $\frac{3}{4}$ **29.** $14\frac{2}{5}$ **31.** \$581.88

EXERCISE 4.1

1. 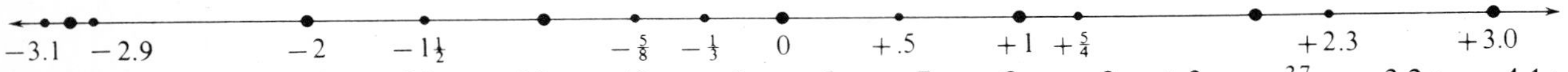

2. $+4 < +6$; $-8 < +10$; $+32 > +23$; $-12 < -3$; $+6 > -7$; $-2 > -3$; $+3 > -\frac{27}{9}$; $-3.2 > -4.1$; $+\frac{1}{4} > -\frac{5}{8}$; $0 > -.1$ **3.** (a) {+3, +4, +5, +6} (b) {+5, +6, +7} (c) {−4, −3, −2, −1, 0, +1} (d) {+1, +2, +3, +4, +5} (e) { } (empty set) (f) {−2, −1, +1, +2} **4.** 12; 3; 8; 91; $\frac{1}{4}$; $\frac{2}{3}$ **5.** $\frac{5}{8}$; $\frac{7}{3}$; 0; 3.06; 3.06; .407 **6.** +19 **7.** −99 **8.** +530 **9.** −103.8 **10.** +170.71 **11.** $+1\frac{13}{24}$ **12.** $-8\frac{7}{15}$ **13.** +48 **14.** +61 **15.** −258 **16.** +9.38 **17.** −578 **18.** +912 **19.** −124.33 **20.** $+\frac{1}{4}$ **21.** $+19\frac{11}{40}$ **22.** $-4\frac{11}{20}$ **23.** $+158\frac{29}{40}$ **24.** −40.85 **25.** −4.035 **26.** −28 **27.** −57 **28.** −59 **29.** +1.3 **30.** +57.4 **31.** −6.08 **32.** $-\frac{1}{12}$ **33.** $+3\frac{17}{24}$ **34.** +6 **35.** −12 **36.** −14 **37.** +.4 **38.** +39.5 **39.** +50.82 **40.** −25.449 **41.** +1449 **42.** +12,571 **43.** −60,508 **44.** $+\frac{4}{5}$ **45.** $+\frac{9}{14}$ **46.** $-\frac{1}{126}$ **47.** $-\frac{49}{6}$ **48.** $-\frac{263}{20}$ **49.** $+52\frac{11}{15}$ **50.** $+87\frac{4}{33}$ **51.** −10.36 **52.** +40.09 **53.** $+40\frac{1}{9}$ **54.** (a) 49°; (b) 14°; (c) 74°; (d) 39° **55.** (a) $47\frac{5}{8}$; (b) $2\frac{5}{8}$; (c) $\frac{3}{4}$; (d) $1\frac{1}{4}$; (e) $24\frac{5}{8}$

EXERCISE 4.2

1. −6 **2.** −48 **3.** +21 **4.** −160 **5.** −4 **6.** +9 **7.** $+\frac{3}{8}$ **8.** −.0486 **9.** +.0000294 **10.** −.18081 **11.** +144 **12.** −70 **13.** $-\frac{18}{55}$ **14.** +.00084 **15.** −24 **16.** +1.62 **17.** .0000012 **18.** +432 **19.** −3892.546 **20.** +2541.2468 **21.** +2 **22.** −2 **23.** +7 **24.** +672 **25.** −562 **26.** +161.5 **27.** $-1\frac{7}{8}$ **28.** $+\frac{3}{4}$ **29.** $-1\frac{3}{4}$ **30.** $+10\frac{2}{3}$ **31.** $+\frac{1}{4}$ **32.** −15 **33.** −24 **34.** −82,400 **35.** +.00162 **36.** +4 **37.** −16 **38.** $+2\frac{1}{2}$ **39.** −3 **40.** +2 **41.** $-\frac{1}{2}$ **42.** −1 **43.** −4 **44.** +7 **45.** $-\frac{4}{5}$ **46.** $-1\frac{1}{3}$ **47.** −2 **48.** (a) +; (b) +; (c) −; (d) + **49.** (a) +; (b) +; (c) −; (d) + **50.** 35 in. **51.** 27

EXERCISE 4.3

1. (a) $7x$; (b) x; (c) $-7r$; (d) $-4d$; (e) 0; (f) $-9xy$; (g) $4k - 2p^2$; (h) $\frac{1}{2}m$; (i) $-3.1s$ **2.** (a) $-2x$; (b) $5.55y$; (c) $3.27x$ **3.** (a) $-6\frac{1}{4}x$; (b) $\frac{9}{100}ap$; (c) $6q - 6q^2$ **4.** $6xy - 8xy^2$ **5.** $4x - 11p$ **6.** $8xy - 2y + 3z$ **7.** $3x + 3y + p^2$ **8.** $3d + 7z$ **9.** $x - 3y + 11z$ **10.** $.1x - 2.7y$ **11.** $\frac{7}{12}x - \frac{7}{3}y$ **12.** (a) $7r$; (b) $-12m$; (c) $-.2k$; (d) $16a + 37b$; (e) $29c$; (f) $-\frac{14}{15}t$; (g) $+34e - 27d$; (h) $+4.4d$ **13.** $-3x + 5y$ **14.** $6x - 4y$ **15.** $3x - 10y - 12$ **16.** $x - 2$ **17.** $\frac{1}{2}x + \frac{1}{4}y$ **18.** $-.1a + .1b$ **19.** $8y - 2p$ **20.** $3.3a - 3.4b - 2.2c - 6d$ **21.** $-2y$ **22.** $-18b - 10c$ **23.** $13.2ab - 1.4b$ **24.** $\frac{29}{30}x - \frac{1}{2}y$ **25.** $1.9g - 4.9h$ **26.** $-13x + 10y$ **27.** $x - 8$ **28.** (a) $1 - 6x + 3y$; (b) $-4x + 3y$ **29.** $6y - 3x + z$

EXERCISE 4.4

1. a^5; b^8; d^{13}; x^{15} **2.** c^4d; a^7b^7; a^9b^9; x^9y^4 **3.** $-10x$; $10n^2$; $21x^2$; $-8xy$; $30x^2y$; $-15x^3y^3$; $24a^2b^2x$ **4.** $-12x^2$; $-21xy$; $-36ax$; $24x^2y$ **5.** $18bx^2$; $84m^4$; $27s^3t^3$; x^5y^5 **6.** $12x^2 + 6xy - 3xz$ **7.** $-4ax + 2x^2 - 3xy$ **8.** $6a^2x - 15ax^2 + 18ax$ **9.** $4a^2 + 2a$ **10.** $-6a^3c + 8a^2c^2 + 4a^3c^2$ **11.** $-24a^2b^2 + 32a^3 + 16a^2b$ **12.** $-3p^2qr - 2pq^2r^2 - 7pqr$ **13.** $-18a^3b^3c^4 + 18a^2b^3c^3 - 18a^2b^2c^4$ **14.** $10x^2 - 31x - 14$ **15.** $5a^2 + 13ax - 6x^2$ **16.** $2pm - 2pn - m^2 + mn$ **17.** $3x^2 - 5xy + 6x + 2y^2 - 6y$ **18.** $9x^3 - 21x^2 + 31x - 35$ **19.** $6x^3 - 15x^2y + 11xy^2 - 10y^3$ **20.** $-10a^2 + 9a^3 - 8a^4 + 3a^5$ **21.** $8d^3 - 20cd^2 + 14c^2d - 12c^3$ **22.** $15x^4 - 19x^3 + 32x^2 - 17x + 7$ **23.** 8 **24.** 5 **25.** -18 **26.** 15 **27.** 5 **28.** 152 **29.** $4p + 7k + 2x$ **30.** $-s$ **31.** $-4a - 2b + c$ **32.** $-2y$ **33.** $13x + 9y$ **34.** $-16a - 7b$ **35.** $6x - 4$ **36.** $20y + 9$ **37.** $2x - (-6y + 3t + 8)$ **38.** $4x - (6y - 3z - 5)$ **39.** $ab - (c - gk + h)$ **40.** $-st + q - (a + c + d)$

EXERCISE 4.5

1. 2^2; $\frac{1}{2}$; $\frac{1}{2^3}$; $2^0 (= 1)$ **2.** x^4; x^6; $\frac{1}{x^2}$; 1 **3.** $\frac{1}{(xy)^2}$; ab; $(-a)^2$ or a^2; $\frac{1}{(-2)^3}$ **4.** 1; $\frac{1}{4^2}$; $\frac{1}{x^3}$; $(-3)^2$

5. $-7a$; $-3x$; $-4a$; $9x^3$ **6.** $-5p$; $12a$; -2; $\frac{5y^2z}{2}$ **7.** $2 - 3a^2$ **8.** $-t + 1$ **9.** $3n - 7k$

10. $2 - 8a^2b$ **11.** $-6xy + 11y - \frac{9y^3}{x}$ **12.** $-4z^3 + 3x - 2xyz$ **13.** $-5a + 4b + 3c^2$

14. $-\frac{6x^2}{y} + 1 - \frac{12}{5x}$ **15.** $4m - 2n - 3mn^2$ **16.** $-\frac{9}{y} + \frac{13y}{x} - \frac{15}{2xy}$ **17.** $2a + 3$ **18.** $3x + 4y$

19. $5a^2 + 21a + 4$ **20.** $2x^2 + x - 1$ **21.** $5x - 2$ (rem 7) or $5x - 2 + \frac{7}{2x + 3}$ **22.** $a^2 - 2a + 3$ (rem -4) or $a^2 - 2a + 3 - \frac{4}{a + 1}$ **23.** $x^2 + 2x + 4$ **24.** $(2x - \frac{1}{2})$ dollars

CHAPTER 4 SUPPLEMENTARY EXERCISES

3. $-\frac{3}{16}$ **5.** $+28$ **7.** $+.5$ **9.** $+\frac{49}{240}$ **11.** $+12$ **13.** $+147.9$ **15.** $-96{,}501$ **17.** $-\frac{5}{36}$ **19.** $-72\frac{7}{40}$ **21.** $+1081$ **23.** $+20$ **25.** $+17.132$ **27.** $+0.000828$ **29.** $+5\frac{1}{2}$ **31.** $-.375$ **33.** $+888\frac{74}{439}$ **35.** $-\frac{13}{21}$ **37.** -3.8 **39.** $-\frac{1}{5}$ **41.** $+3\frac{69}{112}$ **43.** $+$ **45.** $-$, $+$ or $+$, $-$ **47.** $-\frac{1}{40}m$ **49.** $-9.4a$ **51.** $-\frac{37}{120}a$ **53.** $-8.79x$ **55.** $62p - 37q$ **57.** $4\frac{3}{8}d$ **59.** $-650x$ **61.** $6.37x$ **63.** a^{15} **65.** $21a^3b^3$ **67.** $.00036m^3$ **69.** x^6y^4 **71.** x^6 **73.** $-3x$ **75.** $\frac{5}{2}a$ **77.** $\frac{6}{7}x$ **79.** $44xy - 3xy^2$ **81.** $9x - 10y + 7z$ **83.** $8x - 16y$ **85.** $.3a - 4.1b - 4.5c + 3.2d$ **87.** $\frac{5}{4}x + \frac{3}{8}xy - \frac{13}{15}y$ **89.** $-9a + 2b$ **91.** $-12axy + 6x^2y - 21xy$ **93.** $.16a^3b - 6.4a^2b^2 + .48a^2bc$ **95.** $6a^4 - 3a^3 - 2a^2 - 24a^3d + 12a^2d + 8ad$ **97.** $6a^2 - ac - 3ad + \frac{1}{2}cd$ **99.** -18 **101.** $2m - 8t$ **103.** $6a - 11b$ **105.** $8am + 8bm + 6ab$ **107.** $8m + 13$ **109.** $5b - 6a^2$ **111.** $6ab - 4b^2$ **113.** $-12xy^2 - 4x + 5xy$ **115.** $-3ax^5 + 6ax^2 + 9a$ **117.** $7x - 5y$ **119.** $x^2 + 3x + 9$

CHAPTER 4 REVIEW

1. $2\frac{13}{72}$ **3.** $\frac{5}{24}$ **5.** .002268 **7.** $98\frac{1}{3}$ **9.** $1\frac{19}{63}$ **11.** 2.53 **13.** (a) 34%; (b) 15%; (c) 140% **15.** 9.1 **17.** .72 **19.** 300% **21.** 120 **23.** $\frac{2}{3}$

EXERCISE 5.1

1. 4 **2.** 11 **3.** 4 **4.** 12 **5.** -4 **6.** -1 **7.** 5 **8.** 5 **9.** 2 **10.** $\frac{7}{3}$ **11.** 2 **12.** 28 **13.** 16 **14.** $-\frac{1}{9}$ **15.** -125 **16.** 3 **17.** -2 **18.** 6 **19.** 0 **20.** $5\frac{1}{5}$ **21.** 2 **22.** $-28\frac{1}{3}$ **23.** 5 **24.** -24 **25.** $6\frac{2}{3}$ **26.** -70 **27.** -8 **28.** 0 **29.** $\frac{1}{10}$ **30.** $1\frac{2}{3}$ **31.** 3 **32.** 7 **33.** No **34.** (a) No; (b) identity

EXERCISE 5.2

1. $(x + 5)$ or $(5 + x)$ **2.** $x - 17$ **3.** $2x$ **4.** $\frac{1}{4}x$ or $\frac{x}{4}$ **5.** $2x + 3$ **6.** $\frac{1}{2}(x + 3)$ or $\frac{x+3}{2}$

7. $.4x$ or $\frac{2}{5}x$ or $\frac{2x}{5}$ **8.** $1.4x$ or $\frac{7}{5}x$ or $\frac{7x}{5}$ **9.** $x - 6$ **10.** $6 - x$ **11.** $x - \frac{1}{6}x$ or $\frac{5}{6}x$ **12.** $\frac{17}{2x}$

13. $(a + b) - \frac{1}{2}ab$ or $(a + b - \frac{1}{2}ab)$ **14.** $p - x$ **15.** $x + 17$ **16.** px dollars **17.** $\frac{2x - 20}{5}$ or $\frac{1}{5}(2x - 20)$ **18.** $\frac{2x}{5} - 20$ **19.** $60x$ **20.** $\frac{x}{7}$ **21.** $x - 7 = 15$ **22.** $x + 8 = 19$ **23.** $2x - 5 = 16$

24. $3x = x + 8$ **25.** $\frac{10}{x} = 2$ **26.** $60 = 2x + 4$ **27.** $10 + 8 + x = 2x$ **28.** $\frac{1}{3}x = x - 6$

29. $\frac{2}{3}x + 4 = 8$ **30.** $\frac{2}{3}(x + 4) = 8$ **31.** $15\frac{1}{2}$ in., $20\frac{1}{2}$ in. **32.** 45¢ **33.** 16, 17, 18 **34.** 15

35. 12, 19 **36.** Any three consecutive even integers **37.** 33 **38.** 12

EXERCISE 5.3

1. 5, 7, 9 **2.** 438 **3.** 9 quarters, 16 dimes **4.** 15 lb peanuts, 5 lb cashews **5.** 32 × 12 ft **6.** 22 ft **7.** $6\frac{6}{7}$ hr **8.** 18 fouls, 34 goals **9.** 72 **10.** 16 **11.** 90°, 60°, 30° **12.** 24, 40, 30 **13.** \$755 @ 4%; \$1245 @ 6% **14.** $4\frac{2}{7}$ lb **15.** 3 gal **16.** $\frac{3}{8}$ mi

EXERCISE 5.4

1. $\{6, 7, 8, 9, \ldots\}$
2. $\{\ldots, -3, -2, -1\}$
3. $\{4, 5, 6, 7, \ldots\}$
4. $\{\ldots, -5, -4, -3, -2\}$
5. $\{\ldots, 0, 1, 2, 3, 4\}$
6. $\{\ldots, 2, 3, 4, 5, 6\}$
7. $\{1, 2, 3, 4, \ldots\}$
8. $\{\ldots, 6, 7, 8, 9, 10\}$
9. $\{\ldots, -5, -4, -3\}$
10. $\{\ldots, 1, 2, 3\}$
11. $\{12, 13, 14, \ldots\}$
12. $\{\ldots, -13, -12, -11, -10\}$
13. $\{3, 4, 5, 6, \ldots\}$
14. $\{\ldots, -3, -2, -1, 0\}$
15.  7
16.  -5
17.  -2
18.  -5
19.  28
20.  -1
21. 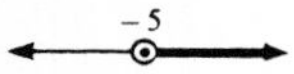 -5
22.  7
23. 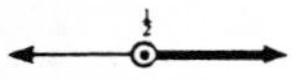 $\frac{1}{2}$
24.  -8
25.  $-\frac{2}{3}$

EXERCISE 5.5

1. (7, 3); (15, -1); (5, 4) satisfy **8 2.** (2, 4)(1, 1); (0, -2) satisfy **3.** (-1, 0); (.5, -1) satisfy **4.** (a) (2, 1); (b) (5, 3); (c) (3, 5); (d) (-2, 2); (e) (-2, -2); (f) (2, -2); (g) (-6, -4); (h) (-4, 4); (i) (5, -2); (j) (-6, 0); (k) (0, -4); (l) (3, -4)

5.

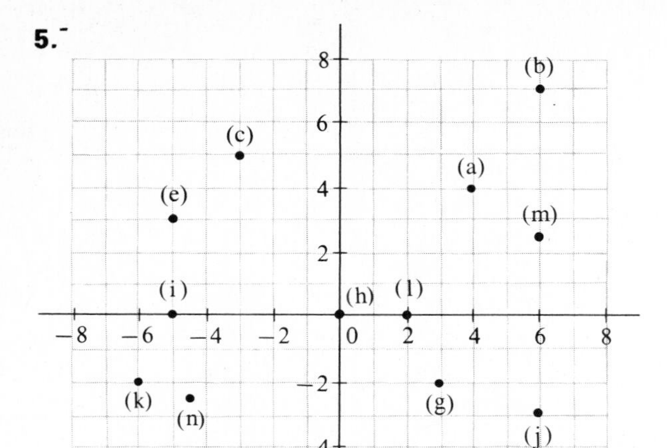

6. (2, 3) **7.** (5, 7) **8.** (2, 1) **9.** (1, −3) **10.** Every point on the two graphs **11.** No point

EXERCISE 5.6

1. (2, 3) **2.** (5, 7) **3.** $(-\frac{1}{2}, 3\frac{1}{2})$ **4.** (−5, 2) **5.** (2, 1) **6.** (1, −3) **7.** (−2, −3) **8.** (7, −2) **9.** $(\frac{5}{7}, -\frac{1}{3})$ **10.** $(3, -\frac{1}{2})$ **11.** Dependent **12.** (5, 1) **13.** Inconsistent **14.** (2, −8) **15.** Inconsistent **16.** 23 × 15 ft **17.** 12¢ daily, 40¢ Sunday **18.** \$1.40 first 3 min, \$.35 each additional min **19.** 30 cc of 20%; 70 cc of 30%

EXERCISE 5.7

1. (5)(3) − (3)(2) = 9 **2.** (4)(3) − (1)(0) = 12 **3.** (2)(−5) − (−1)(4) = −6 **4.** (−3)(3) − (2)(4) = −17 **5.** −11 **6.** 2 **7.** −30 **8.** 11 **9.** 15 **10.** 0 **11.** 28 **12.** 0 **13.** $x = 2, y = 3$ **14.** $x = 0, y = -4$ **15.** (2, 3) **16.** $(2, \frac{1}{2})$ **17.** (−2, −3) **18.** Inconsistent **19.** (0, −3) **20.** Dependent

CHAPTER 5 SUPPLEMENTARY EXERCISES

1. 4 **3.** $\frac{1}{2}$ **5.** .01 **7.** −2 **9.** 24 **11.** $\frac{8}{3}$ **13.** 0 **15.** Identity **17.** −15 **19.** m **21.** $\frac{2x}{9}$ **23.** $1.78x$ **25.** $22 - x$ **27.** $\frac{3x}{x-7}$ **29.** $\frac{x}{12}$ **31.** $x - 3 = \frac{3}{4}x$ **33.** $\frac{2}{3}x = x + p$ **35.** 30 in × 44 in **37.** 41°, 60°, 79° **39.** 52 **41.** −8, −1, 6, 13, 20 **43.** 27 cc/10%; 54 cc/4% **45.** \$40,000 **47.** 6 yr, 12 yr **49.** \$3.40 **51.** {1, 2, 3, . . .} **53.** {4, 5, 6, 7, . . .} **55.** {. . . , 8, 9, 10} **57.** {. . . , 8, 9, 10, 11} **59.** ←○→ (0) **61.** ←●→ (−2) **63.** ←●→ (1) **65.** ←●→ ($3\frac{1}{2}$) **67.** (5, −1) **69.** (−2, −3) **71.** (4, −1) **73.** Inconsistent **75.** $(-3\frac{3}{5}, -\frac{7}{20})$ **77.** (0, −3) **79.** −26 **81.** 0 **83.** −15 **85.** (−4, 5) **87.** (−5, −9) **89.** Dependent **91.** 8 and −2

CHAPTER 5 REVIEW

1. −5 **3.** −.5292 **5.** $-3.6x + 2$ **7.** $-36x^2y^2$ **9.** $2x^4 + x^3 - 2x - 1$ **11.** $-3x^5 + 2x^3 - 2x$ **13.** $y + 6$ **15.** $x + 3$

EXERCISE 6.1

1. $-3a + 18$ **2.** $ax - ay$ **3.** $4a^2c - 4c^3$ **4.** $5ax - 5bx + 5cx$ **5.** $3x^2y - 3xy^2 + 3xyz$ **6.** $-4x + 3y - 2z$ **7.** $-p^5 - p^4 + 3p^3$ **8.** $-a^2b^2 + a^2b^3 - a^3b^2$ **9.** $3mn^2 + 6m^2n - 3m^3$ **10.** $-20a^3 + 15a^2 - 10a - 5$ **11.** $5(x - 2y)$ **12.** $2x(3 - x)$ **13.** $3a(a - 4)$ **14.** $3ap(9ap + 1)$

15. $17p(2q - 3)$ **16.** $6a(1 - 6a)$ **17.** $2axy(7a - x)$ **18.** $x^2(x - 6)$ **19.** $3a(a - 1)$ **20.** $g^2(g^3 + 3g^2 - 2)$ **21.** $2(mn - 2m + 3n)$ **22.** $2(x^2 - 2xy + 6y^2)$ **23.** $2xy(x^2 + 2xy + 8y^2)$ **24.** $a^2(a^4 - 2a^2 + 3)$ **25.** $a^2b^2c^3(ac + b + a^2)$ **26.** $3a(a - 6ab + 9b^2)$ **27.** $xy(9x - 8y + 12)$ **28.** $2a(a^4 - 2a^3 + 4a^2 + 5)$ **29.** $x^2 + 2xy + y^2$ **30.** $a^2 + 2ab + b^2$ **31.** $m^2 + 2mn + n^2$ **32.** $x^2 - 2xy + y^2$ **33.** $a^2 - 2ab + b^2$ **34.** $m^2 - 2mn + n^2$ **35.** $x^2 + 6x + 9$ **36.** $b^2 - 8b + 16$ **37.** $9x^2 + 6x + 1$ **38.** $4y^2 - 4y + 1$ **39.** $9a^2 - 12ab + 4b^2$ **40.** $4x^2 + 12xy + 9y^2$

41. $64x^2 + 112xy + 49y^2$ **42.** $p^2m^2 + 2pmqn + q^2n^2$ **43.** $a^2x^2 - 2abxy + b^2y^2$ **44.** $\frac{x^2}{4} - \frac{3xy}{4} + \frac{9y^2}{16}$

45. $100 + 40 + 4 = 144$ **46.** $100 + 100 + 25 = 225$ **47.** $400 + 200 + 25 = 625$ **48.** $3{,}600 + 120 + 1 = 3{,}721$ **49.** 2 **50.** 24 **51.** 90 **52.** 120 **53.** 84 **54.** 24 **55.** $2ac$ **56.** $20abc$ **57.** $(x + y)(x + y)$ **58.** $(a - b)(a - b)$ **59.** $(x + 2)(x + 2)$ **60.** $2(x - 3)(x - 3)$ **61.** Prime **62.** $(3xy + 1)^2$ **63.** $2(3x + 2y)^2$ **64.** $(2a - 3)^2$ **65.** $(5x^2 + 8y^2)^2$ **66.** Prime **67.** $(5x + 11y)^2$ **68.** $(3 - 2x)^2$ **69.** $5(9x + 4y)^2$ **70.** $(mx - 1)^2$

EXERCISE 6.2

1. $m^2 - n^2$ **2.** $s^2 - t^2$ **3.** $x^2 - 9$ **4.** $y^2 - 49$ **5.** $4x^2 - y^2$ **6.** $9x^2 - 4y^2$ **7.** $64x^2 - 81y^2$ **8.** $a^2b^2 - c^2$ **9.** $400 - 25 = 375$ **10.** $1{,}600 - 9 = 1{,}591$ **11.** $(a - b)(a + b)$ **12.** $(a - 3b)(a + 3b)$ **13.** $(4a - 1)(4a + 1)$ **14.** $(3xy - 8z)(3xy + 8z)$ **15.** $(5p - 2q)(5p + 2q)$ **16.** $(11m - 12)(11m + 12)$ **17.** $(ax + by)(ax - by)$ **18.** $(a^2 + b^2)(a + b)(a - b)$ **19.** $3(2m + 3n)(2m - 3n)$ **20.** $a^2(6b - 1)(6b + 1)$ **21.** $a^3(ab - 1)(ab + 1)$ **22.** $x^2(x - y)(x + y)$ **23.** $x^2 + 5x + 6$ **24.** $5a^2 + 36a + 7$ **25.** $3x^2 + 13x + 12$ **26.** $6x^2 - 19x + 10$ **27.** $8y^2 - 34y + 35$ **28.** $3x^2 - 14x - 5$ **29.** $2x^2 + x - 28$ **30.** $15a^2 - 7a - 4$ **31.** $4m^2 + 16m + 15$ **32.** $40p^2 + p - 6$ **33.** $9x^2 - 42x + 49$ **34.** $15x^2 - 29x - 14$ **35.** $4a^2 - 9$ **36.** $6x^2y^2 - 2xy - 28$ **37.** $49x^2 + 28xy + 4y^2$ **38.** $acx^2 + (ad + bc)x + bd$ **39.** $kmx^2 + (dk - cm)x - cd$ **40.** $2px^2 + (3 - 2p^2)x - 3p$ **41.** $(x + 3)(x + 2)$ **42.** $(y - 4)(y - 3)$ **43.** $(a - 8)(a - 2)$ **44.** $(a + 6)^2$ **45.** $(y + 5)(y - 1)$ **46.** $(x + 5)(x - 3)$ **47.** -8 **48.** $(6x + 7)(3x + 2)$ **49.** $(3x + 5)(x + 8)$ **50.** $(3a - 2)(2a + 5)$ **51.** $(2x - 3y)(2x - 5y)$ **52.** $(3a - 1)(a + 5)$ **53.** $(2x + 3)(x - 3)$ **54.** $(10 - y)^2$ **55.** $(2 - 3b)^2$ **56.** $(2b - 5)(3b - 2)$ **57.** $(x - 2)(x - 1)$ **58.** $(3a - 2b)(4a - 9b)$ **59.** $m(m + 6n)(m + 4n)$ **60.** $(9x - y)(x - 3y)$ **61.** $a(18a^3 + 21a - 4)$ **62.** $(a^2 - 2)(a - 2)(a + 2)$ **63.** $2(5x - y)(3x + y)$ **64.** -80 **65.** $a(2a - 1)(3a - 5)$ **66.** $ap(4x - 9) \times (x - 2)$ **67.** $(a - b)(a + b)(a^2 - 3b^2)$ **68.** -503

EXERCISE 6.3

1. $(a + b)(x + 3)$ **2.** $(c - 4)(x + y)$ **3.** $(d + 1)(x - c)$ **4.** $(c - b)(n - 4)$ **5.** $g(1 + h)(x - 3)$ **6.** $(a + 2)(b + 6)$ **7.** $(a + 1)(a - 1)(x + d)$ **8.** $(x + y)(x - h)$ **9.** $(t - s)(b + 2a)$ **10.** $(2a + b)(y - 3x)$ **11.** $(a + 2b)(5c - d)$ **12.** $(b + 2)(c - 4)$ **13.** $(k - m)(x^2 + y^2)$ **14.** $(x - 2)(x + 2)(a^2 - 3)$ **15.** $[(x + 1) + 1]^2 = (x + 2)^2$ **16.** $(a - b + 2)^2$ **17.** $3(2x + 7)$ **18.** $(2a - b)(2a + b)(b + c)$ **19.** $(2x + 2y - 3)(x + y + 1)$ **20.** $(2m + 2n + 5)^2$ **21.** $(2a + 2b + 3) \times (a + b - 3)$ **22.** $(x - y)(x + y + 3)^2$ **23.** $(4a - 3b)(2a + b)$ **24.** $(y - 1)(y + 1)(1 - 2x)$ **25.** $(a + 3b)(1 - a + 3b)$ **26.** $(a^2 + 3)(a - 2)(a + 2)$ **27.** $100(a + 10b)(a - 10b)$ **28.** $(a - 1)(b - 1)$ **29.** $x^2(5y - 1)(5y + 1)$ **30.** $2(x + y + 5)(x + y + 1)$ **31.** $(a^2 + 2)(b^2 - 5)$ **32.** $(a^2 - 2)^2$ **33.** $(2a + 9)(4a + 1)$ **34.** $2(x^2 - 6)(x^2 + 1)$ **35.** $(4p^2 + 9q^2)(2p + 3q)(2p - 3q)$ **36.** $n(2m + n)$ **37.** $(a + 3b + 2)(a + 3b - 2)$ **38.** $(x^2 + 4)(x + 2)(x - 2)(x - 1)(x + 1)$ **39.** $(2x)(x + 3)(x - 2)(x + 2)$ **40.** $x(x - 2)(x + 4)(x + 6)$ **68.** -503

EXERCISE 6.4

1. $\frac{3a}{5a - 2}$ **2.** $\frac{3}{2}$ **3.** $\frac{b}{5}$ **4.** $\frac{a}{b + c}$ **5.** $\frac{1}{xy - 3}$ **6.** $\frac{x^2 - 4}{x}$ **7.** $\frac{x - 3}{x - 2}$ **8.** $\frac{x + 3}{x - 3}$

9. $\frac{2x - 5}{x - 7}$ **10.** $\frac{2x - 5}{x - 4}$ **11.** $\frac{x - y}{a - 5}$ **12.** $\frac{x - 3}{2(x + 2)}$ **13.** $\frac{x(2x - 3)}{a - b}$ **14.** $\frac{1}{a}$

15. $\frac{x+y-5}{x}$ **16.** $\frac{b-a+3}{c}$ **17.** $-;+;-$ **18.** $+;-$ **19.** $+;-;-$ **20.** $-;+$

21. $+;-;+;-$ **22.** $-a$ **23.** $-\frac{2}{3}$ **24.** $\frac{1}{p}$ **25.** $-\frac{1}{a+3}$ **26.** $\frac{1}{x^2}$ **27.** $\frac{x+y}{x-y}$

28. $-\frac{x+2}{x+1}$ **29.** -1 **30.** $\frac{3x-7}{x+4}$ **31.** $+1$ **32.** -1 **33.** $-(x+3)$ **34.** $\frac{3y-2x}{2x+5y}$

35. $\frac{6-x}{x}$

EXERCISE 6.5

1. $\frac{2(x^2-2)}{x-2}$ **2.** $\frac{b(a-b)}{a}$ **3.** $\frac{3(2x-3)}{xy-2}$ **4.** $-\frac{3y}{x}$ **5.** $\frac{2(a-3)}{(a+2)(2a-3)}$ **6.** $\frac{2a-b}{a-b}$ **7.** $\frac{3y^2}{5a}$

8. $\frac{(x-5)(x-8)}{(x+7)^2}$ **9.** $\frac{x+3}{x-1}$ **10.** $y-b$ **11.** 1 **12.** $y-x-4$ **13.** $\frac{x+4}{3(x-2)}$ **14.** $\frac{cm}{ax}$

15. $\frac{(x+2)^2}{(x+4)(x^2-3)}$ **16.** $\frac{(x+3)(x+y)}{3(3-x)}$ **17.** $\frac{m+9}{2(m-4)}$ **18.** $\frac{9(x+4)(x-4)}{x(x-3)(1-4x)}$ **19.** $\frac{(a+b)^2}{(a-3b)}$

20. $\frac{(a-b)(m-n)}{3am(m+n)}$ **21.** $\frac{x-3}{x-2}$ **22** $-\frac{a^2(a+6)^2}{(a+5)^2}$ **23.** $\frac{2x+3}{x+5}$ **24.** $\frac{(x-y)^2}{3(x+y)^2}$ **25.** $\frac{x-3}{x-4}$

EXERCISE 6.6

1. $\frac{7x+6y}{72}$ **2.** $-\frac{(3t-2s)(t-s)}{42st}$ **3.** $\frac{6a^2-ab-18b^2}{36a^2b^2}$ **4.** $\frac{f_2+f_1}{f_1f_2}$ **5.** $\frac{13t-6r}{rt}$ **6.** $\frac{a^2-3a+5}{a}$

7. $\frac{4x}{2-3x}$ **8.** $\frac{5n-6m+3mn}{mn}$ **9.** $\frac{x^2-5x+17}{x-7}$ **10.** $\frac{x^2-2x-26}{x-7}$ **11.** $\frac{5x^2-12x+17}{x^2-x-12}$

12. $\frac{2q}{q^2-p^2}$ **13.** $\frac{7}{a-b}$ **14.** $-\frac{3x+16}{12x}$ **15.** $\frac{x+4}{x^2-4}$ **16.** $\frac{4}{y+3}$ **17.** $\frac{16}{(4+a)(5-a)}$

18. $\frac{4x^2+14x+5}{(x-3)(x+3)(x+4)}$ **19.** $\frac{d^2-cd-3c^2}{d^2-4c^2}$ **20.** $\frac{4x+7}{2-x}$ **21.** $\frac{5}{(x-3)(x+2)}$ **22.** $\frac{2y-3}{y(x+y)}$

23. $\frac{x^2-x-15}{x^2-5x+6}$ **24.** $2x+3$ **25.** $\frac{x^2+5x+2}{(2x+3)(x-3)(3x-2)}$ **26.** $\frac{2(11-x)}{(x-3)(x+3)(x-1)(x+1)}$

27. $\frac{x^2-20x-11}{(2-x)(1+x)(2+x)}$ **28.** $\frac{1}{x}$ **29.** $\frac{a-1}{a^2}$

EXERCISE 6.7

1. 18 **2.** $\frac{1}{12}$ **3.** -3 **4.** 2 **5.** 2 **6.** b **7.** 3 **8.** -10 **9.** -1 **10.** -7

11. $\frac{a-b}{2}$ **12.** $-\frac{16}{7}$ **13.** $(2, 3)$ **14.** $(\frac{1}{5}, \frac{1}{2})$ **15.** $(-\frac{1}{2}, -\frac{1}{5})$ **16.** 144 mi **17.** 350 mph

18. 2, 3 **19.** $8\frac{2}{5}$ hr

CHAPTER 6 SUPPLEMENTARY EXERCISES

1. $6ab + 9a^2$ **3.** $6ma - 8mb + 2mc$ **5.** $m^2 - 2mn + n^2$ **7.** $4a^2 + 4ab + b^2$ **9.** $9x^2 - 24xy + 16y^2$ **11.** 4 **13.** $30m$ **15.** $2a(a^2 + ab - 3b)$ **17.** $(x - 2y)(x - 2y)$ **19.** $(2m + 3n)(2m + 3n)$ **21.** $2(3 - 2y)(3 - 2y)$ **23.** $4x^2 - 9y^2$ **25.** $x^2 - 10x + 21$ **27.** $12p^2 + 5pq - 2q^2$ **29.** $3m^2n^2 + 10mn + 8$ **31.** $(3x - 7y)(3x + 7y)$ **33.** $(3a + 2)(3a - 1)$ **35.** $(2x - 3)(4x + 5)$ **37.** $(27a - 4) \times (3a + 2)$ **39.** $a(2a - b)(3a + 2b)$ **41.** $(x + y)(a - b)$ **43.** $(2a + 3)(b + c)$ **45.** $a(2b - 3c) \times (x - 2y)$ **47.** $(a - b)(a + b)(1 + 2x)$ **49.** $(x - y)(1 - x - y)$ **51.** $(x + 3)(x - 3)(x - 2)(x + 2)$

53. $\frac{2a - 3}{a - 5}$ **55.** $-\frac{1}{a + 3}$ **57.** $\frac{4 - a}{a - 3}$ **59.** $\frac{a}{2a - b}$ **61.** $-\frac{1}{a + 1}$ **63.** 1 **65.** $\frac{8ax(2x - y)}{7y(a^2 - 3b)}$

67. $\frac{p}{2a + 3}$ **69.** $\frac{(a - 6b)(3a + b - 2)}{5(2a - 3b)}$ **71.** $\frac{7}{5(a - 2)}$ **73.** $\frac{10}{x^2 - 4}$ **75.** $\frac{y - x - y^2}{xy - 2y^2}$

77. $\frac{11a^4 - 22a^2 + 6}{(a^4 - 4)(2a^2 - 3)}$ **79.** -4 **81.** $\frac{1}{2}$ **83.** -12 **85.** $\frac{3}{2}$ **87.** $(2, 3)$

CHAPTER 6 REVIEW

1. $\{\ldots, 4, 5, 6\}$ **3.** $\{\ldots, -9, -8, -7\}$ **5.** $\{\ldots, -6, -5, -4\}$ **7.** 5 **9.** -5 **11.** $-\frac{1}{2}$ **13.** $(3\frac{1}{2}, -2)$ **15.** dependent **17.** -10 **19.** 0 **21.** $(3\frac{1}{2}, -2)$ **23.** dependent

EXERCISE 7.1

1. 8 **2.** 32 **3.** 9 **4.** -27 **5.** -27 **6.** 16 **7.** $\frac{4}{9}$ **8.** 125 **9.** a^8 **10.** a^6

11. x^5 **12.** a^4y^4 **13.** x^3 **14.** 1 **15.** $\frac{b^2}{a}$ **16.** y^2 **17.** $\frac{bc}{a}$ **19.** x^6 **19.** $-b^4$

20. m^{12} **21.** 64 **22.** 729 **23.** 4 **24.** $-a^6b^9$ **25.** $-a^{15}b^{12}$ **26.** $8x^3y^6$ **27.** $81a^8b^4c^4$

28. $-8x^6y^3z^9$ **29.** $\frac{x^{12}}{y^{20}}$ **30.** 1 **31.** $\frac{81}{v^8}$ **32.** a^{2pm} **33.** $\frac{3^{3n}}{8} = \frac{27^n}{8}$ **34.** $\frac{x^8}{y^{4n}}$ **35.** x^{3n}

36. x^{5n-1} **37.** $\frac{1}{x^n}$ **38.** a^{2+2n} **39.** 1 **40.** r^{xy+x+1} **41.** $3x^{2n}$ **42.** m^{pa+2a} **43.** $a^{2r}b^{3r}$

44. $-\frac{a^{14}}{128b^7}$ **45.** $\frac{1}{a^{-4}}$ **46.** $\frac{1}{a^{-2}b^{-3}}$ **47.** y^{-2} **48.** $\frac{p^{-3}}{2^{-1}}$ **49.** $\frac{1}{y^{-3}}$ **50.** 2^{-4} **51.** $(bx)^{-1}$

52. $-\frac{1}{c^{-3}}$ **53.** $-\frac{1}{x^{-2}y^{-3}z^{-3}}$ **54.** $\frac{1}{(a - b)^{-2}}$ **55.** $\frac{1}{a^{-2}} - \frac{1}{b^{-2}}$ **56.** $\left(\frac{1}{a^{-2}} - \frac{1}{b^{-2}}\right)^{-2}$ **57.** $\frac{1}{27x^3}$

58. $\frac{x^2}{4}$ **59.** x^4 **60.** $\frac{y^2}{x^2}$ **61.** $\frac{1}{81y^4}$ **62.** $\frac{x^6}{64}$ **63.** a^2b^2 **64.** $-\frac{8}{27}$ **65.** $\frac{x^8y^3}{ab^7}$

66. $\frac{y - x}{xy}$ **67.** $\frac{x^4}{y^6}$ **68.** $\frac{2y^2 - x}{xy^2}$ **69.** $\frac{y^2}{x^3}$ **70.** $\frac{b^3}{a^5}$ **71.** $3ab$ **72.** $\frac{ab}{b + a}$ **73.** $\frac{6 + x^2y}{x^2}$

74. $\frac{a^2b^2}{b^2 + 4ab + 4a^2}$

EXERCISE 7.2

1. $+2$ **2.** $+6$ **3.** $+2$ **4.** -2 **5.** $+.1$ **6.** $-\frac{1}{2}$ **7.** None **8.** $+.4$ **9.** $-.1$
10. $-\frac{2}{3}$ **11.** None **12.** -2 **13.** $p^{1/2}$ **14.** $s^{1/3}$ **15.** $t^{1/10}$ **16.** $p^{2/3}$ **17.** $x^{2/5}y^{2/5}$
18. $y^{2/n}$ **19.** $p^{-(s/r)}$ **20.** $p^{3m/k}$ **21.** $(c-6)^{1/2}$ **22.** $\frac{c^{3/4}}{b^{1/4}}$ **23.** $x^{1/2}y^{1/8}$ **24.** $\frac{1}{(c+d^2)^{1/n}}$
25. $\sqrt[5]{a}$ **26.** $\sqrt[3]{bc^2}$ **27.** $\sqrt[8]{a^5}$ **28.** $3\sqrt{x}$ **29.** $\sqrt{3x}$ **30.** $2\sqrt[3]{x}$ **31.** $\sqrt{b-c}$
32. $\sqrt[3]{a+2}$ **33.** $\sqrt[3]{(x+y)^2}$ **34.** $\sqrt[4n]{s^8t}$ **35.** $\frac{1}{\sqrt{xy}}$ **36.** 0 **37.** $\sqrt[6]{x}$ **38.** $\frac{1}{\sqrt[15]{m^4}}$
39. a^3 **40.** $\frac{1}{\sqrt[6]{b^5}}$ **41.** $\sqrt[5]{a^3}$ **42.** $\frac{1}{\sqrt[25]{a^{12}}}$ **43.** $\sqrt[12]{m^8n}$ **44.** $\frac{x^5}{64y^6}$ **45.** $\frac{4}{25}$ **46.** 8
47. $\frac{1}{x+y}$ **48.** $\frac{1}{\sqrt{xy}}$ **49.** $\frac{1}{16}$ **50.** $-\frac{1}{16}$ **51.** $\frac{x^2\sqrt{x}}{8\sqrt[4]{8y}}$ **52.** $\sqrt[c]{x^{m+7}}$ **53.** $\sqrt[10]{s}$ **54.** $\frac{1}{\sqrt{3}}$
55. $\frac{1}{\sqrt[16]{a^3}}$ **56.** $\frac{1}{\sqrt{(an)^m}}$ or $\frac{1}{\sqrt{a^mn^m}}$ **57.** $\frac{b^2}{81\sqrt[3]{a}}$ **58.** $\sqrt[24]{x}$

EXERCISE 7.3

1. $\sqrt{3a}$ **2.** $2\sqrt{2ab}$ **3.** $\sqrt[3]{2m}$ **4.** $\sqrt[3]{3m^2}$ **5.** $\sqrt[6]{6s^4}$ **6.** $\frac{1}{2}\sqrt{\frac{3x}{y}}$ **7.** $\sqrt[4]{9}$ **8.** $\sqrt[6]{8x^3y^3}$
9. $\sqrt[6]{16x^2y^2}$ **10.** $\sqrt[3]{64}$ **11.** $\sqrt[8]{169x^4y^6z^2}$ **12.** $\sqrt[6]{(a-b)^4}$ **13.** $\sqrt{18}$ **14.** $\sqrt{75}$ **15.** $\sqrt[3]{48}$
16. $\sqrt{3m^2}$ **17.** $\sqrt{5a^3}$ **18.** $\sqrt[3]{128a^4}$ **19.** $\sqrt{32x^5y^3}$ **20.** $\sqrt[3]{54m^5n^7}$ **21.** $\sqrt[6]{(a-b)^7}$
22. $5\sqrt{3}$ **23.** $2\sqrt[3]{4}$ **24.** $3\sqrt[3]{3}$ **25.** $2\sqrt{2x}$ **26.** $5x\sqrt{2x}$ **27.** $3x\sqrt[3]{xy^2}$ **28.** $3xyz^2\sqrt{2y}$
29. $xy\sqrt[3]{kx^2y}$ **30.** $6x\sqrt{3x}$ **31.** $\frac{1}{4}\sqrt{x}$ **32.** $\frac{s^2}{t^3}\sqrt{5s}$ **33.** $2\sqrt[3]{m^2}$ **34.** $3(1+a\sqrt{5})$
35. $a+(ba+b^2)(\sqrt{a+b})$ **36.** $(a-5)\sqrt[3]{a-5}$ **37.** $\frac{1}{3}\sqrt{3}$ **38.** $\sqrt{2a}$ **39.** $\frac{3x\sqrt{5}}{5}$ **41.** $\frac{7a\sqrt{3}}{6}$
41. $\frac{3\sqrt[3]{2}}{2}$ **42.** $2\sqrt[3]{p}$ **43.** $12\sqrt[3]{x^2}$ **44.** $\frac{\sqrt{30ab}}{5b}$ **45.** $\frac{\sqrt{7m}}{m}$ **46.** $\sqrt{3}+\sqrt{2}$ **47.** $\sqrt{5}-\sqrt{3}$
48. $-(2+\sqrt{6})$ **49.** $\frac{1}{2}\sqrt{3}+\frac{1}{2}$ **50.** $\frac{x+x\sqrt{x}}{1-x}$ **51.** $\frac{a+\sqrt{ab}}{a-b}$ **52.** $-5-2\sqrt{6}$ **53.** $4-\sqrt{15}$
54. $-1-\frac{1}{2}\sqrt{6}$ **55.** $\frac{a-2\sqrt{ab}+b}{a-b}$ **56.** $1-\frac{1}{3}\sqrt{6}$ **57.** $\frac{1}{6}\sqrt{12a-9b}$

EXERCISE 7.4

1. $5\sqrt{a}$ **2.** $2\sqrt{2}$ **3.** $7\sqrt[3]{3}$ **4.** $10(\sqrt{5}-1)$ **5.** $\sqrt{2}$ **6.** 0 **7.** $-\frac{14}{3}\sqrt[3]{6}$ **8.** $5\sqrt{2}-\sqrt[3]{2}$
9. $\frac{5}{6}\sqrt{3}-\frac{3}{4}\sqrt{2}$ **10.** $7\sqrt{6}$ **11.** $6\sqrt[3]{2}-3\sqrt{2}$ **12.** $\frac{17\sqrt{3}}{2}$ **13.** $\frac{3}{5}\sqrt{15}$ **14.** $3-x\sqrt{6}$
15. $(a-b+1)\sqrt[3]{(a-b)^2}$ **16.** $-3\sqrt{b}$ **17.** $2a$ **18.** $48x^3$ **19.** $3a^3\sqrt{2}$ **20.** $2\sqrt[6]{54}$

21. $12 + 2\sqrt{3} - 9\sqrt{2}$ **22.** $72\sqrt{3x} - 18\sqrt{10x} + 252\sqrt{2x}$ **23.** $3x^2\sqrt[6]{32x^5y^4}$ **24.** $42 + 24\sqrt{3}$

25. $a + b - 2\sqrt{ab}$ **26.** $6 - 4x + \sqrt{2x}$ **27.** $2\sqrt{2}$ **28.** $\sqrt[3]{3}$ **29.** $\sqrt{2} - 2\sqrt{5}$

30. $\dfrac{a(1 - \sqrt{a})}{1 - a}$ **31.** $\sqrt{3} + \frac{1}{2}\sqrt{2} + 2$ **32.** $4\frac{4}{7} + \frac{3}{7}\sqrt{15}$ **33.** $\dfrac{3 + 5\sqrt{3}}{11}$ **34.** $\dfrac{28 + x - 11\sqrt{x}}{16 - x}$

35. -1 **36.** $\dfrac{x - \sqrt{ax} - \sqrt{xy} + \sqrt{ay}}{x - y}$ **37.** $7 - 5\sqrt{2}$ **38.** $21 + 14\sqrt{3}$

CHAPTER 7 SUPPLEMENTARY EXERCISES

1. 81 **3.** m^6 **5.** 1 **7.** x^6 **9.** $27a^3b^6$ **11.** 1 **13.** $\dfrac{a^{24}}{b^{16}}$ **15.** y^{3+2n} **17.** d^{an+2a}

19. $a^0 = 1$ **21.** $\dfrac{c^{12}}{a^9}$ **23.** $a^4x^{12}y^5$ **25.** $\dfrac{3x^2 + 3y^2}{x^2y^2}$ **27.** $\dfrac{ab}{a^2 - b^2}$ **29.** $\dfrac{y}{1 + xy}$ **31.** 4

33. -5 **35.** .1 **37.** $x^{1/3}$ **39.** $a^{2/5}b^{3/5}$ **41.** $x^{2/k}y^{1/k}$ **43.** $\sqrt[4]{m^3}$ **45.** $2\sqrt[3]{a}$

47. $\sqrt{m + n}$ **49.** $\sqrt{xy}$ **51.** $a^{1/4}$ or $\sqrt[4]{a}$ **53.** $x^{1/3}$ or $\sqrt[3]{x}$ **55.** $\dfrac{y}{x^{4/5}}$ or $\dfrac{y}{\sqrt[5]{x^4}}$ **57.** $\dfrac{1}{m - n}$

59. $\sqrt[3]{x}$ **61.** $2\sqrt{x}$ **63.** $\sqrt{2x}$ **65.** $\sqrt[3]{2ab^2}$ **67.** $3x\sqrt{7x}$ **69.** $b\sqrt[3]{abc^2}$ **71.** $\dfrac{a}{b^3}\sqrt{2}$

73. $\frac{3}{5}a\sqrt{5}$ **75.** $\frac{1}{9}y\sqrt{3}$ **77.** $\dfrac{2}{b}\sqrt{2b}$ **79.** $1 + \frac{1}{2}\sqrt{2}$ **81.** $\sqrt{6} + 2$ **83.** $\dfrac{m\sqrt{m} + m}{m - 1}$

85. $4\sqrt{3} - 7$ **87.** $\dfrac{3x + 5\sqrt{xy} + 2y}{x - y}$ **89.** $2\sqrt{2}$ **91.** $3\sqrt{3}$ **93.** $9\sqrt[3]{2}$ **95.** $(6a - 2)\sqrt{5a}$

97. $-\dfrac{11\sqrt{3}}{6}$ **99.** $18x^3$ **101.** $2x^2\sqrt{3x}$ **103.** $5 - 2\sqrt{6}$ **105.** $3 - 2x + \sqrt{3x}$ **107.** $4x - 12\sqrt{xy} + 9y$ **109.** $\frac{5}{3}\sqrt{2}$ **111.** $\frac{2}{3}\sqrt{3}$ **113.** $2\sqrt[6]{3} + \sqrt[3]{18}$ **115.** $4 + 2\sqrt{2}$ **117.** $2 - \sqrt{3}$

119. $\dfrac{34 - 7\sqrt{10}}{37}$

CHAPTER 7 REVIEW

1. $(a + 3d)^2$ **3.** $(5 - 3ab)(5 + 3ab)$ **5.** $(x - 7)(x - 3)$ **7.** $5(y - 2)(y - 6)$ **9.** $(a - 2)(6a + 1)$
11. $(m - n)(p + q)$ **13.** $(a - c)(b - d)$ **15.** $2(a + 2c)(b + d)$ **17.** $(x + y + z)(x + y - z)$
19. $(p + q)(1 + p + q)$ **21.** 9% and $4\frac{1}{2}$% **23.** 20 cc of 35%, 30 cc of 20% **25.** 5 and 25

EXERCISE 8.1

1. ± 3 **2.** ± 5 **3.** $\pm\sqrt{5}$ **4.** ± 4 **5.** ± 3 **6.** $\pm\sqrt{6}$ **7.** $\pm 2\frac{1}{2}$ **8.** 0 **9.** $\pm\dfrac{\sqrt{ab}}{a}$

10. $\frac{3}{11}\sqrt{11}$ **11.** $\pm 2\sqrt{3}$ **12.** $\dfrac{\pm\sqrt{z(a + b)}}{a + b}$ **19.** $2i\sqrt{3}$ **20.** $5 + 12i$ **21.** $-2(1 + i\sqrt{3})$

22. $i - 1$ **23.** $7 - 7i$ **24.** $\pm i$ **25.** $\pm i\sqrt{3}$ **26.** $\pm 2i$ **27.** $\pm\frac{7}{2}i$ **28.** $\pm i\sqrt{5}$
29. $\pm i\sqrt{6}$

EXERCISE 8.2

1. 4, 2 **2.** 3, −5 **3.** −4, −6 **4.** $\frac{1}{3}, \frac{1}{2}$ **5.** $-\frac{7}{3}, -5$ **6.** $\frac{1}{3}$ **7.** −5 **8.** $5a, -3a$ **9.** $\frac{2}{5}$ **10.** $\frac{3}{5}, -\frac{3}{4}$ **11.** $0, 2\frac{1}{4}$ **12.** $0, 4a$ **13.** $-\frac{5}{2}, 1$ **14.** 0 **15.** 4 **16.** 12 yd × 26 yd **17.** 4, 6 **18.** 2 × 2 **19.** 12 in. × 12 in. **20.** 3 ft

EXERCISE 8.3

1. 3, 2 **2.** −5, −3 **3.** 3 **4.** $1\frac{2}{3}, -1\frac{1}{2}$ **5.** $1\frac{1}{2}, 1\frac{1}{3}$ **6.** $7, -\frac{2}{3}$ **7.** $\frac{-3+\sqrt{5}}{4}, \frac{-3-\sqrt{5}}{4}$

8. $\frac{6 \pm 4\sqrt{6}}{5}$ **9.** $1 + 2i, 1 - 2i$ **10.** $2 \pm 3i$ **11.** $-\frac{1}{3} \pm \frac{1}{3}\sqrt{3}$ **12.** $\frac{3 \pm i\sqrt{31}}{4}$ **13.** $\frac{1 \pm \sqrt{5}}{6}$

14. +4, −4 **15.** $1 \pm \frac{1}{4}\sqrt{6}$ **16.** $2 \pm \sqrt{2}$ **17.** $4a, 2a$ **18.** $a \pm \sqrt{a^2 + c}$ **19.** 14 **20.** $3 - \sqrt{5}$ or .76 in. (approx.) **21.** Impossible **22.** 6 in. and 8 in.

EXERCISE 8.4

1. $D = \{3, 4, 5, 6\}, R = \{-1, -2, -3, -4\}$ **2.** $D = \{1, 2, 3, 4\}, R = \{0, 1\}$ **3.** $D = \{-1\}, R = \{5, 10, 15, 20\}$ **4.** $D = \{\text{Bill, Sam, Ed, Al}\}$, $R = \{2, 4, 6\}$ **5.** $D = \{6, 3, 2\}$, $R = \{-1, 4, 3, -2\}$ **6.** Exs. 1, 2, 4 **7.** $\{(1, 3), (2, 4), (3, 5), (4, 6)\}$ **8.** $\{(-1, 8), (0, -7), (1, -6)\}$ **9.** $\{(1, 5), (2, 4), (3, 3), (4, 2), (5, 1)\}$ **10.** $\{(-5, 14), (-4, 12), (-3, 10), (-2, 8)\}$ **11.** $\{(-2, 0), (-1, -3), (0, -4), (1, -3)\}$ **12.** $y = x - 1$, x an integer and $0 < x < 5$ (or, $1 \leq x \leq 4$) **13.** **14.**

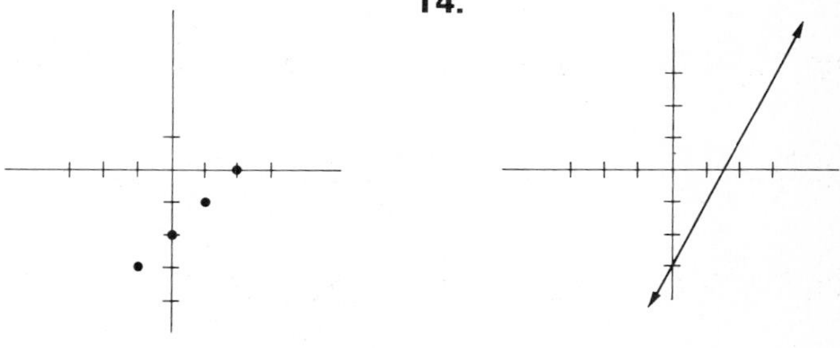

15. Yes **16.** No **17.** Yes **18.** No **19.** −1 **20.** 57 **21.** $f(-3)$ **22.** $2x^2 + x - 5$

EXERCISE 8.5

1. 33 **2.** No solution **3.** No solution **4.** $\frac{1}{2}$ **5.** 2 **6.** 5 **7.** No solution **8.** 2 **9.** +3, −3, +1, −1 **10.** −2, 1 **11.** $\frac{1}{9}$ **12.** $\frac{25}{4}, 4$ **13.** $(-1 + \sqrt{71}) \times (1 + \sqrt{71})$ or 7.4 × 9.4 (approx.) **14.** 4.69 sec (approx.) falling; 350 ft (approx.) deep

CHAPTER 8 SUPPLEMENTARY EXERCISES

1. $6i\sqrt{2}$ **3.** 1 **5.** $9 - 40i$ **7.** $8 + 6i$ **9.** ±4 **11.** $\pm\frac{\sqrt{5}}{2}$ **13.** $\pm 3i$ **15.** $\pm 2ai$

17. −3, 7 **19.** $\frac{1}{2}, -3$ **21.** $\frac{4}{5}, -\frac{5}{4}$ **23.** 0, −14 **25.** −4, −2 **27.** $-2 \pm \sqrt{3}$ **29.** $2 \pm 3i$

31. $\frac{-5 \pm i\sqrt{3}}{14}$ **33.** $2 \pm \frac{2}{3}\sqrt{6}$ **35.** 8 **37.** +4 or −4 **39.** D: {3, 4, 5, 6, 7}; R: {6, 7, 8}

41. D: {A, B, C}; R: {62 lb, 75 lb, 32 lb} **43.** No **45.** (7, −5), (8, −6), (9, −7) **47.** (1, 7), (2, 5), (3, 3), (4, 1) **49.** (−2, 5), (−1, −1), (0, −3), (1, −1) **55.** 10 **57.** 13 **59.** 23 **61.** No soln. **63.** 2 **65.** $\pm\frac{1}{2}, \pm 1$ **67.** $\frac{1}{2}$

1. $\frac{b^6}{a^6}$ **3.** $\frac{a^{12}b^6}{c^3d^9}$ **5.** $\frac{xy}{y-x}$ **7.** $\frac{2}{3}\sqrt[3]{9}$ **9.** $2-\sqrt{3}$ **11.** $14\sqrt{a}$ **13.** $8\sqrt{5}$ **15.** $3\sqrt{xy}$

17. $\frac{2\sqrt{10}-7}{3}$ **19.** $\frac{5-2\sqrt{6}}{2}$ **21.** $-\frac{1}{4}(4\sqrt{2}-4\sqrt{6}-\sqrt{10}+\sqrt{30})$ **23.** Identity, except $\frac{2}{3}$, -4

25. No soln. **27.** ←○→ (3) **29.** ←●→ (2) **31.** $\frac{2a^2-b^2}{b^2}$

EXERCISE 9.1

1. 10 **2.** 12 **3.** 7 **4.** 15.8 **5.** 14.4 **6.** 9.9 **7.** 0 **8.** 7.2 **9.** $\frac{5}{12}$ **10.** $7\sqrt{2}$

11. $\frac{3\sqrt{3}}{2}$ **12.** $18\sqrt{2}$ ft **13.** 10 and $5\sqrt{3}$ **14.** No **15.** 49 **16.** $A=\frac{\sqrt{3}s^2}{4}$ **17.** $2\sqrt{10}$ (or 6.3″) × $6\sqrt{10}$ (or 19.0″) **18.** $\frac{16\sqrt{2}}{3}$

EXERCISE 9.2

1. (a) $\frac{m}{o}, \frac{n}{o}, \frac{n}{m}$; (b) $\frac{o}{m}, \frac{o}{m}, \frac{m}{n}$; (c) tan, cot, csc; (d) sec, sin, cos

2. (a) $\frac{3}{5}, \frac{4}{5}$; (b) $\frac{3}{4}, \frac{4}{3}$; (c) $\frac{5}{4}, \frac{5}{3}$

3. (a) $\frac{3}{\sqrt{34}}, \frac{5}{\sqrt{34}}$; (b) $\frac{3}{5}, \frac{5}{3}$; (c) $\frac{\sqrt{34}}{5}, \frac{\sqrt{34}}{3}$

4. (a) $\frac{1}{2}, \frac{1}{2}\sqrt{3}$; (b) $\frac{1}{3}\sqrt{3}, \sqrt{3}$; (c) $\frac{2}{3}\sqrt{3}$, 2

5. (a) $\frac{1}{2}\sqrt{2}, \frac{1}{2}\sqrt{2}$; (b) 1, 1; (c) $\sqrt{2}, \sqrt{2}$

6. (a) $\frac{1}{2}\sqrt{3}, \frac{1}{2}$; (b) $\sqrt{3}, \frac{1}{3}\sqrt{3}$; (c) 2, $\frac{2}{3}\sqrt{3}$

7.

.6	.6	.6
.8	.8	.8
.75	.75	.75

8. (a) 30°, 45°, 32°; (b) 6°, 73°, 0°; (c) 55°, 35°, 90°; (d) 88°, 62°, 71½°; (e) 41°, 12°, 27°30′

9. (a) 66°, 75°; (b) 1°, cot; (c) tan, sec

10.

(a)	(b)	(c)
.259	5.76	82°
.883	1.11	66°
.754	1.11	38°
1.11	1.33	80°
1.02	7°	57°
2.13	12°	14°
.766	41°	90°

11. .491 **12.** 3.42 **13.** .691
14. 53.8° **15.** 24.7° **16.** 75.1°

EXERCISE 9.3

1. 7.5 **2.** 7.6 **3.** 20.9 **4.** 33° **5.** 46° **6.** 76° **7.** A = 56.0°, a = 43.1, b = 29.1
8. A = 31.9°; B = 58.1°; c = 53 **9.** B = 24.4°; b = 33.5; c = 80.7 **10.** A = 18.9°; B = 71.1°; b = 35
11. 6.1 ft **12.** 225.7 ft **13.** 19°, 71°
(Note: Angle measures may differ slightly, depending upon which function is used.)

EXERCISE 9.4

Solutions will vary.

CHAPTER 9 SUPPLEMENTARY EXERCISES

1. Yes **3.** No **5.** No **7.** 17 **9.** 40 **11.** $3\sqrt{5}$ **13.** 15 **17.** $7\sqrt{2}$ **19.** 9

21. $4\sqrt{3}$ **23.** $\dfrac{10\sqrt{3}}{3}$ **25.** 55 ft **27.** $\sin A = \frac{5}{13}$, $\cos A = \frac{12}{13}$, $\tan A = \frac{5}{12}$, $\cot A = \frac{12}{5}$, $\sec A = \frac{13}{12}$,

$\csc A = \frac{13}{5}$ **29.** 60°; 50°; 20° **31.** 45°; 23°; 82° **33.** .500; .866 **35.** .927; .375 **37.** .875; .488 **39.** 28.6; 1.01 **41.** 16°; 35° **43.** 61°; 67° **45.** 22°; 59° **47.** 37°; 89° **49.** .446; .630 **51.** .665; 2.14 **53.** 42.6°; 78.3° **55.** 45.5°; 79.1° **57.** 143.9 **59.** 57.9 **61.** 50°
63. A = 73.6°; a = 11.5; b = 3.4 **65.** A = 19.4°; B = 70.6°; b = 17.0 **67–75.** Solutions will vary

CHAPTER 9 REVIEW

1. $8i\sqrt{3}$ **3.** $4 + 5i$ **5.** -1 **7.** $(2, -5)$ **9.** $(3, 7)$ **11.** $(0, -4)$ **13.** $(-\frac{3}{4}, -\frac{5}{2})$

15. $-3 \pm \sqrt{13}$ **17.** $\dfrac{-3 + \sqrt{29}}{10}$ **19.** $2 \pm 3i$ **21.** $+11$ **23.** $2a^2 + 9a + 9$ **25.** $\frac{1}{16}$

EXERCISE 10.1

1. $\frac{3}{5}, -\frac{4}{5}, -\frac{3}{4}, -\frac{4}{3}, -\frac{5}{4}, \frac{5}{3}$
2. $\frac{12}{13}, \frac{5}{13}, \frac{12}{5}, \frac{5}{12}, \frac{13}{5}, \frac{13}{12}$
3. $-\frac{15}{17}, \frac{8}{17}, -\frac{15}{8}, -\frac{8}{15}, \frac{17}{8}, -\frac{17}{15}$
4. $-\frac{4}{5}, -\frac{3}{5}, 1\frac{1}{3}, \frac{3}{4}, -1\frac{2}{3}, -1\frac{1}{4}$
5. $\frac{1}{5}\sqrt{5}, -\frac{2}{5}\sqrt{5}, -\frac{1}{2}, -2, -\frac{1}{2}\sqrt{5}, \sqrt{5}$
6. $-\frac{5}{34}\sqrt{34}, -\frac{3}{34}\sqrt{34}, 1\frac{2}{3}, \frac{3}{5}, -\frac{1}{3}\sqrt{34}, -\frac{1}{5}\sqrt{34}$
7. 60° **8.** 15° **9.** 80° **10.** 28° **11.** 48° **12.** 13°
13. 15° **14.** 1° **15.** $17\frac{1}{2}$°
16. 43°; −, −, +, +, −, − **17.** 26°; +, −, −, −, −, +
18. 75°; +, +, +, +, +, + **19.** 30°; −, +, −, −, +, −

EXERCISE 10.2

1. .643 **2.** −.176 **3.** 2.00 **4.** .707 **5.** .839 **6.** 1.02 **7.** −.891 **8.** −1.25 **9.** 0
10. und **11.** 0 **12.** 0 **13.** 25°, 155° **14.** 140°, 320° **15.** 213°, 327° **16.** 54.5°, 305.5°
17. 90°, 270° **18.** 0°, 180° **19.** 321° **20.** 116° **21.** 228° **22.** 180°
23. $\sin x$: 0, .26, .50, .71, .87, .97, 1, .97, .87, .71, .50, .26, 0, −.26, −.50, −.71, −.87, −.97, −1, −.97, −.87, −.71, −.50, −.26, 0
$\cos x$: 1, .97, .87, .71, .50, .26, 0, −.26, −.50, −.71, −.87, −.97, −1, −.97, −.87, −.71, −.50, −.26, 0, .26, .50, .71, .87, .97, 1

$y = \sin x$: $y = \cos x$:

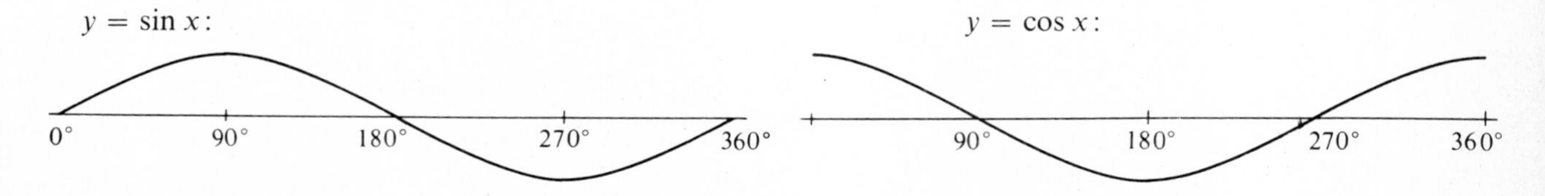

EXERCISE 10.3

1. $a = 24, b = 15, C = 77°$ **2.** $a = 81, A = 128°, C = 29°; a = 11, A = 6°, C = 151°$ **3.** $b = 43, c = 41, B = 83°$ **4.** $c = 31, B = 15°, C = 141°$ **5.** $K = \frac{1}{2}ac \sin B$; $K = \frac{1}{2}bc \sin A$; $K = \frac{1}{2}b^2 \frac{\sin A \sin C}{\sin B}$; $K = \frac{1}{2}c^2 \frac{\sin A \sin B}{\sin C}$ **6.** 175 **7.** 80 **8.** 101 mi **9.** 37°

EXERCISE 10.4

1. $c = 8.2, A = 48°, B = 72°$ **2.** $A = 44°, B = 53°, C = 83°$ **3.** $a = 27, B = 38°, C = 6°$ **4.** $A = 43°, B = 115°, C = 22°$ **5.** 13 **6.** Undefined values for cosine develop. **7.** 20.4 **8.** 4 times **9.** 38 and 55

CHAPTER 10 SUPPLEMENTARY EXERCISES

	sin	cos	tan	cot	sec	csc
1.	$\frac{4}{5}$	$-\frac{3}{5}$	$-\frac{4}{3}$	$-\frac{3}{4}$	$-\frac{5}{3}$	$\frac{5}{4}$
3.	$\frac{2}{5}\sqrt{5}$	$\frac{1}{5}\sqrt{5}$	2	$\frac{1}{2}$	$\sqrt{5}$	$\frac{1}{2}\sqrt{5}$
5.	-1	0	und	0	und	-1

7. 50° **9.** 73° **11.** 36° **13.** 46° **15.** 2°

	17. 132°	**19.** 316°	**21.** 212°
sin	+	−	−
cos	−	+	−
tan	−	−	+
cot	−	−	+
sec	−	+	−
csc	+	−	−

23. .500 **25.** −.602 **27.** −1.35 **29.** −.999 **31.** −1 **33.** 0 **35.** .887 **37.** −1.06 **39.** 158°, 202° **41.** 54°, 234° **43.** 108°, 252° **45.** 236.5°, 303.5° **47.** 90, 270°

49. 322° **51.** 65° **53.** Impossible **55.** $9\sqrt{2}$ or 12.7 **57.** 34° **59.** 68°, 112° **61.** 8 **63.** 21 **65.** 1103 **67.** 16 **69.** 109° **71.** 37 **73.** 90° **75.** 15 **77.** Impossible; not a triangle **79.** 61 mi

CHAPTER 10 REVIEW

1–9. Solutions will vary **11.** $\frac{1}{\cos \theta}$ **13.** $\cos^2 \theta$ **15.** $\cos^2 \theta$ **17.** $8\sqrt{2}$

INDEX